ECOLOGICAL PIONEERS

A SOCIAL HISTORY OF AUSTRALIAN ECOLOGICAL THOUGHT AND ACTION

Whenever the history of ecological thought has been written the contributions of Australian thinkers have rarely been mentioned. Yet Australia as a continent of extreme, rare and complex environments has produced a startling group of ecological pioneers. Across a wide range of human endeavour, Australian thinkers and innovators – whether they have thought of themselves as environmentalists or not – have made some truly original contributions to ecological thought. *Ecological Pioneers* traces the emergence of ecological understandings in Australia. By constructing a social history with chapters focusing on various fields in the arts, sciences, politics and public life, Martin Mulligan and Stuart Hill are able to bring to life the work of significant individuals. Some of the ecological pioneers featured include Tom Roberts, Russell Drysdale, Judith Wright, Myles Dunphy, Philip Crosbie Morrison, Vincent Serventy, Charles Birch, the Gurindji and Yolngu peoples, David Holmgren, Jack Mundey, Bob Brown, Val Plumwood, Michael Leunig, and many more.

Martin Mulligan lectures in the School of Social Ecology and Lifelong Learning at the University of Western Sydney and is the general editor of the journal *Ecopolitics: Thought and Action*.

Stuart Hill holds the Foundation Chair of Social Ecology at the University of Western Sydney, and is the author of many books and articles in the fields of ecology, entomology, soil ecology and sustainable agriculture.

ECOLOGICAL PIONEERS

A SOCIAL HISTORY OF AUSTRALIAN ECOLOGICAL THOUGHT AND ACTION

Martin Mulligan

University of Western Sydney

Stuart Hill

University of Western Sydney

CAMBRIDGE
UNIVERSITY PRESS

CAMBRIDGE UNIVERSITY PRESS
Cambridge, New York, Melbourne, Madrid, Cape Town,
Singapore, São Paulo, Delhi, Tokyo, Mexico City

Cambridge University Press
The Edinburgh Building, Cambridge CB2 8RU, UK

Published in the United States of America by Cambridge University Press, New York

www.cambridge.org
Information on this title: www.cambridge.org/9780521009560

First published 2001

A catalogue record for this publication is available from the British Library

Library of Congress Cataloguing in Publication Data
Mulligan, Martin John.
Ecological pioneers: a social history of Australian
ecological thought and action.
Bibliography.
Includes index.
ISBN 0 521 81103 1.
ISBN 0 521 00956 1 (pbk.).
1. Ecology – Australia – History. 2. Nature – Effect of
human beings on – Australia – History. 3. Landscape
changes – Australia – History. I. Hill, Stuart B., 1943– .
II. Title.

ISBN 978-0-521-81103-3 Hardback
ISBN 978-0-521-00956-0 Paperback

CONTENTS

ACKNOWLEDGEMENTS

It has been a privilege getting to know about the lives and work of some courageous, pioneering Australians. Studying the lives of inspirational people has made some of them feel like old family friends; their inspiration will remain with us forever. We must thank those – Dot Butler, Alex Colley, Vincent Serventy, Banduk Marika, Charles Birch, Val Plumwood, Stephen Boyden, Bob Brown, David Holmgren, Ted Trainer, Isabel Bennett – who gave generously of their time in talking to us directly. Others provided material we could use to incorporate their stories; Jack Mundey provided detailed feedback on what we had written from published accounts; and Allan and Ken Yeomans provided feedback on what we had written about their father P.A. Yeomans. Their brother Neville also gave us a valuable interview about his father's life and work but, sadly, Neville died before this book could appear.

Eleanor Dark's biographer Barbara Brooks provided detailed feedback on what we had said about Eleanor and some valuable comments on the rest of chapter 4. Patsy Jones gave us important insights on the life and work of her late husband Dick Jones. Karen Alexander was able to correct things we had said about major conservation campaigns she had participated in.

Alex Colley and Vincent Serventy were among the first people we interviewed and they both gave us valuable and timely suggestions on other people we should talk to and/or incorporate in the book. Our colleague in Social Ecology at the University of Western Sydney, John Cameron, gave us important early feedback on the structure and nature of the book.

We were very ably assisted in background research by Mary Miller, a graduate of Social Ecology, whom we were able to employ for two years thanks to internal research grants allocated to us by our university. Mary played an active role in helping us to forge the concept of the book and in setting up some of the interviews. Her confidence in our project was heartening.

We have been encouraged at different times by various colleagues at the University of Western Sydney and by students who have read and reviewed draft versions of the manuscript. It has been a real pleasure working with Peter Debus at Cambridge University Press and we are grateful for the very professional approach taken by other members of staff at CUP. We were encouraged to take our manuscript to CUP by Drew Hutton.

Martin would like to acknowledge the patient support given by his wife Nelum Buddhadasa and older son Will and the inspiration provided by twin children Indu and Roshan, whose entry into this world came long after the book was conceived but before it was born. Stuart is indebted to his partner Judy Pinn who also provided patient support and valuable direct feedback on reading sections of the manuscript. It is never easy for those who share the lives of authors struggling to meet deadlines.

There can be little doubt that Judith Wright has been one of our most important ecological pioneers, featuring in two chapters of this book. Her long and influential life came to an end while we were in the late stages of preparing the manuscript and we would like to add our book to the many tributes that were written about Judith when she died in July 2000. Judith had an extraordinary capacity to listen to both the indigenous Australians and the land that had nurtured them for countless generations and she urged all of us to:

> Listen, listen
> latecomers to my country
> eat of wild manna
>   There is
>   there was
>   a country
> that spoke in the language of the leaves.

ILLUSTRATIONS

| # INTRODUCTION

Australia is a new nation in an ancient land. We have a colonial past and an uncertain future. On the cusp of a new millennium – 100 years since feder-ation – Australians are embroiled in debates about our past and future. These are raising challenging questions. How can we come to terms with the terrible things done in the name of settlement and nation building in ways that enable us to deepen our relationships with one another and with the land? How can we reconcile the workings of a modern nation that exists within a global economy and, at the same time, recognise this land's ancient civilisations in ways that are significant to them? What will be our future and role in the southern hemisphere? When will we finally sever our ties with 'Mother England', and what will this mean for our identity? Although some of these debates may reach some level of resolution in the near future, we can be sure that the deeper issues relating to Australian identity will continue to be debated, passionately, for a very long time to come.

The struggle for identity in Australia has been made difficult by the fact that Europeans who migrated here over the last two centuries found themselves in strange and challenging landscapes. It is not surprising that the experience of fragile soils, prickly and pungent bush, seasonal fires and devastating floods combined to create a largely negative frame of mind. Efforts to 'civilise' the land frequently ended in failure (Lines, 1991) and, often as not, those who persisted eventually found themselves being colonised by the land; a very different experience of time and space adding to their insecurity. As a consequence, most immigrants have clustered in coastal towns and cities and in the south-east corner of the continent. Until recently, most white Australians thought of their country as a continent with a 'dead heart' – how extraordinary it is that the 'red centre' has now become a major icon for domestic and international tourism. After a stint in Australia in the early 1920s, the novelist D. H. Lawrence left with a sense of disillusionment, yet he was also sensitive enough to realise that the failing was partly his own, acknowledging in 1923 in his Australian novel *Kangaroo* that:

> The strange, as it were, *invisible* beauty of Australia, which is undeniably there ... seems to lurk just beyond the range of our white vision. (in Griffiths, 1996, p. 103)

The Australian art critic and historian Robert Hughes has made the interesting point that whereas European settlers in North America experienced its spaciousness as a new kind of freedom, in Australia it was perceived by many as a prison.[1] Australia rapidly became one of the most urbanised nations on earth. Yet, even from the beginning, city dwellers seem to have found themselves strangely attracted to the 'outback', as exemplified in the popularity of the 'bush poets', such as Adam Lindsay Gordon and Banjo Paterson. The most popular children's stories focused on furry marsupials (Blinky Bill) and gum-nut 'babies'. Russell Drysdale became famous for his paintings of the outback, when he dared to venture 'beyond the black stump'. Dorothea Mackellar penned a virtual national anthem when she asserted that 'I love a sunburnt country'. Paradoxically, that which initially seemed so strange and mostly unattractive also seemed to encapsulate the soul of the land. Instinctively, we seemed to know that our identity was somehow linked with a better understanding of our 'monotonous' eucalypt forests and our wide-open plains. Many Australians have needed to travel overseas before recognising the smell of a crushed gum leaf as indicative of 'home' and 'belonging'. While we have created a 'hyperseparation'[2] between the city and 'the bush', the latter continues to intrude into our urban mindsets. In literal ways, the nature that we have fought to banish from our cities sometimes comes back to haunt us, as exemplified by the recent re-establishment of colonies of flying foxes in the very hearts of our largest cities.[3]

The Australian literary critic and Jungian psychologist David Tacey has tried to explain why the search for an Australian identity has an ecological imperative:

> The widespread search for identity in Australia is not merely an intellectual or cognitive one that ends when we arrive at comforting national images or consensual conventions about who the 'real Australians' are. Identity cannot be achieved in this pre-programmed way but must arise from within. When society and nature, conscious and unconscious, are organically related (as seems to be the case in most indigenous cultures) there is no more talk about the problem of identity; the emotional depths of existence are filled and a sense of character or personality is assured. 'Alienation and rootlessness,' writes Jung, are the dangers that lie in wait for the conqueror of foreign lands.' (Tacey, 1995, p. 73)

Like it or not, Australian landscapes have had a deep impact on our sense of self. Just as Australian flora and fauna have developed high levels of co-adaptation to survive in tough environments, indigenous Australians, and to a lesser extent white Australians, have developed an ethic of resilience and mutual support. Although the notion of 'mateship' has been over-romanticised, cultural values of white Australia – egalitarianism and the 'tall poppy syndrome', for example – suggest that, from the start, our landscapes represented a challenge to the imported values of elitism and hierarchy. The artist Russell Drysdale, himself a migrant to Australia as a youngster, made the insightful comment to interviewer Geoffrey Dutton that a feeling of loneliness is a common experience when people feel out of place in their environment. People who achieve a sympathetic understanding of the environment within which they are immersed tend to learn to value solitude, rather than fear loneliness. In an earlier interview with Geoffrey Dutton in 1964, Drysdale said that people living in Australian cities lived 'under the shadow of. . . great loneliness' because they feared the vastness of the land, especially in the arid zones. Yet when you spent time in such places feelings of loneliness can evaporate because:

> If you look at the ground you'll find it peopled with insects and life, so much movement that's always going on . . . This is only a matter of adjustment to scale . . .

> The point they [critics] bring up about my paintings is that they depict a lone man in a landscape and therefore it's lonely. I don't, quite frankly, really mean to emphasize that there is a loneliness. There is a loneliness in everybody who lives apart, I quite agree, but this isn't the sole thing . . . You can delineate a landscape, the bones of a landscape, and in this country this age-old thing is tremendously gripping, but if you really want to point up a landscape that is deserted, then you put somebody within it . . . [However], you can try to point out that this is a man . . . unconquered by landscape, because he is a species which has arisen like every other species on this earth and he is not alien to a landscape, otherwise he would have to live somewhere in outer space. (Dutton, 1964, pp. 191–192)

Very late in the day we are beginning to realise that the people who occupied this land for more than 60,000 years before European colonisation had a very deep and detailed understanding of its landscapes. Living close to nature, they developed an intimate understanding of places that most contemporary Australians pass through on the way to 'somewhere else'. You don't have to romanticise Aboriginal society to recognise that, by comparison, white Australians have arrived in this land like plagues of hungry locusts having little apparent concern for the past or future.[4] Only 30 years ago, most white Australians believed that the 'primitive' societies of our indigenous people were in mortal decline; their fate sealed by the relentless 'march of progress'. However, the emergence of the modern Aboriginal land rights movement in the 1960s, and Eddie Koiki Mabo's stunning posthumous victory in the High Court in 1992, has forced many people to think again. Not only were Aborigines and Torres Strait Islanders starting to reclaim their rights, they were reminding the nation that the past would also be part of our future. As Noel Pearson has pointed out, the legal recognition of a long-standing injustice in the High Court's 'Mabo decision' of 1992 has offered the nation an opportunity for redemption from the past and a chance to recreate its future (see chapter 9).

In declaring the land *terra nullius* (see chapter 2), the white settlers in Australia had endeavoured to make the indigenous people invisible in their own land. In doing this, they also turned their backs on the profound ecological wisdom that these people had accumulated through their long associations with the land. Consequently, many of the mistakes that have been made can be traced to arrogance based on ecological ignorance and a false sense of control. White Australians have taken the long route to acknowledging that valuable things can be learnt from the Aboriginal people. Val Plumwood has commented on the extension of the *terra nullius* view to Australia's flora and fauna; seeing an empty land in need of colonisation by more 'valuable' (mainly European) species of plants and animals (see chapter 11). It took the emergence of the 'modern' Aboriginal land rights movement in the 1960s to eventually overturn this *terra nullius* view of the land, its people and its natural history (see chapter 9).

In recognition of the huge shift that has taken place in public perception of Aboriginal people, it is important that we acknowledge and celebrate the enormous achievement of the Aboriginal communities, the Gurindji and Yolngu (see chapter 9), who initiated the modern struggle for indigenous land rights. It is also appropriate to pay tribute to those white Australians – like Eleanor Dark, Judith Wright and Russell Drysdale – who, long before this became fashionable, urged all Australians to consider what we might learn from the indigenous people.

It is significant that painters and writers were among the first to take a less biased attitude towards indigenous Australians. This sensitivity towards the experiences of people on the 'other side of the frontier' probably reached its height in the work of poet and conservationist Judith Wright. She noted that, in addition to our prejudices, the very structure of our language makes it almost impossible for non-Aborigines to be able to understand and relate to the land as they do:

> The very word 'landscape' involves, from the beginning, an irreconcilable difference of viewpoint, and there seems no word in European languages to overcome the difficulty. It is a painter's term, implying an outside view, a

> separation, even a basis of criticism. We cannot set it against the reality of the earth-sky-water-tree-spirit-human complex existing in space-time, which is the Aboriginal world. (Haynes, 1998, p. 17)

In spite of this, people like Wright and Russell Drysdale did find ways to cross this barrier of language. Although many painters contributed to our sense of perceiving ourselves as being separate from the 'landscape' being viewed, they were also the first to point out the limitations of this perceptual framework (see chapter 3). Poets like Wright were particularly effective in giving voice to the growing desire for reconnection.

In her historic novel *A Timeless Land* (1941), Eleanor Dark effectively showed how white Australians have been more influenced by their contact with Aboriginal people than they have been willing to acknowledge. Noticing that the indigenous people were so obviously at home in this foreign land must have made the settlers feel uncomfortable from the start. Although Dark did not use the word 'ecology', she clearly set out to show that people intuitively search for their identity in a better understanding of the land. The word 'ecology' was not in vogue in her time, but her interest in the relationships between people and landscapes could be described as human ecology. It is appropriate to regard her as an 'ecological pioneer' because she encouraged others to think in ecological ways, even before the modern language of ecology was in common use.

The main authors of 'histories of ecology', such as Donald Worster (1994), Robert McIntosh (1985) and Anna Bramwell (1989), acknowledge that ecology-as-science is only one of ecology's traditions. In the preface to the second edition of his classic work, *Nature's Economy*, Worster writes that:

> This account will make clear that ecology, even before it had a name, had a history. The term 'ecology' did not appear until 1866, and it took almost another hundred years for it to enter the vernacular. But the *idea* of ecology is much older than the name. Its modern history begins in the eighteenth century, when it emerged as a more comprehensive way of looking at the earth's fabric of life: a point of view that sought to describe all of the living organisms of the earth as an interacting whole, often referred to as the 'economy of nature'. (p. x)

There is a consensus among such historians that at least one source of development of ecological thinking cannot be *owned* by the biological sciences. That is the 'romantic' philosophical tradition, reflected in the works of writers such as William Wordsworth, Goethe and Henry David Thoreau. Worster has traced this tradition through to the emergence of the organicist school of philosophy that was pioneered by the polymath intellectual Alfred North Whitehead. It is evident also in the more recent 'deep ecology' school of thought. This 'extended' view of ecology is central to the approach that we have taken in this book.

In Worster's account, the famous Swedish botanist Carl von Linné (1707–1778) – better known as Linnaeus – is taken as the starting point of the 'imperial' tradition in ecology, because he sought systemically to categorise and name all of the component parts of nature. According to his beliefs, the 'treasures of nature, so artfully contrived, so wonderfully propagated, so providentially supported throughout her three kingdoms, seem intended by the Creator for the sake of man' (Worster, 1994, p. 36). From this utilitarian perspective, nature exists purely for human benefit; hence the term 'imperial'. Linnaeus' emphasis on the naming of the parts of nature became central to the mechanical interpretation of the natural world that Descartes (1596–1650) had championed a century earlier. Worster contrasts this mechanical/imperial tradition with the 'Arcadian' or romantic tradition exemplified in the writings of the English amateur naturalist and curate Gilbert White (1720–1793). Although White also believed that 'nature's economy' reflected the 'wisdom of God in the creation', he advocated a humble attitude towards this creation, and a more appreciative view of nature (Worster, 1994, p. 8). He also thought that a study of natural systems should have practical outcomes for humanity, but his tone was one of reverence. Although Charles Darwin (1809–1882) was influenced by White's famous collection of letters, *The Natural History and Antiquities of Selborne* (a village in southern England), he was even more strongly influenced by the Linnaean tradition, and his work on evolution and the 'survival of the fittest' helped that tradition become dominant within the biological sciences.[5]

So when the German biologist Ernst Haeckl – a strong advocate of Darwin's theories – coined the word 'ecology' in 1866, he naturally framed it within the imperial tradition. Because of this, his contribution could be viewed as a kind of hijacking of the broader ecological understanding that had been emerging over centuries. In general, we concur with Worster's broader view of ecological thinking, but we go even further by also recognising the emergence of 'ecological wisdom' in non-western cultures and in a wider range of non-biological fields.

The word 'ecology' has achieved strong resonance in the western world because no other term existed for conveying the notion of systemic interrelationships between all living things and their environment. Its colonisation by science has, however, severely limited its value. This has resulted in a failure to acknowledge the contributions of other philosophical traditions that, because of their potential to change our understandings of our relationships with the 'natural' world, could also be called ecological. In our view of ecology we acknowledge the equally powerful, though still commonly ignored, influences of the romantic/Arcadian tradition of poets like Wordsworth and painters like J. M. W. Turner, particularly because of this tradition's roots in aesthetic sensibility and embodied experiences of nature. This integration of humans into the ecological picture contrasts with the scientific/imperial tradition, which emphasises separation and the 'objective' study of the non-human world.

In this book we argue that in Australia the Arcadian ecological tradition emerged with the work of our landscape artists and 'bush poets' (see chapters 2, 3 and 4). The imperial tradition, pioneered in Australia by Joseph Banks and the natural history movement he inspired, became more firmly established between the 1890s and the end of the 1920s and laid the foundation for the development of the scientific study of ecology (see chapter 7). We will describe how the romantic/Arcadian tradition evolved as Australia's artists and writers struggled, early on, to come to terms with our 'foreign' and initially threatening landscapes (see chapters 2, 3 and 4), and more recently as a result of growing criticism of our anthropocentric, western, philosophical traditions (see chapter 11). Meanwhile, Australians have come to play internationally significant roles in the development of ecology-as-science (see chapter 7), in related work in public education (see chapter 5), and in the application of ecological concepts to ways we relate to landscapes (chapter 8). In some fields, the romantic and scientific influences have clearly intermingled, as in the evolution of the nature conservation movement (chapter 6) and in the development of green political thinking (chapter 10).

We have also endeavoured to acknowledge the existence of a third ecological tradition that had its roots in the pre-colonial worldviews of Australia's Aborigines and Torres Strait Islanders. However, only with the emergence of the 'modern' Aboriginal land rights movement in the 1960s, and the consequent Mabo decision made in the High Court in 1992, has this had any impact on the consciousness of white Australians (see chapter 9).

We will argue that artists and writers have been at the forefront of our ecological awakening because they had the sensibility early on to become interested in the *interactions* between people and landscapes, and because some of them became deeply fascinated by what lay 'beyond the frontier' of settlement. Empathy with the other side of the frontier also motivated the preservationists within the nature conservation movement, which gathered force nationally in the early part of the twentieth century. A pressing need to find novel ways to manage 'pests' and pursue agriculture in thin and low-nutrient soils put Australian scientists working for organisations like the CSIR (later CSIRO) at the forefront of developments in applied ecological science. Similarly, landscape designers and managers, who had developed powerful observational skills while working with nature, began to pioneer even more radical approaches to resource and landscape management (see chapter 8). Although environmentalism emerged as a global phenomenon in the 1960s and 1970s, it was particularly advanced in Australia because of the legacy of preservationism and because, paradoxically, the broad notion of 'green politics' had its origins here, and not in Europe or North America as might be assumed (see chapter 10). From the 1970s through the 1980s, Australians have also been at the forefront of international developments in the articulation of ecological philosophies and worldviews (see chapter 11). From the 1960s onwards, the birth of the 'modern' Aboriginal land rights movement enabled writers, artists, philosophers and environmentalists to gain a much deeper under-standing of the ecological philosophies of indigenous Australians.

By taking a broad view of what constitutes ecological thinking we have been able to more fully appreciate that interdependency results in col-laboration more than competition. The imperial/scientific tradition has often lost sight of this in its over-emphasis on the competitive aspects of Darwin's theory of evolution, and in its focus on the survival strategies of individual species. Those who have taken a broader philosophical approach to the exploration of interdependency have been quicker to recog-nise and critique the dangerous illusion of human 'mastery over nature'. We regard the ideas and insights highlighted in this book as *ecological* pre-cisely because they challenge this notion of mastery over nature. As David Abram (1997) explains in his delightful book, *The Spell of the Sensuous*, the consequences of this illusion for our sense of self are profound:

> Human persons, too, are shaped by the places they inhabit, both individu-ally and collectively. Our bodily rhythms, our moods, cycles of creativity and stillness, and even our thoughts are readily engaged and influenced by shifting patterns in the land. Yet our organic attunement to the local earth is thwarted by our ever-increasing intercourse with our own signs. Transfixed by our technologies, we short-circuit the sensorial reciprocity between our breathing bodies and the bodily terrain. Human awareness folds in upon itself, and the senses – once the crucial site of our engagement with the wild and animate earth – become mere adjuncts of an isolate and abstract mind bent on overcoming an organic reality that now seems disturbingly aloof and arbitrary. (p. 267)

We lose sight of ourselves when we lose sight of nature. Ecological thinking can offer ways of re-establishing frayed connections, but what do we really mean by the term 'ecological thinking'. As discussed above, we take a broad view of the term 'ecology' and when we combine that with the word 'thinking' we refer to ways of thinking about the world that emphasise relationships and interconnectedness. We can learn more about ourselves by learning how complex natural systems work and the follow-ing table has been developed over a long period of time by Stuart Hill to demonstrate how this process of learning from nature suggests ecological understandings that conflict with prevailing social assumptions and practices.

Ecological understandings	Prevailing assumptions/practices
• Responsive to early indicators	• Wait for crises
• Cyclical, regenerative relationships	• Linear material flows
• Growth subject to limiting factors	• Unlimited growth (unsustainable)
• Based on solar energy	• Reliant on fossil fuels and nuclear power
• Mutualism favoured	• Competition fostered
• Functional diversity and complexity confer stability	• Simplified, highly controlled systems (dependent and unstable)
• Rich diversity of specialists, generalists, roles and niches within communities	• Few specialists and roles valued
• Most resources used for maintenance	• Production emphasised
• Importance of time and place (uniqueness)	• Events universalised (everything, everywhere, all the time)
• Gradual co-evolutionary structural change, with occasional bursts of creativity	• Fast change with few beneficiaries and many 'casualties'

Source: An earlier version of this table was published in 'A Global and Agriculture Policy for Western Countries: Laying the Foundations', in *Nutrition and Health*, 1:2 (1982), 108–117, AB Academic Publishers, England.

It is clear from this comparison that our so-called civilised society, which has increasingly come to be guided by a short-term economic bottom-line approach to decision making, operates in contradiction to the ultimately more powerful ecological bottom-line, with which the rest of nature has no choice but to conform. The costs of contradicting nature's practices are numerous. Considerably more resources are required, and the risks of harmful side effects are much greater, when solutions to problems are postponed until they reach the crisis stage. Preventative approaches, and early responses to the first signs of problems based on ecological understandings, are clearly the way to avoid such costs and risks.

We used this way of articulating ecological principles in developing the framework for this book and in building a list of pioneers we wanted to include. It also provides us with a point of reference that we can return to in our concluding chapter. However, some of the principles listed on the left of the table need more explanation:

- Nature's cycles are the means by which ecosystems become constantly replenished and renewed. Based on this understanding, it is obvious that our predominantly linear flows of materials can only result in the depletion of 'resource areas', the pollution of 'waste dumps areas', and the degradation of both.
- Whereas other species are subject to immediate limitation by many environmental factors, we humans tend to believe that limits are to be

transcended. Our ability to be able to draw on fossil reserves (until they are exhausted), move resources long distances and make substitutions among raw materials has given us the illusion that growth in our numbers, global production, and consumption per capita can all continue to increase. With ecological awareness, it is obvious that this is only possible because we are daily accumulating an enormous ecological debt, most of which (for example, loss of species and fossil fuels) can never be paid off. Because of this we will be leaving a devastating legacy for future generations.

- Nature operates on a current account, based on daily inputs of solar energy. When the sun goes down, most productive activity stops and the level of production yesterday, last year, or a century ago is about the same as today. This is because the sun is not shining for longer each day, or giving out more light. Again, with such awareness, our obsession with growth and unwillingness to place limits on our numbers and activities are clear indicators of a culture that is still psychosocially immature. Our capacity to mature is one cause for hope.

- Nature is clearly about relationships, especially those in which there are many beneficiaries – what's good for the whole community is usually most beneficial for most of its members. So there are numerous subtle and indirect ways in which the activities of each species help other species to survive. In fact, the more one observes nature and investigates its processes, the more it starts to look like the most beautiful, multi-dimensional jigsaw, which is in the process of continually changing synchronistically over time. Our almost exclusive focus on competition in nature is more a projection of our obsession with this 'expensive' and destructive form of relationship, rather than of its actual importance. Other species devote much more effort to avoiding competition than to engaging in it.

- Nature has been able to continue to put on the most amazing 'play', day after day, for billions of years, because of the rich diversity of actors (the full range from extreme specialists to extreme generalists), and because of their equally diverse, but mostly mutualistic, relationships. Things quickly break down in nature when this diversity is eroded. Put simply, essential jobs that need doing just do not get done, and the system crashes. This applies especially to the complex ways in which systems are maintained in good health.

- In nature it is commonly estimated that over 90 per cent of resource consumption is used for maintenance functions. We naively try to redirect these resources to production functions, and then wonder why systems are continually degrading and breaking down. This is clearly illustrated, over the past century, by the steady decline in response of agricultural production to increases in fertiliser applications to soil. Whereas we have been able to continue to increase the production of fertilisers, this has only compensated for a small part of the increasing degradation of the soil that is associated with our input-output approach to farming.

- Probably more important than all of the above, with respect to efficiency and effectiveness of resource use, is the observation that in nature activities occur at particular locations, only at very precise times and under very restrictive conditions. Trying to make them happen at other, non-ideal, places and times requires levels of inputs that are orders of magnitude more than would be needed if conditions were more supportive. The homogenisation of our world and lifestyles – everything available everywhere, all of the time – is phenomenally expensive to achieve, and also boring to experience! It has been estimated that by living locally, seasonally, and as ecologically informed conservers, we could save over 90 per cent of the resources we currently waste.
- Structural change in nature is also variable in time and space. However, nearly all such change is extremely slow, with only the occasional sudden burst of evolutionary creativity; and probably most of this evolution is, in fact, co-evolution, with more than one species being the beneficiary.

Of course, a willingness to violate or ignore ecological understandings is not unique to Australian society. It is, however, likely to have particularly devastating consequences in a country like Australia where soils are fragile and climatic conditions so unpredictable. Other countries where the soil is rich and climate more predictable may feel the consequences less directly, but the difference is only a matter of degree and timing because the root problem is the unsustainable belief in endless growth. This particular modernist illusion has achieved a global hegemony.

In this economic environment, ecological imperatives are paradoxically regarded by most as far too radical and challenging to respond to, even though they emphasise the very conservative concepts of conservation and the 'precautionary principle'.[6] Although the word 'sustainability' has recently achieved popular currency, and most people would claim to support the concept of 'ecologically sustainable development' (ESD), few would be able to provide much evidence of its application in their lives. Thus, although ecology may have found a place in our language, it is yet to be embraced proactively in most of our lives. Concepts like ESD can so easily be turned into meaningless rhetoric based on naive understandings or, in some cases, on devious manipulations. We can take heart, nevertheless, that there is a growing interest in both nature and the concepts of ecology. The contributions of the ecological pioneers profiled in this book – and of others we were unable to cover – are making a difference. These ideas from the past are starting to influence both present practice and future prospects. Whether we realise it or not, we are all involved in a battle of ideas that will determine the fate of humanity, and if nature loses, so too will we (but not *vice versa!*).

For a small and rather isolated nation, Australia has been able to make a disproportionate contribution to global ecological thought, especially in the latter half of the twentieth century. For example, Australians have been at the forefront of innovative approaches to pest control, land management, bush regeneration, nature education, and environmental action. The Sydney trade unionist Jack Mundey was the first to use the term 'green politics'. P. A. Yeomans was the first to apply comprehensive design principles to landscape management; and David Holmgren and Bill Mollison built on this design approach to create the even more sophisticated system of ecosystem design known as 'Permaculture'. Australians like Val Plumwood, Ariel Salleh, Warwick Fox and John Seed have been prominent in the articulation of ecofeminist and deep ecology worldviews. A central paradox that is explored in this book is that Australians have made contributions like this precisely because the early experience of European settlement in Australia began so badly. Those who were 'unsettled' by this experience often found themselves pushed into the margins of mainstream thought and practice. It has been from this position that many have been able to generate ideas and practices that have not only challenged the *status quo*, but also offered viable alternatives. While some of these pioneers experienced a surprising interest in their work and ideas outside Australia, many more made their main contributions within Australia, in art, literature, film, and more, and were only subsequently recognised internationally.

This book aims to celebrate the achievements of Australian ecological pioneers. Although much has been written about the sorry story of environmental degradation in Australia, much less has been said about the sound ideas upon which we might build better futures. Some of the stories recounted here are well known and some of the people highlighted have had their achievements widely acknowledged. However, their ideas have not always been recognised for their ecological value, and no attempt has been made to integrate the diverse ecological understandings from different fields into a meta-narrative about the overall development of ecological understanding in Australia. Here we have endeavoured to offer fresh interpretations of the better-known pioneers. Our other selections include those that we feel have been either inadequately acknowledged or ignored. Whereas many will be well known to Australian audiences, other 'quiet achievers' will be less familiar.

In researching these people and their stories, we have been impressed by the importance of the cross-fertilisation of ideas. For example, the young Vincent Serventy was inspired by Philip Crosbie Morrison's informed and passionate radio broadcasts of the 1950s (see chapter 5). Eleanor Dark's husband was a colleague of pioneer conservationist Myles Dunphy, and the Darks shared Dunphy's love of the Blue Mountains. Judith Wright was related to Romeo Lahey, a pioneer conservationist in Queensland, and wrote about him in the first edition of her magazine *Wildlife*. When she was working at the Jacaranda Press in 1963, Wright discovered a manuscript by Aboriginal poet Kath Walker (later called Oodgeroo Noonuccal) and urged its publication. The two women became close friends and collaborators, and this encouraged Wright to resume a childhood interest in Aboriginal culture and she continued to deepen her understanding of their perspectives on the land throughout the rest of her life. David Holmgren's development of Permaculture design principles was partly inspired by the innovative landscape design initiatives of P. A. Yeomans (see chapter 8). Tasmania was the site for the launching of the world's first green political party in the early 1970s, and the pioneers of green politics in Tasmania picked up on the ideas of Jack Mundey, who also influenced the shaping of green politics in Europe (see chapter 10).

As well as uncovering examples of such direct influence and collaboration, our study has also highlighted some of the ways in which these separate thinkers have arrived at similar conclusions. Judith Wright, Charles Birch and Val Plumwood have all talked passionately about the need to have compassion for other animals. Eleanor Dark, Judith Wright and Russell Drysdale were all aware of the importance of Aboriginal understandings of the land long before the modern land rights movement brought that to the attention of many more white Australians. Val Plumwood's suggestion of a project in which local communities would be invited to reconsider the relevance of local place names echoes the work that Myles Dunphy did to preserve locally relevant names.

In researching these pioneers we also were saddened to realise that there were so many missed opportunities for collaboration and cross-fertilisation. For example, pioneer animal ecologist Charles Birch was largely ignored, and 'unused', by the nature conservation movement. People like P. A. Yeomans, Ted Trainer and Stephen Boyden often felt isolated, as did many others. Birch found it necessary to segregate his 'two careers' as an animal ecologist and organicist philosopher. Most of those involved in the preservation of wilderness and the pioneers of 'restoration ecology' in urban and rural landscapes seem to have had little interest in each other's work and ideas.

Like Donald Worster, our aim has been to construct a historical narrative around the life and work of exemplary ecological thinkers. We have not set out to produce a comprehensive account of all of Australia's ecological pioneers, or of any particular fields, partly because we did not want our readers to lose the plot by being overwhelmed by the cast of characters.

Also, by being selective rather than comprehensive, we have been able to focus on the development of ideas and practices by individuals over time, taking into account the contextual factors that influenced them. We have also been able to explore the ways in which their ideas began to influence others. We actually began with a much longer list of names before settling on the present combination of people and chapter themes. Although other writers would undoubtedly have selected different casts of characters, we believe that the people we selected and the stories we have told effectively illustrate the main paths of development of ecological thinking in Australia. We have endeavoured to cover a reasonable span of history in rough chronological order and to acknowledge the contributions made in a wide range of fields, without straying too far beyond our own expertise. Within these parameters, we had to leave aside significant areas of endeavour, such as religion, filmmaking, architecture and music. We have not ventured into the complex world of contemporary arts. In constructing a social history of influential ideas we have also been more interested in the creation of ideas than in their separate life as texts after their creation.

We are indebted to a number of people for their feedback on our first attempt to construct a set of chapters around people and themes. In particular, Vincent Serventy, Alex Colley (from the Colong Foundation in Sydney), Stephen Dovers (Australian National University), and Roslynn Haynes (University of New South Wales) provided us with some valuable leads, and we used their feedback to reshape our proposal. In the process of developing the concept of the book, we also were reminded that it would be culturally inappropriate to focus on particular people in writing about the origins and impact of the Aboriginal land rights movement. Aboriginal people prefer to acknowledge the efforts of the Gurindji and Yolngu people as a whole, rather than any of their individual leaders.

Having selected the people and themes to be highlighted in the book, we were determined to maintain an open mind as to what ideas might be uncovered during our research. We did not want to be colonised by the reputations of famous people, nor did we want to miss ideas that had been forgotten or ignored. We struggled with a number of key questions. Which ideas can be classified as ecological, even though they may not have been regarded as such before? What similarities and differences would we find in the emergence of the ideas in different fields? Would the breadth of our approach help to extend the notion of ecology and further a realisation among the population of its relevance today, or would this project prove to be much too ambitious?

Fortunately, we did not have to wait long for confirmation that our approach was working. Starting with chapters on Aboriginal land rights (chapter 9) and landscape art (chapter 3), we discovered a common theme around the idea of learning and communicating stories that are embedded in particular landscapes (introducing the notion of 'storied landscapes'). We found both artists and writers who grappled with the human experience of time and space in an ancient land. We encountered Anglo-European Australians who had developed a deep love and respect for both the 'new country' and its Aboriginal people, and who could not imagine belonging anywhere else. We met scientists who had developed new conceptions at the edges of mainstream science, and political activists trying to create a new kind of green politics at the edges of the established political system. We found various people trying to redesign human systems to better reflect the needs and processes of natural systems. And in a contemporary philosopher, a radical cartoonist and a 'hippie' environmental activist we found common insights concerning the ways in which people can so easily become alienated from their environments (see chapter 11).

In the final chapter (chapter 12), we have attempted to pull together some key understandings that have emerged from this work as a way of showing that ecological thinking continues to provide an important critique of prevailing attitudes and practices. We have endeavoured to show that emergent ecological understandings can help us prepare for a better future. However, such understandings need to be grounded in a deep appreciation of the work of the ecological pioneers. In other words we look backwards in order to look forward with more powerful vision. In constructing the book we have become more conscious of the power of ideas, particularly when ways are found to link them with other, similar, ideas. When we set out on this journey we thought we would end up with more precise conclusions than we have (see chapter 12), but the story became rich and complex and difficult to conclude. However, we became more confident that what we constructed would resonate with a wide variety of readers in a wide variety of ways. We are also confident that it will leave our readers thirsty for more knowledge of the people and ideas that we have introduced. We believe that all of us need sources of inspiration from the past as we head towards our uncertain future.

| THE COLONISATION OF AUSTRALIAN NATURE AND EARLY ECOLOGICAL THOUGHT

INTRODUCTION

Joseph Banks established his reputation in England by being the chief naturalist on board Captain James Cook's *Endeavour* when it sailed around South America, across the Pacific Ocean to the east coast of Australia in 1768–70. He proved himself a master of Linnaean taxonomy by naming many hundreds of species of plants in every land visited by the *Endeavour*. Cook and Banks decided that Australia was essentially unoccupied and Cook claimed what he called New South Wales in the name of the English monarch. It was Banks who initiated the concept of *terra nullius* to suggest that the land was open for claim and he extended this concept to Australian plants and animals because he saw little commercial potential in them. Despite his gloomy report on the character of Britain's new colony, Banks went on to play a significant role in the affairs of the colony. He encouraged explorers to continue collecting new specimens to be sent back to museums and gardens in England.

Charles Darwin made virtually the same journey as Banks some 65 years later in a similar role as naturalist on board the *Beagle*. Whereas Banks had been essentially a scientist at the service of empire, Darwin used his experiences on the *Beagle* to revolutionise the way biologists conceive the world. Darwin was not impressed aesthetically by Australian landscapes but saw them as ecosystems rather than as simply home to exotic 'new' species.

Among the early white settlers were some who took the time to study the natural world they had arrived in and some of them came to love what they found. More commonly, however, they were driven by the desire to 'conquer' the wilderness; to 'tame' the 'new' land. To their surprise, some of the common soldiers in the war on the wilderness found that they were being changed by the land they had come to conquer. A subtle shift in attitudes towards Australian landscapes can be detected in the work of the romantic poet Adam Lindsay Gordon and of landscape painters working in Victoria in the 1860s and 1870s. The lead-up to federation of the Australian nation in 1901 stimulated wider public interest in the nature of Australian identity. Although ecology as a scientific discourse did not reach Australia until the 1920s ecological understandings of the relationships between people and the land began to emerge in various fields in the last decades of the nineteenth century.

FIRST CONTACT: JAMES COOK AND JOSEPH BANKS
When Captain James Cook first set sail for the South Seas on board the *Endeavour* in 1768 he knew there was a large landmass – known only as *Terra Australis Incognita* – which, as the name suggests, was largely unexplored by Europeans. Dutch and English sailors had started to map the west coast of this 'unknown' continent; enough to know that it was a very large land mass. But they were not impressed with the arid landscapes they encountered. With the eastern side of the continent unexplored, the assumption was that a significant landmass centred largely in the temperate zone would have to be an attractive proposition for an expanding empire. It might be home to natural treasures or to a society producing tradable goods. Even if the west coast had disappointed, the east coast might be a different proposition altogether. Cook left with instructions from the Royal Society of London, which sponsored his voyage, to approach the continent from the east; map the east coast; and, if the land was not already claimed by others, claim it in the name of the king (Lines, 1991, pp. 21–23).

Cook was to reach the great south land by sailing around the tip of South America and then across the Pacific. On the way he was to stop in Tahiti so that an astronomer on board could observe the transit of Venus across the face of the sun; data that would be used to calculate the distance between the earth and the sun (Lines, 1991, p. 21). As far as the Royal Society was concerned this was primarily a scientific expedition. But if Cook had the opportunity to add another colony to the British empire then it was his patriotic duty to do so. Otherwise it would be claimed by some other European power.

James Cook was a brilliant mariner, renowned for his attention to detail and willingness to embrace new ideas (Mackay, 1985). Charts that he made of the rugged coastline of Newfoundland were still in use well into the nineteenth century. He experimented with new navigational techniques and discovered that access to fresh fruit could curtail outbreaks of scurvy among his crew. He was said to possess powers of intuition that enabled him to detect hidden reefs (except the one the *Endeavour* collided with on the Great Barrier Reef in 1770) and the early signs of a change in weather. He became the darling of the Royal Society, being awarded the society's prestigious Copley Medal, because he was a skilful sailor who also had an interest in developments in various domains of science. The society sent him out in vessels with such worthy names as *Endeavour* and *Resolution* and he pioneered a tradition of exploration that continued to, and beyond, the voyages of *HMS Beagle*, which carried Charles Darwin to the same parts of the planet in the 1830s.

One of those who benefited from sailing with Cook on the voyage of the *Endeavour* was Joseph Banks who, at the tender age of 25, was chosen to head the ship's party of scientists. By then he was a graduate of Eton and Oxford and had been as far as Newfoundland on a naval vessel that made regular patrols of the North American fisheries. Although not trained as a botanist, he first discovered an interest in plants when he made the acquaintance of the head gardener of the Physick Garden in London (Mackay, 1985, p. 16). He learnt the Linnaean system for classifying plants from Dr Daniel Solander, a student of the system's inventor – Swedish botanist Carl von Linné (better known as Carolus Linnaeus). Solander is the man credited with introducing the Linnaean system to England and Banks became Solander's star student in London. The voyage that Banks took to Newfoundland in 1766 seemed to intensify his interest in exotic plants and he applied for the job on the *Endeavour* soon after returning to England. No doubt his application was helped by the fact that he was a man of 'independent means', having collected his inheritance on the early death of his father. He was from the rural gentry and his forebears had established a reputation for being innovative farmers. So, Banks was seen to be of 'good stock'. However, he must have impressed Captain Cook with his enthusiasm because he was given an authority on board the *Endeavour* that was much greater than that of his teacher Daniel Solander, who also joined the expedition. Cook invited Banks to work closely with him as a general adviser and as overseer of the expedition's scientific work. Banks was in charge of a team that included Solander, an astronomer, and two natural history artists.

Banks and his party of naturalists tackled their task with enthusiasm. At the first stop – the Madeira Islands in the Atlantic – they listed 230 species of plants; 25 of which had not been previously named by the Linnaean system (Lines, 1991, p. 21). They also described 18 species of fish. At Tierra del Fuego, Banks compiled a list of 104 species of flowering plants, six ferns, and 34 mosses and lichen. When they reached the Pacific his team warmed up by collecting specimens of 255 plant species in the Society Islands without venturing beyond the coast. They also confined themselves to the coast in New Zealand but managed to name some 400 species.

Banks first sighted the coast of Australia on April 19, 1770, but was not very impressed with what he saw (Lines, 1991, p. 22). He set foot on the continent when the *Endeavour* put into what Cook called Botany Bay just over a week later. His landing boat was challenged by an irate resident who hurled several spears at the alien invaders. The rather hostile group of Aborigines only withdrew when some of their number were wounded by musket fire. Banks recorded that he came across a hut in which some children were huddled in fear. He threw some 'beads, ribbands, and cloth' into the hut and took away some spears to add to his collection. His party also collected so many specimens of plants that he struggled to preserve them all. As the *Endeavour* sailed north from Botany Bay, Banks spent most of his time working with the material already collected. Emerging from this work every now and then, he became convinced that '[f]or the whole length of coast we saild along there was a sameness to be observed in the face of the countrey very uncommon' (citations from Banks' journal are from Lines, 1991, pp. 22–23). He next set foot on land in the tropical north, after the *Endeavour* had collided with a reef. While the ship was being repaired, his team collected a further 300 plant specimens. Although Banks and Solander were the first European botanists to visit this part of the world they seemed to tire of their task. Less than three months after their experience at Botany Bay, almost a month before the *Endeavour* parted company with the coast of Australia, Banks wrote that he was:

> Botanizing with no kind of success. The Plants were now entirely compleated and nothing new to be found, so that sailing is all we wish for if the wind would but allow us.

Banks had quickly reached the conclusion that the new land had little to excite the interest of Europeans. The animals were exotic and the plants very unusual but there seemed little that could be eaten or even used by humans. Despite his close encounter with Aborigines, he thought they held 'a rank little superior to that of monkeys'. He suggested that the land was essentially unoccupied, thus coining the notion of *terra nullius*. He wrote that the soil was 'intirely void of the helps dervd from cultivation' that it was unlikely to 'yield much to the support of man'. Cook also decided the land was essentially unoccupied and, without any indigenous people present, he staked Britain's claim to all he had seen in a little ritual carried out on Possession Island off the northern-most tip of the coast on August 21, 1770. He called the coast New South Wales and specified that his sweeping claim included all the 'Bays, Harbours, Rivers and Islands Situate upon the said coast' (in Lines, 1991, p. 23).

Although Cook and Banks were jointly responsible for declaring the 'new' land a *terra nullius*, it seems that Cook had a more charitable view of the indigenous people than Banks. It was a view akin to the Rousseauian notion of the 'noble savage' for, in his journal, Cook wrote that they seemed:

. . . far happier than we Europeans . . . they live in a tranquillity which is not disturbed by the inequality of the condition. The earth and sea of their own accord furnishes them with all the things necessary for life. (in Grove, 1995, p. 317)

However, Cook knew that European settlers would not even try to live in such 'tranquil' ways and, as already noted, Banks was pessimistic about the prospects for establishing agricultural practices on the sandy soils of the 'new' land. So the reports they both submitted to the colonial authorities in England discouraged thoughts of establishing a new colony so thoroughly that the idea was not seriously considered until the House of Commons set up a committee of inquiry in 1779 to explore the prospects for sending off shiploads of surplus 'felons' to other parts of the empire. By this time Cook had met his death in Hawaii, and Banks suggested that Botany Bay could be a good site for a penal colony because the prisoners could do the hard work required to grow food in such difficult circumstances and the settlers could anticipate little or no serious resistance from the existing inhabitants.

By 1779 Banks' opinion carried some weight. His work on the *Endeavour* and his association with Captain Cook first established his reputation but it was enhanced in the years that followed. Although he never again made a long journey, he was seen as an established expert on the colonies and their potential. He became a virtual director of the Kew Gardens, which aimed to research and propagate plants collected from all parts of the world. He acted as adviser to the Board of Trade and to the East India Company. Banks added to his reputation by experimenting on his family farms. He introduced merino sheep to England and experimented with new strains of wheat and other grains. He published research work conducted on his farms on the problem of blight in corn. He had made his name as a man of science; a characterisation that could improve one's social standing, if not financial status, in his day. He rose to become Sir Joseph Banks, Privy Councillor. He was president of the Royal Society of London from 1778 until his death in 1820.

Banks and his peers in England were essentially scientists at the service of empire. They did little to disguise the fact that their scientific work was aimed at unlocking the commercial potential of the colonies. Through his association with the East India Company, Banks became particularly interested in expanding the base of agricultural industries in India (Mackay, 1985, chapter 7). He was involved with long-running efforts to introduce tea from China into northern India and in establishing a silk and cochineal industry at Madras. He was convinced that plants could be easily transported from one tropical country to another and saw potential in introducing 'useful' species from the West Indies to India. For this purpose he helped establish botanical gardens in India that would explore ways of propagating such species in Indian conditions.[1] Banks was certainly not unaware of the fact that European colonisation of foreign lands could have disastrous consequences for local plants and animals. He had, for example, personally witnessed the sorry state of the Atlantic island of St Helena that had been logged to desolation by short-sighted settlers (Grove, 1995). His interest in establishing botanical gardens was not solely to explore the commercial potential of 'exotic' plants, but also to preserve species that might be lost due to the impact of colonisation (Grove, 1995). He was very sure that scientific knowledge would enable colonial authorities to avoid past mistakes or rectify any that might be made in the future.

When the decision was made to establish the colony at Botany Bay, Banks was asked to equip the First Fleet with food plants suitable for the climate and conditions. He chose plants that would thrive in a climate 'similar to that of southern France' (in Malouf, 1998, p. 54), leaving room for some trial and error by including several varieties of most plants. As David Malouf has pointed out (1998, p. 53), many of these plants originated in places like China and Persia before 'travelling westward to Europe'. Others – such as tomatoes, corn and potatoes – had been taken to Europe from the Americas in a west to east trajectory. To survive their earlier travels they had to be highly adaptable plants, but in Australia they struck soils that lacked important minerals and the manure of large herbivores. They also struck unpredictable weather patterns. It took the First Fleeters some time to find niches where some of their food plants could survive – almost running out of food in the process. While still depending on supply ships from England, they managed to work out which plants and animals would do well enough in a land that had little in common with southern France. They did not explore the potential of native plants that had been used by indigenous people for thousands of years. They were bent on carving out familiar farms in unfamiliar settings; radically transforming landscapes into approximations of the Arcadian visions they had in their minds' eyes.

What they failed to understand; what scientists like Banks failed to predict, was that the predominantly skeletal soils were easily leached of their nutrients and easily disturbed by hard-hoofed animals. So even when the pioneer farmers had some initial success there was no guarantee that it could be sustained. This was no simple process of pushing back the frontiers of wilderness, but rather the ebb and flow of successes and failures – a colonisation of the land by the people and of the people by the land. The land was not well prepared for this battle. As writer/farmer Eric Rolls has put it:

> No wheel had marked it, no leather heel, no cloven foot – every mammal, humans included, had walked on padded feet. Our big animals did not make trails. Hopping kangaroos usually move in scattered company, not in

damaging single file like sheep and cattle . . . Every grass-eating mammal
had two sets of teeth to make a clean bite. No other land had been treated so
gently. (in Malouf, 1998, p. 54)

Yet it also proved to be more resistant to transformation than many had
imagined.

Although Banks never returned to Australia after his experience on
the *Endeavour* he continued to play a major role in the affairs of the colony
well after the First Fleet had been despatched. He had a say in the appoint-
ment of every governor from Arthur Philip to William Bligh in 1806 (Lines,
1991, p. 25). He was also seen as the patron of all efforts made to collect
scientific information about the new land. In 1801 he named the ship
Investigator that would carry its commander Matthew Flinders and his crew
on a circumnavigation of Australia and he ensured that Flinders took with
him a party of qualified scientists and the best equipment available (including a greenhouse for transporting live specimens). Banks tried to instil a
passion for collecting among the explorers and pioneers in the colony and
was rewarded for his efforts by being sent the pickled head of Pemulwuy, a
troublesome leader of Aboriginal resistance to the colony at Sydney. Much
of what was collected was simply sent back to England to be added to the
bulging collections of museums and other institutions; often to be left in
boxes without being sorted or analysed, let alone displayed. Flinders had
with him a fine artist called Ferdinand Bauer who made an exquisite set of
drawings and paintings of plants and birds that were collected. However,
his work remained largely unknown in Australia until put on display by
the Sydney Museum in 1998.

As a scientist, Banks was little more than a collector and cataloguer.
As Tom Griffiths (1996) has said, European approaches to natural history
and anthropology in the eighteenth and nineteenth centuries grew out of a
mentality of hunting and collecting – a desire to fill up endless museums
with 'exotic' trophies of the hunt. Despite all the collecting and naming
done by Banks and all those he stirred into action in Australia, he was able
to make little serious contribution to an understanding of how Australian
ecosystems work. His concept of *terra nullius* extended to the flora and
fauna of the land. If there was nothing of commercial value in it, the land
held little lasting interest. He obviously felt that 'men of science' had
nothing to learn from indigenous Australians, whom he saw as being little
above the rank of monkeys. The habit of discounting the ecological knowledge of people who had lived in the land for countless generations
subsequently meant that explorers and pioneers frequently met their death
without seeking the advice of people who lived in the landscapes where
they perished. Ironically, the discounting of Aboriginal knowledge meant
that Banks and his successors failed to notice the commercial potential of

Australian plants and animals. Instead, the colonists introduced the hard-hoofed animals that destroyed ground-covering plants and caused extensive soil erosion. They introduced plants and animals that quickly became pests in the absence of predators of other forms of 'natural control'. In short, they acted out of ecological ignorance.

CHARLES DARWIN IN AUSTRALIA

Charles Darwin was a much more important scientist than Banks. He arrived in Sydney on board the *Beagle* on January 12, 1836, near the end of a long journey during which he had collected information that would not only transform his own understanding of nature but revolutionise biology as a science. The *Beagle* had followed a route very similar to that of the *Endeavour* and Darwin had been employed to perform a role very similar to that of Banks (Brent, 1981). But that is where the similarities end. Darwin, a devout Christian, was disturbed by the contrast between his experience of rural England and his observations of 'untamed' nature in South America and on the Galapagos Islands (Worster, 1994). He became less convinced that nature behaved according to some divine plan and more convinced that a struggle for survival was leading to adaptations that had become increasingly refined over time. This, of course, led to the theory of natural evolution outlined in the book that Darwin published more than 20 years after the voyage of the *Beagle – On the Origin of Species* (1859).

While the *Beagle* was in Sydney, Darwin hired 'a man and two horses' to take him on an excursion over the Blue Mountains to Bathurst, during which he met a party of 'good-humoured' Aborigines and participated in a kangaroo hunt.[2] During his journey he was struck by the 'extreme uniformity' of the vegetation but noticed that the forests had a very different structure to those of Europe:

> The trees nearly all belong to one family, and mostly have their leaves placed in a vertical, instead of, as in Europe, in a nearly horizontal position: the foliage is scanty, and of a peculiar pale green tint, without any gloss. Hence the woods appear light and shadowless: this, although a loss of comfort to the traveller under the scorching rays of summer, is of importance to the farmer, as it allows grass to grow where it otherwise would not. (*Voyage of the Beagle*, p. 280)

This passing comment showed a level of perception of the unique systemic characteristics of dry woodland forests in Australia that was largely absent from the comments of earlier naturalists who had concentrated on collecting and naming species. Darwin made it clear that such forests did not appeal to his aesthetic tastes because they fail to produce the 'glorious spectacle' of new foliage bursting forth in spring and the peeling bark of the eucalypts gives the forest a 'desolate and untidy appearance'. Earlier he had written with much more reverence about standing in tropical forests in Brazil. For him Australia was a bit of an anticlimax. When he arrived in Sydney he wrote that he felt comfortable being back in an English colony but when the *Beagle* finally departed King George Sound on its way across the Indian Ocean, he wrote:

> Farewell Australia! You are a rising child, and doubtless someday will reign a great princess in the South; but you are too great and ambitious for affection, yet not great enough for respect. I leave your shores without sorrow or regret. (*Voyage of the Beagle*, p. 289)

Reaching Australia at the end of a long and emotionally draining journey, Darwin may not have been in the frame of mind to tackle another set of challenges and his time in Australia was limited. He sensed a fairly desperate struggle for identity for the English migrants who were trying to make their home in foreign and challenging landscapes. He made it clear it was not a challenge he was willing to tackle. He wanted to retreat to the familiarity of his homeland; he was feeling homesick. However, his lack of enthusiasm for Australian landscapes also suggests that the most perceptive of Englishmen struggled at this time to appreciate their unique attractions.

Darwin's comments about Aborigines were fairly typical of the time. After his trip to Bathurst he bemoaned the fact that they were rapidly decreasing in number and noted that 'Wherever the European has trod, death seems to pursue the aboriginal' (*Voyage of the Beagle*, p. 282). At King George Sound he had the privilege of witnessing a large corroboree and, although he noted that the 'natives were in such high spirits and so perfectly at their ease', he thought the performance was a 'rude and barbarous scene' and a 'festival of the lowest barbarians' (*Voyage of the Beagle*, p. 283). In this case, his observations were no more perceptive than those of Banks. The indigenous people of the land once again were arrogantly dismissed by another 'man of science'.

While Darwin was not interested in tackling the huge challenge of settling in Australia, there were, of course, many who were and some of them learnt to love what they found. An interesting example was the painter John Glover who arrived to join his son in Tasmania in 1830 after a long and successful career as a landscape artist in England. Glover made a detailed study of the flora he found in the new country and made impressive efforts to interpret the ambience of the forests in his work. He painted some rather idyllic scenes that included images of Aborigines recreating happily in their natural settings (well after they had been all but annihilated by the genocidal policies of the colony's authorities). While he presented a rather romanticised interpretation of his new home, he clearly felt inspired by what he had found.

A GARDENER IN A WILDERNESS: GEORGIANA MOLLOY

An even more interesting story is that of Georgiana Molloy, who arrived with her husband Captain John Molloy to settle in the Swan River colony (Perth) in 1830 (Lines, 1994). Caught up in 'Swan River fever' generated in England by a press campaign launched by the colony's chief advocate and founder, Captain James Stirling, the Molloys joined other 'free settlers' who saw an opportunity to build a prosperous life in an 'empty land'. As a man of military rank, John Molloy was accorded a significant status in the fledgling colony and he and his wife soon became friends with Stirling and his wife Ellen. However, they felt unexpectedly hemmed in by the 'strange' 'bush' that seemed to keep the settlement pinned to the coast. While Georgiana Molloy found the native trees to be intriguing, her husband could not muster much enthusiasm for becoming a farmer in such landscapes and he complained bitterly about the flies, fleas and mosquitoes in the Swan River settlement. Recognising that his friend was not happy, Captain Stirling suggested to Molloy that he and another settler who had arrived on the same boat, John Bussell, should lead a party to establish a new settlement at Flinders Bay, on the southern coast just around from Cape Leeuwin. Stirling was not one to waste time and so, even though Georgiana Molloy was heavily pregnant at the time, he accompanied the Molloys and Bussells and their support staff on the journey south. The Molloys were certainly not unhappy with what they found. The climate was much milder than that of Perth, with much more reliable rainfall. Although thick forest – predominantly jarrah (*Eucalyptus marginata*) with pockets of giant karri trees (*Eucalyptus diversicolor*) – grew right to the edge of the sea, the luxurious growth suggested that farming would also be easier here. The settlers set about building the township of Augusta.

The Molloys' initial enthusiasm for their new home was severely tested when their new baby, born in a tent just weeks after their arrival, died nine days after birth. Feeling terribly isolated in her grief, Georgiana Molloy first turned to the Bible for solace and then decided that she would feel more comfortable if she could establish a splendid garden like those she had enjoyed working in back in Scotland and England. So, while still living in a tent, she threw herself into the task and soon had nice displays of peach blossom and orange nasturtiums as she eagerly awaited the arrival of more seeds from England. From her growing patch of familiarity, Molloy gazed towards the surrounding forests containing karri trees so big that it took a dozen men three days to fell just one of them. From the outside, such forests looked inaccessible and she wrote:

> This is certainly a beautiful place – but were it not for domestic charms the eye of the emigrant would soon weary of the unbounded limits of thickly clothed dark green forests where nothing can be described to feast the imagination and where you can only say there must be some tribes of natives in those woods and they are the most degraded of humanity. (in Lines, 1991, p. 62)

By April 1836, Georgiana Molloy had established a much-admired garden. In that month she gave birth to a fourth child, a son John, to join her two surviving daughters, and in the same month she received a surprise present from a stranger in London that would change her life. It was a box of English seeds for her garden sent by a cousin of Captain Stirling's wife Ellen, a horticulturalist named Captain James Mangles. What was unusual about the gift was that it came with a note asking Molloy to return the box filled with Australian seeds and pressed flowers. This presented a big challenge for Molloy because she had rarely ventured into the nearby forest and was unable to identify distinct native species. However, she decided to tackle the challenge, and in the spring of 1837, when the flowers were at their peak, she made her first foray into the bush. After her first tentative mission she found she looked forward to each outing, taking along her daughters who went 'running like butterflies from flower to flower'. Tragically the growing joy of her new work was cut short when her son fell into a well and died. Again she sought solace in the Bible, but this time she felt less isolated and resumed her collecting about a month later. The trips became a daily event and her knowledge about the diversity of life in them grew enormously.

Captain Mangles was delighted by the collection he received and he and Molloy maintained an active correspondence about her new discoveries. In one of her letters to Mangles, Molloy thanked him

> for being the cause of my more immediate acquaintance with nature and variety of [Australian plants] . . . but for your request, I should never have bestowed on the flowers of this Wilderness any other idea than that of admiration. (in Lines, 1991, p. 63)

The Molloys left Augusta in 1839 to take up a land holding at Vasse, on the other side of the peninsula between Cape Leeuwin and Cape Naturaliste to the north. The isolated settlement at Augusta had been faltering, particularly after a bitter feud between John Molloy and the Bussell brothers, John and Charles. Molloy continued botanising around her new home and her reputation for knowing about native species spread but her heart remained at Augusta. Tragically, she died in April 1843, aged just 37, after a prolonged period of ill-health that followed the birth of her seventh child. As she approached the certainty of her own death, she wrote about her changing religious convictions, criticising the rather vengeful and brooding perception of God presented in Milton's *Paradise Lost*; a God that tested religious devotion by inflicting pain, particularly on women (Lines, 1994, pp. 329–331). As a result of her joyful experiences in the Australian bush, Georgiana Molloy had come to see life as a precious gift and an opportunity to revel in the extraordinary and unexpected beauty of God's creation without trying to conquer it.

While Molloy had approached Australian plants with the same classificatory tools as Joseph Banks and his disciples, she had been led to a feeling of reverence rather than a desire to exploit nature's 'bounty'. This put her more in line with Gilbert White's Arcadian notion of 'nature's economy' rather than the imperial tradition represented by Banks. However, her passion and reverence were not widely shared in colonial Australia. Most early settlers thought it was their manifest destiny to tame the wild and 'primitive' land to make it fit for 'civilisation' (Lines, 1991). Few allowed themselves to be diverted from the task of chopping down trees long enough to absorb the beauty of what they were destroying. They were driven by the values of the Enlightenment, which suggested that human 'progress' must involve a 'mastery over' nature. They came to conquer wilderness,[3] but were often left feeling disappointed and deluded. By the 1870s it had become a fashion to write romantically about the 'bush experience' (see chapter 4), but for many foot soldiers in the war on wilderness, the experience was not a pleasant one. A poem published in a Western Australian newspaper in the 1870s[4] expresses this frustration clearly:

> The poets, and writers of stories
> And sylvan refrains
> All sing of the bush and its glories
> In rapturous strains . . .
>
> Now, many years I've been a rover
> O'er bush and o'er plain,
> Yet little I've seen of its clover
> And much of its pain . . .
>
> The fruits of our toil which we seize on
> Expectant and glad,
> We find like fruits grown out of season,
> Small, shrunken and sad.

Despite the frustrations involved in trying to 'tame the great south land', generations of settlers pushed on with the project because they deeply believed there was no other way. As David Horton has said, those who continued to bring to the task a colonising mentality saw Australian 'scrub' as 'rubbish vegetation' waiting to be cleared (2000, p. 13). They tended to believe that the indigenous people had wasted their opportunity to do something with the land and felt that it needed 'hands of industry with a strong work ethic to buckle down and lick it into shape and make up for lost time'.

David Malouf (1998) has argued that the very idea of Australia began as 'an experiment in reformation using the rejects of one society to create another' and he cited an entry in Charles Darwin's diary to suggest that this experiment 'succeeded to a degree unparalleled in history' (1998, p. 13). Given opportunities they never would have had in the Old World – especially to own land – these 'vagabonds' proved to be extremely determined and very creative. According to Malouf, 'Before long, and well within the first two decades, all the amenities of an advanced society had been conjured up' with very few of the appropriate resources. What was created drew heavily on the traditions of the Old World – not just England but Europe more broadly – but it was also different. As Malouf puts it, the settlers displayed

> . . . inventiveness and industry beyond the mere making do; . . . a determination to create a world here that would be the old world in all its diversity, but in a new form – new because in these new conditions the old world would not fit. (1998, p. 21)

Yet, for all their determination and inventiveness, European colonisers had never before found themselves so 'out of place'. 'Unlike the Americas', Malouf writes, 'we found ourselves in the opposite hemisphere to Europe with opposite seasons, different plants and animals and birds, and disorienting stars overhead' (1998, p. 32). This disorienting experience left a strong mark on the culture that was created and Malouf puts a positive spin on this by saying that 'our uniqueness may lie just here, in the *tension* between environment and culture rather than what we can salvage by insisting either on the one or the other' (1998, p. 32). However, we must never forget that the environment has suffered grievously under the clumsy hands of those who acted out of ecological ignorance.

Of course, those who came to conquer the land were transformed by it as well. The 'vagabonds' and 'misfits' from Europe tried to create the landscapes of their Arcadian dreams, but they were also determined to create a more open and egalitarian society than the one that had rejected most of them. Ancient and resistant landscapes colluded with this aspect of their dreams because the difficult conditions undoubtedly had an equalising effect on the settlers, especially in the 'frontier' regions. As the indigenous people had learnt long before, and encoded into their culture, the land demanded collaboration rather than competition. Alongside the myths surrounding the 'daring' adventures of failed explorers, new myths built around the now cherished notion of 'mateship' began to emerge. These were the myths that would gradually replace those reflecting a deep fear of hostile environments. And with them was born a new mythological character, the Australian 'bushman'.

THE BUSHMAN'S POET: ADAM LINDSAY GORDON

The romantic poet Adam Lindsay Gordon may not have been the first to write this new character into his work, but he probably did it more successfully than any other writer who was around to witness its 'birth'. At the height of his influence, in the 1860s, his published poems were often referred to as 'the Bushman's Bible'.

Adam Lindsay Gordon (1833–1870) was born into a family deriving from Scottish gentry that had settled in the Cotswolds region of England. In 1853, when he was 20 years old, he was sent to South Australia by a father who thought that he needed more discipline and responsibility in his life. An accomplished horseman, he first joined the mounted police and then went into business as a horse-breaker, visiting remote farms across South Australia and into Victoria. He joined steeple-chase clubs and eventually established a reputation as Australia's finest steeple-chaser (Sladen, 1934). Having come into some of his inheritance, he was a man of 'independent means' before he was 30 and could follow various interests. He tried his hand at creating a new cattle station in Western Australia but this did not last long. In 1865 he became a member of parliament in South Australia but a few years later was living in Melbourne as a member of the horse-racing fraternity. While living in Melbourne he learnt that his claim to an estate in Scotland had been denied and soon afterwards he shot himself near the coast at Brighton. People who met him on the day of his suicide said there was nothing to suggest he was deeply depressed. His main concern had been to make arrangements for delivering the corrected proofs of a book of poems to his publisher.

Gordon's own life had been one that might fire the imagination of a romantic. Coming from 'good stock', he gained a reputation as being courageous (especially on a horse), skilful, and charming. Even the circumstances of his death seemed to add to the legend.[5] His poetry was fast-moving and stirring; sharing his love of horses and horsemanship. He wanted to celebrate the achievements of those who had come to conquer the land – the volunteers in the war on the wilderness. Yet he also found that he was touched by the landscapes that his characters moved within. According to Brian Elliott (1967), Gordon helped his readers appreciate landscapes by making them part of the story he told. Instead of writing about the splendours of sunset and dawn, he was at his best 'when he describes (or merely evokes) some other time of day: hours of sunshine, or glimpses of starlight . . . when he is capturing routine impressions . . . when he records what everybody can see . . . [and] the fact that everyone does see just these things makes them memorable' (Elliott, 1967, p. 79).

Gordon's poetry does not rate very highly in reviews of Australian literature written in the second half of the twentieth century. If he is referred to at all, it is generally only in relation to his influence on later writers. Yet, he did receive warm popular acclaim in his lifetime (Humphries and Sladen, 1912) and for many years afterwards was considered to be Australia's greatest poet. As late as 1934 a bust of Gordon was added to Poets' Corner in Westminster Abbey, alongside those of poets such as Tennyson, Coleridge, and Wordsworth, and at the time he was called the 'National Poet of Australia' (Sladen, 1934). Clearly, there was something in his poetry that resonated with his readers. Perhaps it was the fact that he was interested in the interaction between people and the land; this was certainly a theme picked up by some of the later writers who paid tribute to him (see chapter 4).

After Gordon's suicide, the established writer Marcus Clarke was asked to write a preface for a volume of his poems, published in 1873. Clarke chose as his central theme the strong and disturbing influence that Australian landscapes can have on those who venture into them. Perhaps he was influenced by the pathos surrounding Gordon's death in suggesting that those who wrote about Australian landscapes would be unable to write 'airy' or 'freshly happy' pieces because 'the dominant note of Australian scenery' is a 'Weird Melancholy' (in Thornhill, 1992, p. 146). The mountain forests, he wrote, are 'funereal, secret, stern . . . Their solitude is desolation. They seem to stifle, in their black gorges, a story of sullen despair' (p. 146). 'Hopeless explorers' venture into landscapes of 'frightful grandeur' to be overwhelmed by 'their sentiment of defiant ferocity' (p. 146). Clarke's comments about the 'Weird Melancholy' of the Australian bush have been widely quoted in reviews of Australian literature and art to suggest that most writers and artists of his time perceived the bush as being a depressing environment. Many critics have suggested that this view of the Australian bush reflected a feeling of alienation caused by a longing for the distant, romanticised landscapes of 'mother England'. The renowned art critic Bernard Smith has suggested it was more a product of fear and guilt – fear of the unknown and guilt stemming an unspoken suspicion that the bush was haunted by the ghosts of indigenous people slaughtered by settlers (Griffiths, 1996, pp. 3–4). According to Smith's analysis, writers like Clarke were projecting the pain and anxiety associated with the settlement experience, conflated with a guilt associated with the conviction that the Aboriginal people were now facing certain extinction.

However, the widespread analysis of Clarke's passage has been very selective and one-sided because he went on to say that the very weirdness of the bush was the thing that made it so intriguing. When you read the passage in full it is clear that Clarke did not see the melancholy evoked by the bush as being something that should be avoided, but rather an emotional response to be engaged with because an exploration of it could be deeply rewarding. He continued:

> In Australia alone is to be found the Grotesque, the Weird, the strange scribblings of nature learning how to write. Some see no beauty in our trees without shade, our flowers without perfume, our birds who cannot fly, and our beasts who have not yet learned to walk on all fours. But the dweller in the wilderness acknowledges the subtle charm of this fantastic land of monstrosities. He becomes familiar with the beauty of loneliness. Whispered to by the myriad tongues of the wilderness, he learns the language of the barren and uncouth, and can read the hieroglyphs of haggard gum trees, blown into odd shapes, distorted with fierce hot winds, or cramped with cold nights, when the Southern Cross freezes in a cloudless sky of icy blue. (in Thornhill, 1992, p. 146)

THE EARLY LANDSCAPE ARTISTS

Clarke's more sensitive appreciation of Australian landscapes had been reflected even earlier in some landscape painting, particularly in Victoria. From the late 1850s through to the 1870s, for example, Eugene von Guérard was able to produce some landscape representations that seemed to stir unexplored emotions in their viewers, including some who were themselves aspiring artists. Von Guérard, a migrant to Australia from Austria, managed to convey in his work his unambiguous admiration for the unique qualities of the scenes he painted and he travelled widely in Victoria and in New South Wales. An 1857 painting he made of a scene inside the forest at Ferntree Gully in the Dandenong Ranges east of Melbourne had such an impact that the art critic of the *Argus* newspaper, James Smith, stepped up to lead a public campaign against a proposal to log the area when that threat emerged in 1861 (Bonyhady, 1994, 2000b). In the following years Smith continued to be inspired by von Guérard's work in becoming a strong public advocate for nature conservation. Von Guérard's influence on the next generation of Australian artists increased when he subsequently became the teacher of landscape painting at the National Gallery in Melbourne.

In the 1860s, the globe-trotting artist Nicholas Chevalier also fetched up in Australia for a while and he produced some fresh and sensitive representations of what he saw. At a similar time a Swiss migrant photographer, Louis Buvelot, turned his hand to painting and produced some evocative images of scenes around Melbourne, which also inspired the next generation of artists, including Tom Roberts and Frederick McCubbin (see chapter 3). Clearly a shift in attitudes towards Australian landscapes began in the 1860s and gathered force during the 1870s. In a sense, the work of artists like von Guérard, Chevalier and Buvelot represented a departure in Australian art but it would be an exaggeration to call it a movement. It took the Heidelberg painters, working together under the leadership of a strong personality in Tom Roberts, to turn the proto-movement into something more coherent and significant (see chapter 3). While artists like von Guérard and Buvelot in Victoria and John Glover in Tasmania managed to capture more 'authentic' representations of Australian landscapes they were still trapped within the legacy of the picturesque style of painting that depicted landscapes from an outside perspective (for discussion of the picturesque style, see chapter 3). Roberts returned from a stint in Europe with new ideas and techniques that had challenged picturesque painting in its European heartland, and he encouraged his colleagues to join him in bush camps where they would attempt to paint what they saw from the inside. Thus, the Heidelberg painters were responsible for a deeper shift in perception about the relationships between people and landscapes in Australia. Soon after the Heidelberg painters began to make their mark in the late 1880s, a group of writers, following in the tradition pioneered by Adam Lindsay Gordon, began to cluster around the *Bulletin* magazine in Sydney (see chapter 4). The *Bulletin* writers were determined to tackle the cultural cringe towards all things English in order to assert a new and proud Australian identity. For both the Heidelberg painters and the *Bulletin* writers the emerging Australian identity was rooted in the experience of distinctive, and challenging, landscapes.

Whilst trying to take a fresh look at the relationship between Australian identity and the land, the *Bulletin* writers maintained the dominating *terra nullius* attitude towards the indigenous people. Their notion of an Australian identity was a very narrow one and the *Bulletin* magazine remained at the forefront of campaigns to keep Australia 'for the white man' (a slogan actually used for a time on the magazine's masthead), in other words free of a 'takeover' by immigrants from nearby Asia. This notion of creating and preserving a white Australia was aimed more at Asia than at the indigenous people, because conventional wisdom of that time was that the latter would surely die out in due course. However, it exacerbated the 'erasure' of a people and a history that could have taught the white settlers much about land and identity. It is rather ironic that in the 1890s – a time when white Australians invariably believed that they had nothing to learn from the indigenous people – Aboriginal images and icons began to appear in the world of advertising.[6] At one level this might be seen as nothing more than a rather token effort to locate distinctive Australian imagery, but it might also suggest that, subconsciously at least, white Australians were beginning to engage with the thought that Aboriginal people did find a sense of belonging in a land that they wanted to feel part of. However, it would be left for writers like Eleanor Dark, Xavier Herbert and Judith Wright and painters like Russell Drysdale and Arthur Boyd to articulate this thought overtly. Even more importantly, it took a revitalisation of Aboriginality – in the form of the 'modern' land rights movement launched by Aboriginal communities in the Northern Territory in the 1960s (see chapter 9) – to make a significant number of white Australians want to grapple with this realisation. Of course, issues related to the identity of a nation established by an act of seizure will always be contentious. Yet, since the last decades of the nineteenth century there has been a growing realisation that Australian identity must be connected to the land itself and the people who have understood it best.

Although the word 'ecology' was coined by the German biologist Ernst Haeckl in 1866, seven years after the publication of Charles Darwin's most important work, *On the Origin of Species*, the application of ecological ideas within science did not occur in Australia until the 1920s (see chapter 7). However, ecological concepts were beginning to emerge in other language and in other fields in the last decades of the nineteenth century, largely because of the emerging discourse on land and identity. David Malouf (1998) has suggested that by the 1870s Australia had become less colonial and more 'confidently provincial'. However, he argues, this confidence was steadily eroded by the economic depression of the 1890s, the death of so many young men in World War I, the impact of the Great Depression between the wars, and the shock of coming under direct attack during World War II. By the 1940s and 1950s, the nation had become timid and inward-looking; self-confidence and a 'sense of play' only re-emerging after the 1960s. Whether or not this is a fair characterisation of changing national moods, it is a useful reminder that the notion of steady 'social progress' is a delusion born of Enlightenment thinking. Social change is more complex and paradoxical than that.

In arguing that new ecological understandings began to 'emerge' in Australia from the 1860s onwards, we are not suggesting that they were readily embraced or even widely understood. What we are arguing, more modestly, is that it is possible to detect in this period the stirrings of a new sensibility, or attentiveness, that could challenge the characterisation of Australia's natural heritage as being inferior and low in value. This change was not without its contradictions, of course, and so we argue that a fascination for the melancholy that could be triggered by contact with the Australian bush was at the same time an acknowledgement of fear of the unknown, a sense of guilt about what had been done to the land and its indigenous inhabitants in the process of 'settlement', and an engagement with the emotions triggered by the emerging sensibility. Such contradictions are manifest in Marcus Clarke's 'discovery' of 'the subtle charm of this fantastic land of monstrosities'.

| CHAPTER 3 | SEEING THE LAND IN A NEW LIGHT: *People and Landscapes in Australian Art* |

INTRODUCTION

In the last two decades of the nineteenth century, Australia moved inexorably towards the goal of becoming a single nation rather than a set of rather disparate colonies. As discussed in chapter 2, the stirrings of a new sensibility towards Australian landscapes started to emerge in the 1860s and 1870s through the work of artists like von Guérard, Chevalier and Buvelot and writers like Gordon and Clarke. In the 1880s and 1890s, this new attentiveness towards the Australian bush began to dovetail with attempts to define a distinctive Australian identity, which, in turn, was fuelled by a growing desire to overcome an exaggerated deference towards English culture (the so-called 'cultural cringe').

While some painters, like those mentioned above, had begun to produce more authentic representations of Australian landscapes, they remained faithful to the picturesque style of painting that was strongly European in its origins and inspirations. Ironically, it was a trip back to Europe by the budding artist Tom Roberts that created the impetus for a new movement in Australian landscape painting because he brought back to Australia the techniques of impressionism and a desire to paint *en plein air*. Together with Frederick McCubbin he organised artists' camps in the bush that gave birth to the 'Heidelberg school' of painters – the first real movement in euro-Australian art. Roberts was the leader of the Heidelberg painters because he had returned to Australia with some fresh ideas, but also because he had the force of personality and an interest in fostering the talents of some younger artists like Arthur Streeton and Charles Conder. He also had some entrepreneurial ability that served the group very well. Like his friend and colleague McCubbin he had an interest in depicting human stories unfolding in bush settings (narrative painting) but unlike McCubbin he travelled widely in Australia to explore the interactions between people and a variety of landscapes.

Once he moved beyond the influence of Roberts and McCubbin, Streeton moved also away from a focus on particular stories to a reinterpretation of the Arcadian vision of landscapes that seemed to hold the promise of a peaceful coexistence of people and the land. In a later period he used his work to campaign for the protection of forests endangered by human greed but at the height of his influence his 'pastoral' paintings had become a very conservative influence in Australian art. It was not until artists like Hans Heysen, Sidney Nolan and Russell Drysdale began their treks to arid, inland landscapes that some new perspectives began to emerge. With his strong interest in story-telling and in the colonisation of people by the land, Drysdale followed closely in the footsteps of Roberts, while Nolan began a new trend by focusing on stories that had become part of a national mythology. It has been suggested that Nolan was the first white Australian artist to emulate Aboriginal artists in being able to visualise landscapes 'from the inside out', although Margaret Preston had come close to this with her earlier interest in the design-like qualities of Aboriginal representations of patterned landscapes.

In the 1950s, artists like Drysdale, Nolan, Arthur Boyd and Albert Tucker began to attract international attention for their distinctly Australian paintings. They were seen as the leaders of a modern movement in Australian art yet, ironically, their interest in looking more deeply at the land and the stories embedded in particular landscapes meant that they were moving closer to an Aboriginal perspective drawn from ancient cultural sources. However, it was not until a group of artists from the Aboriginal settlement of Papunya, north-west of Alice Springs, began working with acrylic paints on board and canvas in the 1970s that the artistic potential of the Aboriginal perspective began to properly reveal itself and reach a broad 'audience'. Beginning with traditional iconography, but becoming increasingly abstract, the 'dot paintings' of the desert communities came to constitute an even more distinctly Australian school of art that combined ancient and modern influences in an ecological interpretation of the inter-relationships between people and their non-human environments.

'THE HEIDELBERG SCHOOL': ROBERTS, McCUBBIN AND STREETON
Towards the end of 1886 three Melbourne-based artists – Tom Roberts, Frederick McCubbin and Louis Abrahams – set up camp on Houstens' farm, just outside the tiny village of Box Hill, which was then in the rural fringe east of Melbourne. These three men were not experienced campers. A painting of the scene by Roberts (*The Artists' Camp*) shows McCubbin and Abrahams looking rather awkward as they prepare food beside the wall of their tent, which was pitched so that it would have sucked in smoke from a

nearby fire. Roberts also recalled later that when they were carrying supplies into the camp, Abrahams had slipped on a log crossing a creek, delivering some of their food to a watery grave (McQueen, 1996). But the camp was a significant event in Australian art history because it brought together artists who wanted to get out into the bush and experiment with the new European fashion of painting *en plein air*. They sought to challenge their own perceptions as to what constituted paintable subjects. Instead of seeking out the grand vistas and pastoral scenes that were popular with English landscape artists they sought to paint among the trees and, in so doing, they found a new kind of beauty in a neglected environment. As Roberts later fondly recalled, the experience was a pleasant one:

> the land sylvan as it ever was . . . tea-tree along the creek . . . young blue gums on the flat bit alongside, and on the rise our tent. The evenings after work . . . the chops perfect from a fire of gum twigs . . . the 'good night' of the jackies as the soft darkness fell . . . then talks round the fire, the 'Prof' [McCubbin] philosophic – we forgot everything, but the peace of it. (in Moore, 1934, p. 70)

As the paintings of Roberts and McCubbin show (Abrahams confined himself to drawings), they saw beauty in the peeling bark of Victorian bluegums, set among the dry sclerophyll forest that had been an anathema to earlier generations of artists in Australia. Because they pitched their tent *among* the trees and came with a determination to paint whatever they saw, they painted *inside* the forest rather than from an outside perspective. They began a new movement in Australian painting that was eventually dubbed the 'Heidelberg school' by visiting American art critic Sid Dickerson.

Tom Roberts, in particular, became a colourful and influential figure in Australian art. In addition to his craftsman-like skills as a draughtsman and portrait artist, he injected an enthusiasm for the impressionistic style of painting he had witnessed during a stint in Europe and his entrepreneurial skills created opportunities for a range of other artists. After arriving in Australia in 1869 as a 13-year-old from England he found ways to love his adopted country and he subsequently travelled widely in his efforts to capture the essence of the land and its people. He had been given the nickname 'Bulldog' by his younger brother Dick when they were still boys, but he was happy to carry that name into adulthood, probably because it hinted at the steely determination that enabled him to become a leader.

'Bulldog' Roberts first met Fred 'The Prof' McCubbin when they were both attending an art class at the Artisans' School of Design in Carlton in the early 1870s (Bonyhady, 1985, p. 136). In 1874 they were together in a painting class run by Austrian-born landscape painter Eugene von Guérard at the National Gallery in Melbourne. For Roberts the art classes started as a hobby because he had a responsibility to support his mother and two younger siblings (their father had died before they migrated from England). But his interest grew as his talent blossomed and, by the end of the 1870s, he had resolved to become a professional artist. McCubbin, one year older than Roberts, was born in Melbourne in 1855. He was the first Australian-born artist to break into the professional ranks.

Thanks to the presence of painters like Nicholas Chevalier, Eugene von Guérard, and Louis Buvelot – all of them born and trained in Europe – Melbourne had become the centre of Australian landscape painting by the 1870s. However, art classes that Roberts and McCubbin attended (other than those run by von Guérard) were still dominated by a narrative painting style, that primarily took subjects from the Bible or classical literary sources. This style of painting, popular in England, was made popular in Australia after the National Gallery purchased a work by the English artist George Folingsby in 1864. Folingsby's reputation in Melbourne was such that he decided, in 1879, to accept an invitation to succeed von Guérard as the master of the National Gallery's painting class. So, early in their careers, Roberts and McCubbin were influenced by accomplished European-trained landscape painters, like von Guérard and Buvelot, and by the narrative style popularised by Folingsby. Both influences would remain with them in their later work.

By 1880, Roberts had decided that he needed to further his art education by going to Europe and in the following year he left for the 'old country'. Over the next four years he attended art classes in London and Paris, visited all the galleries he could find and participated in intense discussions that were taking place about contemporary art trends. Like all artists interested in landscapes he was influenced by the work of J. M. W. Turner, but he was equally fascinated by the controversy over the work of emerging American artist James MacNeil Whistler. On a trip to Paris, he was deeply impressed by an exhibition of work by French artist Jules Bastien-Lepage who, as a pioneer of the emerging *plein air* fashion, was seeking to capture 'truth to nature'. Artists like Bastien-Lepage became critical of carefully constructed landscape paintings and wanted to capture more fleeting impressions of what they saw on their outdoor excursions. According to Robert Hughes (p. 53), 'Bastien-Lepage worked with a square headed brush, laying his areas directly and accurately without painting "wet into wet" or retouching. This directness attracted Roberts: it was an adaptable way of working, suitable for depicting objects for which "tradition had not yet prescribed a way of handling and regarding . . . " '.

Near the end of his time in Europe, Roberts accepted an invitation from a Melbourne acquaintance Thomas Maloney (who would later become a prominent Australian socialist) to join him and another Australian artist John Peter Russell on a trip to Spain. During their journey, Roberts and Russell were able to witness the work of two Spanish *plein air* artists, Barrau and Casas,[1] and dabble in the style themselves. But, just as importantly for Roberts, they visited the Prado in Madrid and saw the work of Diego Velásquez. Velásquez had been a forerunner of the 'truth to nature' school, and his work had influenced both Whistler and French impressionist Edouard Manet. Roberts was interested in his focus on people at work in the outdoor environment, a theme that Roberts would turn to repeatedly in his own work.

The trip to Spain also stimulated Roberts' desire to return to Australia. He told Maloney that the sunshine and scenery often reminded him of his adopted country. On one occasion he found a gum tree and crushed its leaves to remind himself of the smells of the Australian bush. He was beginning to feel surprisingly homesick and ready to tackle the challenge of painting in Australia again.

Roberts arrived back in Australia in 1885 and obtained work as a photographer's assistant (useful for learning about the effects of light on subjects). He renewed his friendship with 'Prof' McCubbin who was fascinated to hear about the new trends in European art. McCubbin told Roberts that the area around the village of Heidelberg, on the fringe of Melbourne, had become popular for day-tripping landscape painters. Following his experience in Spain, Roberts suggested the idea of a camp where they could work together without distraction. According to Moore (1934),[2] they chose Houstens' farm near Box Hill because Roberts had been there as a 14-year-old (not long after migrating from England), and fondly remembered the hospitality of the Houstens. It was an inspired choice. As inexperienced campers they were able to rely on the generosity of their hosts, yet they found themselves embedded in the bush.

During this first camp, McCubbin created one of his most famous paintings, *Down on his luck*, inspired by a poem by Adam Lindsay Gordon. Roberts asked McCubbin's sister, a visitor to the camp, to pose for *Summer Morning Tiff*, which, together with its sequel *Reconciliation*, would also become famous. Although the paintings were finished when they eventually returned to their city studios, they used the brush-stroke techniques that Roberts had learnt in Europe and a wide palette to try to capture the effect of the light among the trees at specific times of day. These were narrative paintings in the tradition that the two artists were taught, but a starring role for the young, peeling, bluegums gave them a distinctive character. From being a backdrop to a story, the environment had become more clearly a part of the story. And the new techniques helped the artists capture the ambience of warm and sunny Australian summer days.

The experience of working together was obviously a good one because, when they had had their fill of Box Hill, Roberts, McCubbin and Abrahams moved together to a rented cottage near the seaside resort of Mentone on Port Phillip Bay. From here they were able to make forays back into Melbourne and plan painting excursions on the weekends when there were more people about. During a solo Sunday walk, Roberts happened across Arthur Streeton, a precocious young art student at the National Gallery painting school, 'out on the wet rocks, doing a very free sketch of sea and shore' (Moore, 1934). Impressed by his obvious talents and energy, Roberts invited the younger artist to join him and his colleagues at the cottage. According to Moore's account, Roberts and Streeton began their life-long association in a chance encounter, but this story bears the character of a myth. By this time McCubbin was the master of painting at the National Gallery school where Streeton was a student and he had been nurturing the young artist's talents for some years. Of course, it is possible that Roberts met Streeton independently but, more likely, 'The Prof' engineered a meeting between the two men. Whatever the circumstances, Streeton quickly became a devoted member of the Roberts 'gang', earning the nickname 'Smike' after a Dickens' character who liked to follow.

Streeton may have taken immediately to Roberts and his entourage, but he was not so impressed with their seaside cottage. He later wrote disparagingly about its 'vile hammocks' and 'puce-coloured walls' (Moore, 1934, p. 75). But he did enjoy the outings, later recalling that:

> On Sundays we took a billy and chops and tomatoes down to a beautiful little bay which was full of fossils, where we camped for the day. We returned home during the evening through groves of exquisite tea-trees, the sea serene, the cliffs at Sandringham flushed with the afterglow.

When the 'golden summer' of 1886–87 was over the artists retreated to Melbourne. Roberts had set up a studio in a chamber in Collins Street and he invited 'Smike' to join him. There they could complete works begun *en plein air* while Roberts also earnt a living as a portrait painter. The work was quite lucrative and Roberts was able to plan a trip to Sydney to renew his acquaintance with the founding editor of the *Bulletin*, J. F. Archibald, who had sailed to Australia on the same ship as Roberts in 1885. During his 19 days in Sydney he was also able to meet Julian Ashton, founder of a famous Sydney art school, and another precocious young artist, Charles Conder, with whom he spent some time painting around Coogee (where Conder was staying with relatives). Conder's father had sent him to 'the colonies', in the company of an uncle, so that he would be obliged to find a job rather than indulge his passion for art. Instead he relied on the support of his relatives and continued to paint until finding a job as an illustrator on *Illustrated Sydney News* in 1887. Conder used his charm to make friends in the Sydney art circle and, according to Moore's account (1934, p. 75), he met Roberts at the home of an artist called Madame Roth.

Whatever the circumstances in which they met, Roberts and Conder established a quick rapport and spent time together while Roberts was in

Sydney. Roberts later recalled that on one occasion they met for a drink at a wine bar overlooking Mosman wharf. According to Roberts' account, the conversation was so engrossing that the hourly ferries seemed to be arriving every ten minutes and 'three hours seemed to last but half an hour'. As they talked, Roberts – an 'incurable sketcher' (Moore, 1934, pp. 71–72) – produced a likeness of Conder and gave him the nickname 'K' so that he might be equipped for joining the Melbourne gang (where nicknames seemed to be mandatory). Conder accepted Roberts' invitation to come to Melbourne later that year where he could work in a new, purpose-built studio in the Grosvenor Chambers in Collins Street.

Back in Melbourne, Roberts and Streeton began to plan for a new summer camp. They were interested in returning to the Heidelberg area and they heard that fellow artist David Davies could get access to a large, empty house on an estate known as Eaglemont. Davies' brother was the secretary of a company that happened to own the title to the estate and it turned out that the house had never been completed or occupied. But it was spacious – eight bedrooms – and much more comfortable than the Box Hill tent of the previous year (although the beds were made from saplings and flour sacks (Moore, 1934, p. 72). The estate had landscaped gardens and a view across the Yarra Valley to the blue ranges in the north and east (a view made famous by Streeton in *Still glides the stream and will forever glide*). Roberts, Streeton and Conder began the camp where they worked on 'glare paintings' that depicted landscapes saturated with sunlight. Later, they were joined by other artists like Davies and Walter Withers, who can be regarded as members of the Heidelberg school even though their painting styles were quite different to those of the famous four – Roberts, Streeton, McCubbin and Conder. The Eaglemont camps were repeated over two summers and Streeton, in particular, spent a lot of time there. In this conducive environment he was able to produce works like *Still glides the stream*, which was bought by the Art Gallery of New South Wales soon after it was finished, and *Golden Summer*, which he sent to Paris where it received a 'Mention Honourable' at the famous Paris Salon of 1892.

Because they were no longer camped among the trees, the Eaglemont paintings were different to those painted by Roberts and McCubbin at Houstens' farm. Streeton, in particular, was less interested in trees and more interested in capturing the effect of bright sunlight on more open landscapes. Roberts' influence can be seen in a focus on men at work in the outdoors, but Streeton led the way in developing the 'glare painting' style that became strongly associated with the Heidelberg school.

While they were working at Eaglemont, Roberts turned his fertile mind to the task of how to win more public acceptance for the new form of art they were producing. Since arriving back from Europe in 1885 he had been impressed by the growing profile that art was achieving in Melbourne.

In October 1885, Buxton's artistic stationery warehouse moved into a prime location opposite the Melbourne Town Hall in Swanston Street and the building, with a gallery on the first floor, became a mecca for artists and art followers. Around the corner in Collins Street, new commercial galleries began to open and, as mentioned, purpose-built art studios were opened in Grosvenor Chambers in Collins Street in 1888. Roberts decided that they should take advantage of all this to put on an eye-catching exhibition in Buxton's gallery and so he began discussing the idea with Streeton and Conder. Between them they brainstormed ideas for capturing the spontaneous character of *plein air* painting and they came up with the idea of displaying works that would appear to have been painted on the lids of cigar boxes – popular, portable, surfaces for artists making outdoor sketches. Their colleagues were asked to submit 'mood' works that displayed the quick brushwork of impressionistic painting. For their own works, Roberts, Streeton and Conder ordered planks of red kauri from a local timber yard to make bulky frames that would emphasise the small size of the paintings. This idea became the now-famous '9x5' exhibition, so named because of the size (in inches) of most of the works exhibited.

The theme of the exhibition may have been spontaneity, but Roberts left nothing to chance. He managed to convince a drapery firm to contribute some lavish cloth to decorate the hall and he organised a group of Melbourne girls to serve afternoon tea. The exhibition included 182 works by seven artists (not all of them in the small format). Roberts' old partner, McCubbin, contributed just five works (possibly reflecting some doubts about the bold venture). Most of the works were by Roberts and his two young associates.

The exhibition opened on August 17, 1889, and, predictably, it was greeted by a very hostile review in the *Argus* by Melbourne's foremost art critic James Smith. Smith had already made it plain that he was not a fan of the 'unfinished' works that were beginning to emerge from Europe and he was determined to prevent a spread of this disease into Australian art. In his review he wrote:

> ... like primeval chaos, 'it is without form and void'. To the executant it seems spontaneous and forcible. To the spectator it appears grotesque and meaningless ... of the 180 exhibits catalogued on the present occasion, something like four-fifths are a pain to the eye. (Moore, 1934, p. 75)

Smith had earlier been an admirer and promoter of Eugene von Guérard and Louis Buvelot. As mentioned in chapter 2, he had also made his mark at the *Argus* in the early 1860s by writing editorials condemning plans to log the forest at Ferntree Gully, made famous by an 1857 painting by von Guérard. He admired Australian artists who strove for authentic representations of Australian landscapes and he was well aware of the intrinsic beauty of the Australian bush. However, he was artistically conservative and he failed to see that impressionistic techniques held the potential to make artists more sensitive to their surroundings. Certainly it is true that none of the individual works in the '9x5' exhibition would become enduring masterpieces, although they would eventually be sought after by collectors and galleries for what they represented in the development of Australian art. The point of the exhibition, which Smith missed, was to celebrate a style and 'mood' of painting that the artists found fresh and stimulating. They wanted to share their enjoyment with a broader public, hoping to create a new receptiveness to more enduring works that would follow. They wanted to exhibit works that reflected an engagement with changing moods and a sensitivity to places rather than polished works of art. In their catalogue for the exhibition, the artists explained that when you are painting outdoors, no two hours are the same because the light conditions are constantly changing. To highlight this they included works painted at the same location but at different times of day and evening and in different weather conditions. Impressionistic painting is interested in capturing the unique and emergent qualities of a particular scene, just as ecologists are interested in the particularities of specific ecosystems and their changing qualities over time.

Not fazed by Smith's predictable response to the exhibition, Roberts actually used it to the artists' advantage by having a copy of the review pasted on a board outside the gallery to create more public interest. The ploy worked as the exhibition attracted significant public attention and most of the works were sold before it was over.

The Eaglemont camps ended when the summer of 1888–89 came to a close. In April 1890 Conder left for Paris where he achieved considerable success (never returning to Australia) and later that year Roberts and Streeton moved to Sydney where they linked up with another artists' camp at Little Sirius Cove, near Mosman. The Heidelberg movement had effectively ended, but a new era in Australian painting had been born. When they later reflected on what was special about the experience of the camps, both Streeton and Conder emphasised the spirit of collaboration that had been engendered (Bonyhady, 1985, p. 136). It was a rare opportunity for Australian artists to collaborate rather than compete and Streeton, in particular, gave the credit to Roberts for being an inspirational leader.

Because of his role in the Heidelberg movement, Roberts has sometimes been dubbed 'the father of Australian landscape painting'. This is rather misleading because there were, as already mentioned, accomplished landscape painters before Roberts and his colleagues who were finding 'authentic' ways to represent what they saw. For example, John Glover was a highly accomplished English landscape artist who arrived in Tasmania in 1830 at the age of 64 and used all his experience to produce some very striking interpretations of what he saw. Another artist trained in England, Conrad Martens, was able to create a domestic market for his landscapes soon after he arrived in 1835 and he survived as a professional artist for nearly 40 years. The globe-trotting Nicholas Chevalier (who was born of Swiss parents in St Petersburg and studied art in Lausanne, London and Rome) spent 13 years in Victoria and produced the first local painting – *The Buffalo Ranges* (1864) – to be bought by the National Gallery in Melbourne. Eugene von Guérard migrated from Vienna in 1852. Roberts himself said that Louis Buvelot, who gave up portrait photography in 1866 and started painting around Melbourne, was the first artist to capture the essence of the Australian landscape.

A panel of artists and art critics who joined Julie Copeland for a discussion on Australian landscape art on Radio National in August 1998[3] agreed that Glover and von Guérard, in particular, deserve more credit than they have been given for pioneering distinctive ways of painting Australian landscapes. One of the panel, Andrew Sayers, suggested that these two painters 'start to paint an Australian landscape that really couldn't be anywhere else in the world'. According to another panellist, Christopher Allen, both von Guérard and Buvelot had success in Australia that they could not have had if they stayed in Europe and he suggested that Glover produced his finest work in Tasmania. Apart from his direct influence on Roberts and McCubbin, von Guérard is also said to have inspired the work of W. C. Piguenit, whose painting *The Flood in the Darling* (1888) was described by Julie Copeland as 'one of the most sublime landscapes ever painted in Australia'. In an earlier article, Tim Bonyhady said that Piguenit can clearly be called a conservationist even though his concept of conservation, true to his time, was concerned with the wise use of natural resources rather than the conservation of nature of its own sake ('The Art of the Wilderness', 1994).

Clearly, artists like Glover, Martens, Chevalier, von Guérard, Buvelot and Piguenit all found inspiration in Australian landscapes and some, like von Guérard and Piguenit, used their work to promote a conservation message. Several of these artists might well be called the 'father' of Australian landscape art (or maybe it is a case of multiple births). However, what the Heidelberg painters, led by Roberts, were able to do was to break out of the picturesque mode of representing landscapes in order to explore the *interactions* between people and their environments (an ecological idea).

The picturesque genre in landscape art was first pioneered in Europe by the seventeenth-century French artist Claude Lorraine. This genre, which remained a dominant influence in all of Europe throughout the eighteenth and nineteenth centuries, grew into a formula for constructing landscapes with trees used to frame more distant views and people used only to emphasise the scale of the scene. While working in England, Glover had said that he wanted to become known as 'the English Lorraine'[4] and he continued to use the genre in his Australian work. Popular watercolours produced by Conrad Martens during his long career were clearly formula-driven. When the honorary secretary of the New South Wales Academy of Art, Eccleston du Faur, organised a series of camps for artists and photographers in the Grose River Valley in 1875, one of the first things the participants were asked to do was to participate in felling trees that blocked more distant views of 'some grand crags with broken foreground and picturesque skyline' (see Bonyhady, 1994, pp. 170–180). Piguenit participated in one of these camps without, apparently, participating in the axe work. However, he did remain faithful to the picturesque genre, as did his mentor von Guérard and his Victoria-based colleagues Chevalier and Buvelot.

It has been argued that the picturesque genre lingered so long in England because it represented a longing for a disappeared ideal of pre-enclosure rural life (Taylor, 1999). In Australia, by contrast, the picturesque represented the construction of a future ideal in a new Garden of Eden. As Ken Taylor has put it:.

> The representations of the picturesque were part of the imaginative occupa-
> tion of the landscape prior to the physical occupation ... [I]t was the
> manifest destiny of European settlers to turn awaiting nature into an Eden,
> a pastoral Arcadia. So people saw what they wanted to see. Representations
> of an Arcadian ideal resulted from a state of mind where memory and
> allusion played primary roles. (1999, p. 51)[5]

So, even if artists from Glover to Buvelot and Piguenit managed to achieve more authentic representations of what they saw they were still, perhaps unwittingly, engaged in a kind of aesthetic imperialism – the imposition of an ideal hatched in the Old World onto the colonised land. Their style of painting did not encourage them to explore the transformative power of the land on those who had come to 'tame' it.

Of course, the Heidelberg painters did not pioneer the break from the picturesque genre. This has already been achieved in Europe as Roberts discovered during his travels. Roberts and his colleagues simply borrowed the *techniques* of impressionism; they certainly did not pioneer a new genre of art. However, new techniques and a commitment to paint *en plein air* enabled them to revel in their surroundings without worrying about what would come out of that experience. Much more than their pre-decessors they were able to paint scenes from the inside out and were able to pay more attention to the feelings evoked by the experience of certain landscapes. The kind of breakthrough that this achieved can be seen in the way in which the so-called 'glare paintings' of the Heidelberg group captured the ambience of landscapes that are commonly saturated by sunlight.

Roberts and McCubbin, in particular, were still interested in the human dramas that were being acted out in natural settings, but, like their contemporaries in Europe and North America, they were sensitive to the influence on such dramas of their settings; a theme of interaction. It has been noted that McCubbin reflected the melancholic themes popular in writings about interactions between people and the Australian bush. This can be seen in paintings like *Down on his luck* and *Bush burial* and he also picked up a popular theme of the day – the plight of a child lost in the bush. However, it should also be noted that many of the Heidelberg paintings display a spirit of joy and, even, light-heartedness. It was as if the darkness of the strange forests had been penetrated and the therapeutic effects of sunlight had been noticed.

While von Guérard had drawn public attention to the beauty found in patches of rainforest, or wet sclerophyll forest, in places like Ferntree Gully and the Illawarra district of New South Wales, it was a new departure to find beauty *inside* the dry sclerophyll forests that cover much of south-eastern and south-western Australia (the so-called 'coastal fringe'). Even the earliest artists had been intrigued by the twisted shapes of gum trees and by some highly 'exotic' species of plants and animals. However, the open woodland eucalypt forests of the coastal fringe were generally seen as being monotonous and dull and the arid inland devoid of any aesthetic merit.

Long before von Guérard, Augustus Earle, a globe-trotting English artist who would be offered a job on the famous survey ship the *Beagle* soon after his stint in Australia, was attracted by the diverse and lush forests of the Illawarra district (below the escarpment). When he eventually returned to England he did quite well selling a series of paintings inspired by a visit to that region in 1827. Yet, during that visit he wrote a poem that betrayed very mixed feelings about the landscapes he passed through:

> Such tangled thickets now we passed, such mighty trees we saw,
> Such giants of Australia of Australian growth, now fill'd my mind
>   with awe,
> They seemed to say in future times, we'll guard our native shore
> Such Natives shall grow out of us, as ne'er were seen before.

> We now had reach'd a lovely spot, by Farmer call'd his Farm
> And hop'd to get our bellies fill'd, with a drop to keep us warm
> But O what horror we all felt – when wide we gaz'd around
> To find a barren wilderness of gum trees most profound.
> (in Bonyhady, 1994, p. 170)

For Earle, the sight of old growth trees in wet sclerophyll forest filled him with awe but a large stretch of open eucalypt forest filled him with horror.

In his time the word 'wilderness' was used to refer to wild and uncultivated places. As biblical references to the word suggest, it has long been associated with barrenness and only recently has the concept of wilderness as a repository of extraordinary biodiversity and great beauty been popularised.[6] So, in his poem, Earle used the term to express his distaste of the prevailing eucalypt forests. He chose to paint more 'lush' scenes, featuring plants like the cabbage-tree palm that might have seemed more familiar to the European eye.

While recent attention has focused on the contribution of people like von Guérard, the contribution of another nineteenth-century euro-Australian artist, Samuel Thomas Gill, remains rather neglected. Gill arrived in Adelaide in 1839 at the age of 21 and, ambitiously, set out to become a professional artist. Struggling to survive in his Adelaide studio, he decided to join an ill-fated expedition led by explorer John Ainsworth Horrocks in 1846 as an unpaid artist (Haynes, 1998). Five weeks into this journey Horrocks was badly injured while trying to fire a gun from the back of a feisty camel and Gill was left to look after him while others went for help. Sadly, there was little Gill could do for him and Horrocks died before help could arrive. However, while he waited, he completed two paintings that showed the scene with Horrocks' tent in the foreground and a vast space behind. Gill included himself gazing expectantly into space while the camel that caused the problem is the only thing that does not seem to be out of place. In the 1850s Gill headed for the Victorian goldfields where he painted cartoon-like scenes of hopeful people seeking wealth in the soil and in the hastily constructed villages and towns. While often working in the picturesque genre, he became interested in the notion of an interactive struggle between people and the land. There was little romanticism in what he portrayed.

The renowned art critic Geoffrey Dutton (1974) has described Gill as 'the most brilliant artist of ordinary life in Australia's history'. Along with von Guérard, he was probably the first white Australian artist to produce relatively unbiased images of Aboriginal people and their cultural practices. One of Gill's Victorian paintings, *Native Dignity*, was a direct assault on prejudicial stereotyping of Aboriginal people because it depicted a virile young Aboriginal couple dressed in rather formal European clothes striding rather arrogantly down the street while some rather scruffy looking white people look on mournfully. Another perceptive painting from the same period, *The Colonized*, depicted a white settler standing proudly next to his house, being attended to by Aboriginal servants and with his sad, stooped, wife standing beside him.

Gill's contribution was probably undervalued in his time because his work often resembled caricature. He was much less influential in art circles than von Guérard or Buvelot. Yet his interest in stories being acted out in particular landscapes was something that Roberts, in particular, would later pick up as well. In fact, Roberts might be better described as a narrative painter than a landscape artist. Although he introduced impressionist techniques into Australian painting, he was not particularly comfortable in that mode of painting. When the younger, more intuitive, painter Charles Conder shared a studio with him in 1888 he complained that the older man worked to a strict 9am to 5pm schedule like a clerk in the public service. Roberts, on the other hand, found the younger man messy and unreliable. However, what Roberts may have lacked in spontaneity he made up for with hard work and determination and he certainly contributed some classics of Australian art. For example, when he first travelled inland in 1890 to capture scenes remote from the coastal fringe he spent eight months camped in a shelter constructed of canvas and corrugated iron to make all the sketches he needed to be able to complete the classic *Shearing the Rams* (1890).

In his travels, Roberts developed an interest in depicting scenes and stories that were passing into history. For example, when he visited the New England district in 1894 he heard stories about the 'gentleman bushranger' Captain Thunderbolt which inspired one of his most famous paintings, *Bailed Up.* When this painting was exhibited it was criticised for showing a crime as a rather relaxed and non-threatening affair. Yet this was exactly the way a former driver of Cobb and Co. coaches had described the routine hold-ups he had experienced and that was what interested Roberts the most. Part of the problem was that it took the meticulous Roberts nearly 30 years to finish the painting (thus encouraging a perception that he was romanticising the past). However, the work has endured with the background, widely considered to be a masterful depiction of open rangeland forest, being added a long time after the completion of the story shown in the foreground. Despite his rather laboured approach, Roberts' work often displayed a freshness of vision.

Roberts was the only one of the Heidelberg painters to venture out beyond the coastal fringe. His interest in doing so was stimulated by a trip to the Riverina district in December 1886 when he attended the marriage of his cousin, Ethelwold Burchill, to one of the owners of the Brocklesby station near Corowa. When he subsequently returned he not only completed the sketches for the shearing paintings for which he is well known, but also for the painting *A break away!* (see page 49*)* which was a significant reinterpretation of prevailing settler myths. At the time this painting was created it was popular to depict Australian stockmen as skilful horsemen totally in control of the animals they were working with. *A break away!*, by contrast, depicts a stockman who has lost control of a mob of thirsty sheep who are plunging recklessly down a slope towards a small waterhole. Beyond the dust-storm the sheep have created, a silent plain stretches out to the distant horizon where a whirlwind threatens to create an even bigger dust-storm. This painting suggests that the 'forces of nature' are beyond the control of humans. It presents a sharp contrast between the stillness of the plain and the frenetic activity of the sheep.

As well as travelling to the Riverina district and New England, Roberts went to north-western New South Wales and as far as the Torres Strait in Queensland. He was equally at home as a dinner guest of the New South Wales governor and in the pub at Cooktown in north Queensland. And, like his contemporary Henry Lawson, he was fascinated by the characters he met during his travels. His striking portrait of the 80-year-old owner of Yugilbar station on the Clarence River of New South Wales, Edward Ogilvie, shows a face full of strength and good humour. While he was at Yugilbar working on the portrait of Ogilvie, Roberts experimented with some portraits of Aboriginal people working at the station. Although they were not particularly successful, they foreshadowed later efforts made by Russell Drysdale to capture the combination of dignity, patience and sadness that can be reflected in such faces.

Paintings by Roberts and his colleagues have often been criticised for presenting a romantic and outdated view of Australian identity. If this is the case, it would be difficult to argue that they gave Australians some fresh interpretations of their relationships with the land. The critics say, for example, that in *Shearing the rams* and *Golden fleece: shearing at Newstead*, Roberts depicted shearers using hand shears at a time when they were being rapidly replaced by mechanical shears. *Bailed up* (begun in 1895 and released in 1927) depicted a scene with bushrangers even though the last bushranger – Ned Kelly – was hanged in 1880. McCubbin's paintings of colonial figures in the bush are said to represent a romanticised view of the triumph of humans over nature yet he swung between prevailing fears about the bush, as in his 'lost child' series (1886, 1892 and 1907), and an interest in the stoicism of those who had largely conquered such fears without ever being in control.

The harshest critics have even suggested that paintings like *Shearing the rams* and Streeton's pastoral scenes (like *The land of the golden fleece*) were a part of the conscious construction of an Australian nationalism based on rural mythology (see Astbury, 1985). Such critics point out, for example, that by the 1890s many more Australians were living in the cities than in the country and the 'mateship' of the pioneers was about to be shattered by the bitter national shearers' strike of 1891–92. These critics suggest that the relaxed scene in the shearing shed at Brocklesby station depicted in *Shearing the rams,* with the station owner quietly watching the men at work, belied the tension that existed on the eve of the strike. As already noted, Roberts was related to the owner of Brocklesby station by marriage and he needed his consent to paint the scene. But Astbury has also noted that there is some

A break away! by Tom Roberts (1891); a challenge to prevailing myths.
(Art Gallery of South Australia)

evidence to suggest that relations between the shearers and the owner at Brocklesby were, in fact, very good. It is certainly unfair to suggest that Roberts would not be unsympathetic to the cause of the shearers. Following the influence of Velásquez and the English writer Thomas Hardy, he was strongly interested in the dignity of ordinary people at work and among his friends he counted Victorian socialist Thomas Maloney and J. F. Archibald, editor of the radical, pro-union *Bulletin*.

No doubt the Heidelberg painters were caught up in a growing tide of Australian nationalism, but that is hardly surprising in the context of their time. The economic depression of the 1890s, and the big industrial disputes that went with it, fuelled the debate about the formation of a national government. After its formation in 1891, the Australian Labor Party became the first national political party. A decade later, a national constitution was adopted and the federal parliament was created. In the second half of the 1880s, radical nationalism –a form of nationalism that promoted egalitarianism – gained its voice, particularly through Archibald's *Bulletin* magazine. Henry Lawson had his first work, a poem titled 'A Song of the Republic', published in the *Bulletin* in 1887.

Eventually Streeton's pastoral paintings may have been used to promote an image of Australia far removed from the sprawling cities and emerging class distinctions; an image that may have been convenient for those who wanted to depict a prosperous and harmonious nation. But for a long time the Heidelberg painters could not get any serious patronage; they were not embraced by the rich and powerful. When it was completed Roberts could not convince a public gallery to buy *Shearing the rams* and McCubbin's most famous work, *The Pioneers*, was rejected by the National Gallery in 1903 and only purchased in 1906 following sustained public pressure on the directors. This work was radical in its time.

Although Streeton got his early break by hitching himself to the Roberts' star, he went on to overshadow his mentor and all the other Heidelberg painters. He was the first to have success with public art galleries when the Art Gallery of New South Wales bought *Still glides the stream* soon after it was completed in 1890. He had been so annoyed when the National Gallery in Melbourne rejected his earlier work *Golden Summer* (1889) that he sent it to Paris where it was hung in the prestigious Paris Salon. The Art Gallery of New South Wales bought two more Streetons, *Fire's on! Lapstone tunnel* (1891) and *Cremorne Pastoral* (1895), before the National Gallery finally acknowledged his growing reputation by purchasing *The purple moon's transparent might* in 1896. With Conder in Europe from 1890 and Roberts returning to England in 1904 (where he stayed for 19 years), Streeton carried the legacy of those who had first gained a public profile through the controversial '9x5' exhibition. He became the first Australian artist to be knighted (in 1937) and, by 1930, his domination of Australian landscape painting was such that Professor W. K. Hancock wrote that his style had become a 'national habit' (Bernard Smith, 1979, p. 146).

Streeton was more specifically a landscape painter than either Roberts or McCubbin. In most of his paintings – particularly the later ones – any people present were either small or incidental. While working with Roberts and Conder he explored the theme of men at work and he did produce one of the great narrative paintings of the Heidelberg era in *Fire's on! Lapstone tunnel* (1891). It was painted while he was staying at Glenbrook in the Blue Mountains west of Sydney and, in a letter to 'The Prof' McCubbin, he delighted in describing the scenery in which he was living and painting. He found a spot from which he could paint a scene of men working on the construction of a railway slashing its way through some steep country. Unexpectedly, he witnessed a huge drama when two of the men were killed in an explosion inside a tunnel under construction and his painting depicts the scene when one of the bodies is being carried out of the tunnel on a stretcher. Despite the enormity of the tragedy for the men involved it seems dwarfed by the powerful images of the landscapes in which they are working.

Yet, as Streeton moved away from the influence of Roberts and McCubbin his focus moved from the potential for conflict between people and nature to Arcadian representations of pastoral landscapes that seemed to hold the promise of peaceful coexistence between people and the land.

Critics have suggested that all of his best work was completed before the turn of the century and that, thereafter, he became a rather mechanical painter turning out recipe pictures in blue and green for a well-established market (Astbury, 1985). Writing in 1979, Bernard Smith suggested that the 'national habit' Streeton established with his style of painting had become a 'national vice' because 'the Streeton tradition has prevented or thwarted the natural development of original artists who paint in quite a different manner from the disciples of Streeton'. Critics like Smith also point out that Streeton used his highly influential column in the Melbourne *Argus* to condemn modernism in art, much like James Smith had earlier condemned the impressionists in the same newspaper. Such critics say that Streeton became backward-looking and narrow in his thinking.

This criticism is much too harsh. Streeton himself cannot be blamed for his extraordinary popularity and nor can he be held responsible for the lazy art critics and market-oriented art dealers who exaggerated his achievements and encouraged imitators. More importantly, those who criticise his much-imitated pastoral paintings also ignore the fact that Streeton had a strong interest in the preservation of native Australian bush. Tim Bonyhady (2000b, pp. 325–327) has pointed out that he joined a campaign against construction of a coal mine on Bradleys Head (near the Sydney artists' camp at Sirius Cove) in 1895 and in the 1930s Streeton used both his

Fire's on! Lapstone tunnel by Arthur Streeton (1891); a narrative painting.
(Art Gallery of New South Wales)

paintings and his column in the *Argus* to campaign for the protection of forests in the Dandenong area (Bonyhady, 1993, pp. 8–12). A 1934 painting of his beloved Sherbrooke forest in the Dandenongs was titled *Our vanishing forests* and two works included in his final exhibition of 1940 – *Silvan Dam and Donna Boang, AD 2000* and *Gippsland Forests for Paper Pulp* – had a sharply critical edge to them. According to Bonyhady,

> Far from simply repeating himself Streeton increasingly imbued his paintings with his environmental concerns – trying to awaken his contemporaries to the importance of conservation. No Australian modernist of his day was more political in either their prose or their paintings.

Some of the contradictions apparent in Streeton's work reflect the fact that he was strongly influenced by romantic philosophers and writers. For example, he was clearly influenced by the poetry of William Wordsworth; shown by the fact that an early work took its title from a line in a Wordsworth poem (*Still glides the stream*, 1890) and by the fact that a well-worn volume of Wordsworth's poetry was found among his possessions on his death. Like Streeton, many of the romantics have been criticised for harking back to an idealised interpretation of what existed in the past, when their real aim was to highlight the dangers of 'Enlightenment' thinking that suggested humans could achieve a mastery over nature. Like other romantics, Streeton can be rightly accused of presenting, at times, a rather simplistic and idealised image of peaceful coexistence between people and nature, but his aim was to inspire people with the visions of a coexistence and, in this sense, he belongs to the romantic ecology movement that has been partially 'rediscovered' (Mulligan, 2001b) in recent times.

HANS HEYSEN AND MARGARET PRESTON

The Heidelberg painters have also been criticised for not looking beyond the coastal fringe for their images of Australian identity.[7] As we have already seen, this criticism is wide of the mark when it comes to Roberts. Indeed, it might be said that he began a trek to the inland that was resumed by later artists. Perhaps the next to go on that journey, and to take it further, was Hans Heysen, who arrived in Adelaide from Germany in 1877 at the age of six. In 1899 he returned to Europe to attend art classes in Paris but he soon returned to Australia to pursue his fascination with gum trees. Even more than the Heidelberg painters, he converted trees from framing devices into subjects, with particular attention paid to the massive and exquisitely textured trunks of his favourite gums. He showed his hand with the title of an early painting, *A lord of the bush* (1908), which depicted one giant eucalypt standing defiantly while its felled mate was being dragged away by a man with a bullock. Heysen certainly fell in love with the landscapes of the country his family had adopted. In a letter to Lionel Lindsay in 1912, he wrote that 'Australia has got *something* in its light that colours local substances with just something that another country has not got, and it is this something out of which an Australian school has got to come . . .' (in Pearce, 1983, p. 53). He also commented that he thought 'Streeton has got something, but he didn't explore beyond its very surface . . .' He was disappointed that Streeton had decided to go to Europe rather than stay to probe more deeply into the 'mysteries of our forms – and light.'

For Heysen the 'mysteries of our forms' became even more intriguing when he discovered the 'ancient' landscapes of the Flinders Ranges in 1926. In another letter to Lionel Lindsay he wrote that the rock formations in this area gave the impression of being 'the bare bones of the landscape' and he explained his artistic interest in the following way:

> My first impression upon arrival was that of expanse, of simplicity and beauty of contours – the light flat and all objects sharply defined; distances very deceptive, and no appreciable atmospheric difference between the foreground and the middle distances; indeed, hills at least four miles away . . . appear to unite, and scale becomes an important relative factor. (in Haynes 1997)

Heysen became increasingly aware that he was gazing at ancient terrains that had passed through some extraordinary transformations and, in his letter to Lindsay he wrote:

> It was in the Flinders Ranges that I was made curiously aware of a very old land where the primitive forces of nature were constantly evident. The barren hillsides incised and torn by Nature's forces, hold peculiar attraction . . . great masses of stone are piled layer on layer in regular formation, as if built up by some very ancient people, their appearance is given an architectural order. (in Haynes, 1997)

Heysen's art was seen as a resistance to modernism, and, ironically, it took something ancient to stimulate his interest in the sculptural abstraction of the modern school. He wrote to painter Sydney Ure Smith: 'here is the very thing you moderns are trying to paint' (Haynes, 1997). Heysen himself became so engrossed in the arid zone of northern South Australia that he made 10 trips into the region between 1926 and 1933. However, his invitation to the 'moderns' to follow suit was not taken up until the 1940s when artists who had been influenced by surrealism, including Peter Purves Smith, Russell Drysdale and Sidney Nolan, began to turn their attention to the desert.

Modernist influences in art had only really reached Australia in the late 1920s and it was a loose circle of painters in Sydney, including Roland

Wakelin, Roy de Maistre, Grace Cossington-Smith and Margaret Preston, who were the first to emulate post-impressionist European painters such as Paul Cézanne. Of this group it was Margaret Preston who made the biggest public impact, especially with her bold representations of previously neglected Australian flora.

Born in Adelaide as Margaret McPherson, Preston studied drawing with Frederick McCubbin at the National Gallery in Melbourne before returning to the School of Design in Adelaide in 1894 (Butel, 1985). Her interest in decorative art attracted her to the efforts being made by people like Cézanne to highlight the inherent geometry in their subject matter. However, she did not believe that art should be cosmopolitan and, in 1925, she began a personal quest to identify a distinctive form of Australian art. Instinctively, she turned to Aboriginal art – probably because it had strong design qualities – writing that it offered:

> . . . an escape from the South Kensington dullness which has pervaded and perverted the decorative impulse of this country. (Butel, 1985, p. 50)

Whether Preston was leading or responding to a new interest in Aboriginal art is unclear because 1925 was also the year when the University of Sydney first established a chair in anthropology and people like Professor A. P. Elkin and Charles Mountford were attracting interest for their descriptions of Aboriginal cultural practices. However, Preston was certainly a pioneer in being an established artist willing to make direct use of Aboriginal motifs in her own work. Her biographer Elizabeth Butel has said that her views on Aborigines were rife with racist assumptions (pp. 51–69) but she was interested in the fact that for Aboriginal people art was not so much an exercise in aesthetics, but more an effort to represent meanings:

> . . . which endow everyday events and familiar features of the landscape with a cosmic significance, by referring to and conjuring up images of the ancestral past. (Butel, 1985, p. 56)

During the 1940s, most of Preston's art showed strong Aboriginal influences, but this work did not achieve the enduring popularity of her earlier work. Perhaps her deliberate efforts to forge an integration of western and Aboriginal art was too laboured or too tokenistic to reproduce the 'freshness, grace and vigour' (Butel, 1985, p. 69) of her earlier work. There were similarities between what she was trying to do and the efforts of the 'Jindyworobak' poets based in Adelaide because they also believed that 'authentic' Australian cultural expression must find its inspirations in Aboriginal sources. The Jindyworobaks, led by Ian Mudie, sought out Aboriginal words and place names to use in their poetry and they made excursions to Central Australia. However, their understanding of Aboriginal culture remained very superficial. Preston maintained a correspondence with Mudie and they probably influenced each other but neither of them really succeeded with the cultural fusion that inspired them both. In view of prevailing white attitudes towards Aboriginal culture, it was never going to be an easy project and in hindsight it was probably fortunate that people like Preston – who were ahead of their time in some ways only – did not succeed in appropriating aspects of Aboriginal art. The first successful fusion came from the other direction when Albert Namatjira used a style of painting introduced to him by the white artist Rex Battarbee to capture some distinctive representations of his tribal homelands.

ALBERT NAMATJIRA AND ABORIGINAL ARTISTS

Born at Hermannsburg Mission outside Alice Springs in 1902, Namatjira was introduced to watercolour painting when Battarbee visited the mission in 1936 (Hardy, Megaw and Megaw, 1992) and he used the style of painting to create images of places that were of significance to his Aranda people. He used a wide range of colours to depict the red earth, purple ranges and ghostly white gums and those who had not been to Central Australia, or those who simply failed to notice the range of colours in its landscapes, thought that Namatjira had painted a fantasy world. However, his paintings helped white Australians to remove their blinkers and notice the extraordinary beauty in the country's 'dead heart'. The vibrancy of his colours highlighted the clarity of light that is a feature of arid regions.

Namatjira held his first exhibition in Melbourne in 1938 and this was followed by an exhibition in Adelaide the following year, when the Art Gallery of South Australia endorsed his talent by purchasing one of his works. Ironically, the influx of troops into Central Australia during World War II increased the market for Namatjira's work and others at Hermannsburg began to emulate his style. Between 1941 and 1959 there were several highly successful exhibitions of Aranda watercolour paintings in major Australian cities, but the success also drew Hermannsburg painters into fringe camps around Alice Springs, where alcohol consumption became a major problem. A ready market encouraged the production of poor quality imitations. Namatjira himself was invited by the Menzies government to meet Queen Elizabeth II when she visited Australia in 1954, yet, as an Aborigine, he was not an Australian citizen when this meeting took place (Aborigines did not get citizenship rights until after a referendum in 1967). Because of his high profile, Namatjira was given individual citizenship rights in 1957, yet his own children would remain as wards of the state and could not stay overnight in Alice Springs. Citizenship brought personal grief to Namatjira because he was entitled to buy alcohol for himself but prevented from supplying it to other Aborigines. He was jailed in 1958 for supplying alcohol to his relatives and he died, broken-hearted, the following year.

Meanwhile, more traditional forms of Aboriginal art were beginning to be noticed by non-indigenous Australians. In the 1930s, the anthropologist Baldwin Spencer had recognised that there might be a market for the 'X-ray' paintings on bark that were common in Arnhem Land (Bonyhady, personal communication) and, at a similar time, the Methodist Mission at Yirrkala on Gove Peninsula started to encourage Yolngu artists in the area to use modern paints to reproduce their images on bark (Hardy, Megaw and Megaw, 1992, p. 6). As these paintings began to find the market that Spencer anticipated, a north Australian style of Aboriginal painting was popularised. Much later, in the early 1970s, a white arts and craft teacher at Papunya Aboriginal settlement in Central Australia, Geoffrey Bardon, encouraged local artists to translate images they had previously made in sand mosaics or in body painting onto canvas or board using acrylic paints (Bardon, 1979). The movement had an inauspicious beginning when Bardon invited his school pupils to participate in painting a mural on the school building and they chose to use icons and images that they were familiar with. This excited the interest of the old men, in particular, who were encouraged to see that the children had maintained an interest in traditional iconography. As the community discussed the mural, Bardon heard stories from the old people about what some of the traditional icons represented and he encouraged the old men to experiment also with modern paints and materials. This led to the creation of a grouping of artists who adopted the name Papunya Tula and, subsequently, to the emergence of the now-famous 'dot paintings' that have enabled Aboriginal artists in desert communities across Northern Territory, South Australia and Western Australia to offer fresh new interpretations of their stories of land and identity.

Some of those who pioneered the Papunya Tula movement had recently been encouraged, or forced, to move into the Papunya settlement from their traditional homelands further to the west and their early art reflected a nostalgia for places they had left. While many of the early works were a fairly direct translation into the new medium of traditional sand mosaics and body art, the artists also enjoyed the freedom of expression that the new technology facilitated and subsequent work tended to become more abstract and more subjective in its interpretation of the stories embedded in the land. Through the 1980s the dot paintings gained recognition as an internationally unique art tradition. As with the paintings produced by Namatjira and his imitators, the rather sudden creation of a ready market led to the production of many hastily constructed paintings with limited artistic merit. However, the tradition has also thrown up some new masterpieces in Australian art and a movement that is more distinctively Australian than any created by non-indigenous Australian artists. Furthermore, the artists of this movement were treated with more respect than was Namatjira.

The success in Australia of the dot painting movement reflects both a growth of interest in the ecological worldviews of the indigenous people (see chapter 9) and a reappraisal of the landscapes that were once considered dead (see Haynes, 1998). Of course, it has become much easier to visit the desert in relative comfort (with air travel, air-conditioned buses and motels). Australians have also been encouraged to look again at desert landscapes because visitors from outside the country have been impressed by what they have seen. However, it is also important to give credit to those white Australian artists who did find ways to engage with such landscapes. As Roslynn Haynes (1998) has pointed out, the first European artists to venture into the desert were poorly equipped to make sense of what they saw because they carried with them the picturesque mindset discussed earlier in this chapter.

As Haynes puts it, the picturesque movement encouraged the artist to think that a 'good' landscape required at least some of the following elements: framing trees, water as a focus, historic ruins, and/or distant people to provide a sense of scale. The foreground was used to create a perspective and to draw the eye into the more distant view. Such paintings celebrated the human domestication of nature, with just a hint of the 'wild'. Everything appeared as God surely intended it. However, this way of constructing landscapes left an artist paralysed and depressed when confronted with the desert. When they looked across the desert with a lateral perspective they could see none of the elements that would contribute to a 'good' landscape. Instead they were affronted by what seemed to be a direct challenge to notions about the 'divine order'. The land appeared to be of no possible 'use' to humans. As Haynes has put it, these early artists were confronted with what appeared as a 'ghastly blank'[8] on the maps of the continent. Desert landscapes mocked the iconography of European landscape art because their 'vastness and emptiness . . . magnified the [western interloper's] sense of dislocation, exile, alienation and insignificance' (Haynes, 1998, p. 89).

In her review of the struggle by nineteenth-century artists to come to terms with the desert, Haynes highlights three artists in particular: Edward Frome, Surveyor-General of South Australia from 1839 to 1849; S. T. Gill and his paintings of the ill-fated Horrocks expedition in 1846; and Ludwig Becker, nature artist with the ill-fated Burke and Wills expedition of 1860–61. She picked out Frome because of a single painting of his called *First View of the Salt Desert – called Lake Torrens* 1843), which depicted a lone figure on a camel gazing towards a vast and shimmering space. What this painting managed to capture was the sense of space and the intimidating power of the desert; nearly three-quarters of the image is of the bright sky and the figure on the camel appears to be hypnotised by the eerie stillness of the scene. Becker is highlighted because he not only left behind some exquisite drawings of the flora and fauna he encountered, but also because he produced some striking watercolours of the party of explorers as transient (almost translucent) interlopers passing through timeless landscapes.

Haynes identifies a gap of 60 years between the death of Becker (during the expedition) and the emergence of Heysen's paintings of the Flinders Ranges in the 1930s as a period in which the barren landscapes were again neglected. The 1930s also saw the emergence of Namatjira, the publication of a book titled *The Red Centre* (1935) by H. H. Finlayson, the availability of air travel (that enabled people to get wider horizontal views of desert landscapes), and the emergence of colour photography. One artist who was strongly influenced by such developments was Sidney Nolan who first gained public recognition for works he produced after flying over Central Australia in 1949. Nolan changed the perspective by looking down rather than across. Aborigines did not have the advantage of seeing their land from a plane, but they had long known that the desert has an extraordinary diversity of shapes, textures, colours and living things appearing in and around the surface of the land and their dot paintings could only be appreciated once white Australians understood the importance of looking down rather than across.

As Haynes has pointed out, it is quite extraordinary to think of the shift that has taken place from the time when the desert was seen as a 'ghastly blank' until a time when it is seen by many as the source of inspiration. The 'dead heart' has become the 'red centre' and desert images – particularly those of Uluru – have become major icons for promoting tourism. As mentioned, it has become much easier to 'experience' the desert from the comfort of air-conditioned buses and motels or by simply watching promotional films shot from the air. Such images can also become marketing icons. 'The Rock', for example, has been used to promote things as diverse as insurance and tropical drinks (!)[9] According to Haynes (1999), the advertising industry uses the desert as a symbol of immense size and tough conditions requiring endurance and innovation to survive (just like the products being advertised!). It is unlikely that this commercial exploitation of desert symbolism has increased ecological understandings of arid landscapes, but it does illustrate how far popular perceptions have shifted from the time when the 'dead heart' appellation was in vogue. Sidney Nolan probably had more to do with this shift than any other individual.

SIDNEY NOLAN

Sidney Nolan first emerged as a member of a group of Melbourne artists that included Albert Tucker, Arthur Boyd and John Perceval. According to Bernard Smith, Tucker was the intellectual leader of the group. Whereas Tucker had his first public exhibition in Melbourne in 1936, Nolan did not make his public debut until 1939. These painters were part of a broader circle of writers and artists who reflected their discourse into a magazine titled *Angry Penguins* and art critic Robert Hughes has said the magazine provided a focus for a powerful creative movement that found its voice in the 'angry decade' beginning in 1937. The Angry Penguins were alarmed by the rise of fascism in Europe in the 1930s and the outbreak of World War II. Nolan and Tucker were both conscripted into the army against their will. Tucker, who was strongly influenced by surrealism in European art, vented his anger with some highly cynical paintings of the wartime experience and he was followed in this mode by both Boyd and Perceval. The ending of the war did not necessarily improve Tucker's mood and he fired off some critical comments about the treatment of artists in Australia before leaving for Europe in 1947 (not returning until 1960).

Nolan spent his military service at a base in Victoria where he found time to renew his interest in landscape art. He deserted the army in 1944, but was able to find refuge at the Heidelberg home of John Reed, the wealthy patron and editor of *Angry Penguins*, and his wife Sunday. John and Sunday Reed enjoyed the company of artists like Tucker and Boyd but Nolan was their favourite and their patronage enabled him to continue working from the time he walked out of the army base in 1944 until he got a dishonourable discharge under a government amnesty of 1948. His time in the army had not been wasted artistically, because, while based at an army camp in rural Victoria, he found time to experiment with landscape paintings and his interest in non-urban themes was stimulated. In 1947 he produced his first series of Ned Kelly paintings that featured the image of the armoured bushranger in stark and colourful landscapes. In working with the popular Kelly story, Nolan was, in one sense, reviving the narrative approach to landscape painting pioneered by artists like Roberts and McCubbin. But he was more explicitly interested in mythology and iconography than in a 'realistic' recording of history. His exploration of myths embedded in landscapes had strong, but unwitting, echoes of Aboriginal art and, together with his aerial interpretations of the desert painted in 1949, the Kelly series began to stimulate a new interest in 'storied landscapes'.

Following the Kelly series in 1947, Nolan travelled to Queensland, where he got interested in the story of Eliza Fraser, a white woman who survived an 1836 shipwreck near what later became known as Fraser Island. She survived on Fraser Island by living with Aborigines who had killed other, male, ship survivors, including her husband. Again Nolan was attracted by a story that had become mythological and he produced a series of paintings that showed distorted images of Fraser in strangely beautiful, yet threatening, landscapes. Later explorations featured myths surrounding the ill-fated Burke and Wills expedition and the equally ill-fated invasion of Gallipoli during World War I. His most influential works, however, were those that combined references to popular Australian myths set in his fresh interpretation of colourful, yet confronting, landscapes.

Robert Hughes has said that Nolan's 1947 Kelly series was a major turning point in Australian art and these paintings are probably what he became best known for. Just as important, however, was his exploration of the changing mythology surrounding the ill-fated Burke and Wills expedition of 1865. In saying that Nolan produced 'parodies of nineteenth-century grand historical canvases', Roslynn Haynes has compared his image of a lonely Burke setting out on a camel from a 'deserted two-dimensional stage set of Melbourne' (*Departing Melbourne* 1950) with Nicholas Chevalier's early representations of the explorers departing Royal Park amidst a huge crowd of cheering citizens (*Memorandum of the Start of the Exploring Expedition* 1860) (Haynes, 1998, p. 214). Ironically, Burke and Wills were turned into heroes after the failure of their mission and their lonely deaths in the desert because this was the only way that coast-clinging Australians could cope with their own fears of the vast spaces of the inland. It was necessary, Haynes points out, to simultaneously demonise the 'treacherous' environment that was so difficult to 'tame' and at the same time lionise the explorers, especially those who paid the 'supreme' sacrifice.

Tim Bonyhady's study (1991) of the mythology surrounding the Burke and Wills expedition makes it clear that the motivating idea of the expedition had less to do with exploration – for it actually covered little territory not already traversed or even settled – than a desire on the part of the emerging Melbourne elite to demonstrate its capacity to play a leading role in the forging of a nation. The whole thing had been hyped up from the planning stages onwards and when the bodies of the fallen 'heroes' were brought back to Melbourne a huge public funeral procession was organised and commemorative statues were erected in Melbourne and in all the Victorian towns that the expedition had passed through on its way north. However, the enthusiasm declined steadily with the passage of time and, as early as 1888, an 'official' history of exploration commissioned by the New South Wales government to mark the centenary of the New South Wales colony described the Burke and Wills expedition as a 'perverse absurdity' (Bonyhady, 1991). Even in Victoria the monuments were gradually removed from public places and annual commemorations of the explorers' deaths had petered out by the turn of the century.

So, when Sidney Nolan set out, from 1950 onwards, to depict Burke as an 'obsessed, arrogant and self-satisfied' man, he was not forging a new

Ned Kelly by Sidney Nolan (1946); mythological icon placed in iconic landscapes. (National Gallery of Australia)

myth but rather drawing attention to the extent to which the earlier hero myth had turned around. Nolan enhanced the revised mythology by depicting Burke as an anxious and lonely figure fused to the back of his camel as it picked its way through landscapes of intimidating power. The self-satisfied leader of THE expedition, seen prancing around on a white horse in Chevalier's image of the departure from Melbourne (see above), had been replaced by a frightened individual who could not even control the movement of the camel which moved as if in a trance. Perhaps the implied message is that only a fool would enter such country with an attitude of mastery rather than humility or at least the kind of stoicism represented by the camel. Much discussion of the folly of Burke and Wills has centred on their lack of willingness to seek or accept help from Aborigines living in the landscapes where they perished. However, as Bonyhady has said, the hero to fool transition also reflects a growing criticism of their 'more general inability to understand the land' (p. 31).

Nolan's work on Burke and Wills encouraged some of his colleagues, notably Albert Tucker and David Boyd, to also explore the changing mythology surrounding 'heroic' explorers in their work. Nolan's work probably also encouraged Patrick White to take up the explorer theme in his classic novel *Voss* (1957) because it is known that White was an admirer of Nolan and asked him to provide an image for the dust-cover of the first edition of *Voss* (see chapter 4).

As mentioned, during his career Nolan moved from one area of Australian mythology to another – from Ned Kelly to Eliza Fraser to Burke and Wills to Gallipoli. He was interested in the interplay between the stories and the places in which the stories unfolded – noting that stories can gain the status of myths when they have an ability to transcend the specific places in which they were born. He liked to return to the image of Ned Kelly and celebrated the mythological status of the Kelly icon by placing it in vibrant arid landscapes (far removed from the Victorian rangelands where the real Ned Kelly operated) because he also saw the colourful desert as an icon of our struggle to establish a sense of belonging in a challenging but seductively beautiful land. Nolan himself has said that he was continually attracted by desert images because 'I wanted to deal ironically with the cliché of the 'dead heart' . . . I wanted to paint the great purity and implacability of the landscape' (Haynes, 1998, p. 60).

Nolan left Australia in 1951 and spent most of his remaining career based in London. Perhaps he found it easier to deal with the 'implacability of the landscape' from a great distance. However, unlike some earlier exiles, he continued to work on his Australian stories and he succeeded in attracting attention to both himself and Australian art in both London and New York. His interest in the relationship between landscapes and an emergent mythology continued to resonate with many Australians who followed his career with great interest. An art critic on the Melbourne *Sun*, Harry Tatlock Miller, probably put it best when, in a review of a 1953 exhibition of the Burke and Wills paintings, he said that they had a 'curious hypnotic power' because they gave the viewer:

> ... the sensation of seeing and knowing our country, both in its landscape and its legend, for the first time. (Bonyhady, 1991, p. 302)

ARTHUR BOYD AND RUSSELL DRYSDALE

Another of the Angry Penguins, Arthur Boyd, also made his mark as a landscape painter in the period following World War II. An accomplished 'straight' landscape painter early in his career, Boyd came to be influenced by Tucker's surrealism and by his own interest in European narrative painters like Pieter Brueghel. His postwar paintings were emotive and disturbing (reflecting the influence of the Angry Penguins), with common biblical references. He reinterpreted the scene of Adam and Eve being cast out of the Garden of Eden; placing it in an Australian landscape. In *Lovers by a creek* (1944) he developed a style of painting in which people and trees seem to metamorphose into each other. In the 1950s, Boyd turned his attention to the situation of Aborigines living in rural Australia and, again, he produced a series of disturbing, haunting, images. He maintained the critical 'social realism' of the Angry Penguins, but channelled it into evoking emotional responses from the viewer. According to Allen (1997), his temperament was totally different to that of either Tucker or Nolan. Whereas Tucker had painted with cynical anger and Nolan had been detached, Boyd showed a 'passionate involvement' in his themes:

> His early work creates an hallucinatory world of anguish, pity, loneliness, terror and frustrated desire. Nothing remotely approaching this emotional intensity had been seen in Australian art before.

In 1971 Boyd decided to move his Australian base from Melbourne to near Sydney,[10] and he settled on a property on the Shoalhaven River, west of Nowra. He subsequently attracted attention for some rather lyrical paintings of the region surrounding his new home. Any people that made their way into his landscapes were mostly depicted as clumsy intruders or lost souls. He particularly resented the noise and riverbank erosion caused by speedboats with waterskiers in tow and some satirical paintings from the mid-1980s depicted the skiers as clumsy-looking, lobster-red sunbakers who were too large for their own boats. As Tim Bonyhady (2000a) has commented 'No other artist has made Australians appear so like blights on the landscape'. Like Streeton before him, he became an ardent conservationist and, in 1993, he gave his property, Bundanon, to the federal government for protection as a national heritage site.

Following Boyd's death in April 1999, tributes highlighted the fact that he was a modest and reserved man with an immense and intense imagination. His subjects ranged from the depiction of a vagina with snapping teeth to a peaceful scene on the Shoalhaven. However, it was an endless fascination with Australian landscapes, maintained even during lengthy stays in England, that sustained him the most. In a late interview he said 'If I was whisked away . . . I think I could put up with anything, except not seeing the Australian landscape. It would be torture to have it cut off' (Hawley, 1999).

Russell Drysdale was another painter who began his career in Melbourne in the 1930s but he was never associated with the Angry Penguins. He studied at an art school run by George Bell in Melbourne in 1932 before spending two years on a tour of art galleries and painting classes in Europe. He returned to Europe in 1938 with a former schoolmate from Geelong Grammar and graduate of George Bell's school – Peter Purves Smith. When Drysdale returned to Australia he spent time in the Riverina district (like Roberts before him) where his own family and that of his wife, Elizabeth 'Bon' Stephen, owned several properties. He moved to Sydney in 1941.

In Sydney, Drysdale became a leader of a group of painters that has been dubbed, rather disparagingly, the 'charm school' (Hughes, 1970); a group that also included Donald Friend, William Dobell and Sali Herman. According to critics such as Robert Hughes, the Angry Penguins in Melbourne were more interesting than their contemporaries in Sydney because they were driven by passion, whereas the Sydney artists were interested in producing less challenging, quirky images that would go down well with the public. Critics have pointed to the upper middle class social background of artists such as Drysdale and Friend to explain why they were not angry like the Melbourne group.

It is true that Drysdale came from a privileged background; his family owned properties in the Riverina and the Pioneer sugar mills in north Queensland and he attended the elite Geelong Grammar school. He certainly won the heart of the public more easily than his Melbourne contemporaries and he was the first Australian artist to have a solo exhibition in London (in 1950). He quickly gained wealthy backers in newspaper owners Keith Murdoch and Warwick Fairfax. But it would be foolhardy to dismiss his influence on Australian art because he had some lucky breaks. It must be acknowledged that his public acclaim reflected his capacity to combine raw talent, extraordinary sensitivity and perception, and a good sense of humour. He produced accessible, yet challenging, images of life on the land. Before Nolan, he had turned his attention to the strange beauty of the arid inland and he found his own way to paint seemingly desolate scenes that had been an anathema to earlier generations of artists.

Drysdale also had to cope with the extraordinary disadvantage for an artist of permanently losing the sight in one eye before he had learned to paint. Ironically, it was while he was in hospital for an unsuccessful operation on the damaged eye (when he was aged 20) that he got his lucky break in art. His doctor, a friend of artist Daryl Lindsay, was impressed with the cartoons Drysdale churned out while confined to his hospital bed and he arranged for a meeting between Lindsay and the patient. As a result of this meeting, Drysdale enrolled in George Bell's art classes where his new mentor honed his drawing skills and introduced him to European artists such as Cézanne, Van Gogh, Matisse, Picasso, and Modigliani, whom he studied further during his trips to Europe (Dutton, 1964; Hawley, 1997; Geoffrey Smith, 1997).

Drysdale was strongly influenced by his friend Peter Purves Smith and his interest in surrealist art. When they returned from Europe in 1938, it was Purves Smith who first experimented with the elongated figures that became a feature of Drysdale's work. Unfortunately, Purves Smith's own career was tragically interrupted, first by military service during the war and then by a debilitating illness that ended his life in 1949.

Drysdale could not be considered for military service because of his partial blindness. Instead he turned his attention to documenting the wartime experience in Australia. His wartime images lacked the biting cynicism of the wartime paintings of Tucker, Perceval and Boyd, but he used humour to question Australia's involvement in the war when he produced an image, *Local V.D.C. Parade*, of a ragtag bunch of volunteers turned out in a country town for military training. Most of the men are dressed in their 'Sunday best' (complete with ties). There are only two guns among the lot of them and none are paying any attention to a demonstration of how to handle a rifle. The opportunity for a chat seemed more important than preparing for a distant war.

However, the wartime paintings were much less important to Drysdale's career than a series of images started in 1941 of 'normal' life in the inland. Two of these early works – *Man feeding his dogs* and *Man reading a paper* show a strong influence of surrealism. More successful, perhaps, was *Sunday evening* in which the gaunt figures of a family are gathered around as a mother prepares to bathe a baby in a pan. Seemingly hypnotised by their surroundings, the family members are almost overpowered by the vast space and brilliant sky behind them. Drysdale achieved the effect by lowering the horizon to include a vast expanse of sky and by reversing the normal gradation of tone to make the sky very bright at the horizon.[11] A similar effect was achieved in *Going to the pictures* (1941), which depicts a rather comical scene of a family preparing to leave their remote home for a long journey to the movies.

In 1942, Drysdale started another theme in his work with the painting *Joe's Garden of Dreams.* The title of the painting reflects the name of a cafe that has been opened in a country town by an Italian migrant and the proud owner and his family stand in the deserted street outside their cafe dressed in smart clothes and aprons. An apparent lack of customers does not seem to dim their enthusiasm for life in the new country. Later paintings of *Joe* and *Maria* – both painted in 1950 – show strong-faced Greek migrants gazing, trance-like, towards brilliant skies and open space beyond the confines of their cafes; both hypnotised by this new experience of space/time.

Drysdale received early recognition with the purchase of one of his early works by the prestigious Metropolitan Museum in New York in 1941 and he had solo exhibitions in Sydney in 1942 and 1943. But his big break in terms of public recognition came in 1944, when the *Sydney Morning Herald* (owned by Warwick Fairfax) commissioned him to travel through drought-stricken areas of New South Wales and send back images that could be published in the paper. Images he sent back of dead cattle rotting into the earth and twisted trunks of dead trees seemingly 'tortured to death by thirst'[12] not only brought the drought home to city-dwellers, they also drew attention to Drysdale's capacity for capturing a sense of space and time not seen in earlier art. Drysdale developed a rather melancholic fascination for the images of the dead and dying in the dry inland but he depicted death and dying as part of a process of transformation by which everything becomes part of the land. His images of people suggested that it is possible to feel at home in such environments but it takes stoicism, patience and good humour to reach that goal. Like Henry Lawson and Tom Roberts before him, he was fascinated by the characters that seem to be created by tough environments and those he painted included *Tractor-face Jackson* (1950) and *Brandy John* (1965). *The Drover's Wife* (1945) has the same title as a famous story by Henry Lawson.

The melancholy that is evident in many of Drysdale's images is also tempered by his strong sense of humour. The art historian Bernard Smith (Smith and Smith, 1991) has said there are a number of interesting parallels in the careers of Russell Drysdale and Henry Lawson. Both had a sense of the absurd and a strong sense of humour. They were both humanitarians with a tendency towards sentimentality. But, above all, both were keen to explode romantic myths about life in the bush. In their own ways, they each became interested in the ways in which the rhythm of the land seems to get into the bones of the people who live there. There are also similarities in the interests of Drysdale and Tom Roberts. Both men followed their artistic noses inland to explore the interaction of people and landscapes. Both got interested in the faces of Aborigines living on the margins of colonial bush society, although Drysdale took this interest much further.

Drysdale first became interested in Aborigines while visiting north Queensland to attend board meetings of his family's business – Pioneer sugar mills. He began by painting of out-of-place Aborigines visiting Cairns, *Mother and child, north Queensland* (1950) and *Shopping day* (1953) and he depicted groups of Aborigines dressed in their 'Sunday best' for a trip to town, yet still refusing to wear shoes so that they might feel connected to the earth – see *Shopping day* and *Station blacks, Cape York*. According to Roslynn Haynes (1998), these particular paintings suggest that 'civilisation' was imposed on the Aborigines from the head down but it failed to disturb their connection with the land. One of Drysdale's most powerful works, *Mullaloonah Tank* (1953), shows a group of Aborigines seemingly at home in an arid landscape, but separated from their land by a flimsy dwelling that acts as a kind of screen. Drysdale later pursued his interest in Aborigines by making trips to the Northern Territory and Broome, in the Kimberley district of Western Australia. In the Northern Territory in 1959 he was invited to attend a funeral and mourning ritual by the Tiwi people on Melville Island because, according to tradition, artists were invited to such ceremonies to record the event. The ceremony is called *pukumani* and the invited artist normally creates a carved and painted pole – a *tutini*. Instead Drysdale created several paintings that merged together the images of people and the *tutini* poles. The faces of the people in these paintings appeared much more positive than those of the people visiting Cairns in earlier paintings.

A wonderful story surrounds one of Drysdale's paintings of an Aboriginal man, simply titled *Man in a landscape*. This story begins with a visit to Australia by the relatively young Queen Elizabeth II in 1963. According to royal protocol, visiting monarchs might expect a gift fit to take back and display in Buckingham Palace and, before this visit, Prince Philip let it be known that the appropriate gift might be a work of art by a leading Australian painter. Drysdale was, by then, Australia's most popular artist so Prime Minister Robert Menzies took it upon himself to visit the artist in his Sydney studio to see what work he had for sale. There were two pieces that were ready for sale but one of them (from the Tiwi series) featured a naked woman, so Menzies settled on *Man in a landscape*, in which the man in the foreground has his arms draped over some boulders. It is unclear what the Queen thought of the painting because when Drysdale was later invited to attend a dinner with the royals in Canberra, she asked the artist what the painting 'means'. Drysdale replied that it is a picture of 'a man trying to hold onto his land'![13]

Unlike Streeton, whose popularity he inherited, Drysdale was essentially a studio painter. He was a keen photographer and his paintings were based on sketches and photos taken during extensive travels. However, they were always compositions put together from a range of images, rather than impressions of what the artist actually saw. Like Roberts, Drysdale was a story-teller, but he was less interested in authenticity than Roberts. He wanted to capture a sense of space and time that was totally different to anything that people living in the cities could imagine. He became fascinated with an ancient land and its old and new inhabitants.

The example of the two Arthurs – Streeton and Boyd – shows that landscape painters in Australia have sometimes felt a need to speak out for the cause of nature conservation. However, an even closer relationship between art and activism has shaped the development of Australian wilderness photography, because skilled photographers have sometimes sparked public interest in conservation campaigns and because campaigns have sometimes encouraged photographers to take their work more seriously. An important pioneer in the field was the Tasmanian photographer John Watt Beattie who led a public campaign in 1908 to greatly expand the area set aside by the state government for 'preservation of the scenery' along the Gordon River in the south-west corner of the island (Bonyhady, 1994). Beattie began his campaign by giving an illustrated lecture to the Royal Society of Tasmania to coincide with the opening of parliament that year and one of his images was subsequently published in the Launceston *Weekly Courier*, the largest circulation weekly in the state. After Beattie showed his slides to a large public meeting in Queenstown, the government responded by promising the expansion of the reserve that Beattie was demanding. Unfortunately, ill-health on the part of the premier prevented the government from acting on this promise and so Beattie had to begin campaigning once again in 1910, when he learnt that a new Gordon River Sawmilling Company had already been given access to part of the area he wanted reserved.

It is ironic that Beattie's efforts had to be repeated sixty years later by another accomplished photographer, Olegas Truchanas, who tried to build public opposition to government plans to flood the beautiful Lake Pedder in the same region by presenting a slide show at the Hobart Town Hall titled 'Tasmania's Unique Southwest' (Bonyhady, 1994). Truchanas also took his slides to Canberra where he showed them to an audience of federal parliamentarians in the hope that they would be moved to intervene against the Tasmanian government's plans. Unfortunately, his efforts – and the efforts of many other Lake Pedder campaigners – failed to save the lake and Truchanas tragically drowned during a canoe trip on the Gordon River in 1972. Truchanas had acted as a mentor for another migrant from Lithuania, Peter Dombrovskis, who took up a similar role in the subsequent, successful campaign to prevent the flooding of the Franklin River (see chapter 10). One image by Dombrovskis, *Rock Island Bend*, became an icon of the Franklin campaign and his work was subsequently featured in posters and calendars sold nationally by the Wilderness Society. Dombrovskis raised the art of wilderness photography in Australia to a new level. Like his mentor Truchanas he died in the wilderness he loved; during a solo walk in 1996.

Although wilderness photography was pioneered and developed into a fine art in Tasmania, it also played an important role in conservation campaigns in other states as well. An important pioneer in New South Wales, was Henry Gold, who migrated from Hungary in 1956. Soon after arriving in Australia Gold joined the Sydney Bushwalkers and fell in love with the Australian bush.[14] He got so interested in photographing wild places that he took himself to the USA to study the work of renowned landscape photographers like Anselm Adams. Gold's work came to public prominence in 1968 when his images were used extensively in the public campaign against plans to mine the Colong Caves in the Blue Mountains district (see chapter 6). To capture some of the images used in that campaign Gold camped alone in the area for weeks at a time, sometimes rising early on bitterly cold mornings in order to capture the ambience he wanted for particular shots.

Virtually by definition, wilderness photographers have focused their work on 'pristine', people-free places on the other side of the frontier from human settlement (see chapter 6 for a discussion of some of these concepts). This put them into a very different tradition to that pioneered by artists like Tom Roberts and Russell Drysdale, which was interested in interactions between people and landscapes. From the early 1970s, some film-makers participating in a renaissance of Australian film-making picked up the latter tradition and a notable feature of 'second wave' Australian films is that different landscapes often feature quite prominently in the shaping of the story – from early films like Peter Weir's *Picnic at Hanging Rock* to the mid-90s classic *Priscilla, Queen of the Desert*.

Visual artists – from the early painters through to contemporary film-makers – have played a significant role in ongoing public discourse about land and identity. In the first instance, ecological pioneers in this field have challenged the inappropriate perceptual frameworks that were brought from Europe, causing a deep sense of not belonging among the settlers. Perhaps it needed the sensibility of artists to recognise that there was a major problem of perception, even if the early euro-Australian artists had contributed to that problem. Sensibility may also explain why individual artists like Margaret Preston and Russell Drysdale were among the first public figures to say that white Australians have much to learn from Aboriginal perceptions regarding people and the land.

Artists who have gained popularity and influence in Australia have generally engaged in story-telling and myth-making. Perhaps there is a natural progression from an interest in local stories embedded in particular landscapes to broader themes of land and identity that are reflected in the constant creation and recreation of a prevailing cultural mythology. When a group of artists – that included Arthur Boyd and John Perceval – held a joint exhibition of their work in Melbourne in 1959 they wrote in the preface to the catalogue that:

> We live in a young society still making its myths. The emergence of myth is a continuous social activity. In the growth and transformation of myths a society achieves its sense of identity. In the process the artist may play a creative and liberating role. (Philips, 1967, p. 15)

Of course, it is only white Australian society that can be considered young, because indigenous Australians have very ancient roots in this land. Nevertheless, the point is well made that artists have helped white Australians engage more creatively in the process of myth-making in order to find a more grounded sense of identity in the 'new' land. It was probably Sidney Nolan who engaged most consciously and playfully in the art of myth-making. But he was certainly not alone and a mutual interest in mythology is probably one of the things that has brought about a certain convergence in the work of indigenous and non-indigenous Australian artists. This convergence has been less the product of conscious intent (because that did not work well when Margaret Preston tried it) and more the result of mutual interest in landscape stories and myths. Of course, it is important to note that this has not been a one-way convergence, because the now-famous 'dot paintings' of the desert communities were significantly inspired and facilitated by the availability of non-indigenous technology and traditions. The dot paintings have also moved from a fairly direct use of traditional iconography to more abstract images aiming to capture the land in its transient moods. At the dawning of the twentieth century the Heidelberg painters were intent on forging a distinctive school of Australian landscape art that would grow out of increased attentiveness to Australian landscapes. The subsequent cross-fertilisation of euro-Australian and indigenous Australian art has given that project much more momentum by the dawning of the twenty-first century.

CHAPTER 4 | OF DROVERS' WIVES
AND A TIMELESS LAND:
*Land and Identity in Australian
Literature*

INTRODUCTION
In the lead-up to the federation of the Australian nation in 1901, both the
Heidelberg painters and the *Bulletin* writers were engaged in the nation-
building task of tackling the 'cultural cringe' towards all things English. For
the painters the task had been to unsettle the perceptual framework, which
meant that the settlers had gazed at Australian landscapes with European
eyes. Painters like Roberts and McCubbin were also interested in the stories
that were being embedded in particular landscapes (see chapter 3), drawing
inspiration from 'bush poets' such as Adam Lindsay Gordon. In their exploration of an emerging Australian identity, the *Bulletin* writers were interested
in similar stories, but, of course, their job was to embellish them; to turn local
stories into legends of the 'Australian experience'. While Banjo Paterson, following in the footsteps of Gordon, picked up the tradition of romanticising
the experience of the bush pioneers, Henry Lawson revisited the melancholic
mood of earlier writers and moved from there into an exploration of stoical
responses to defiant landscapes (foreshadowing the later work of the painter
Russell Drysdale). Lawson's approach took him more deeply into the cultural
roots of a distinctively dry Australian sense of humour and intolerance of
rank and privilege (the 'mateship myth' that pointedly excluded the indigenous people). While Paterson's work was more popular, Lawson's was more
influential; confronting as it did the dark side of the settler experience. It was
Lawson who had more influence on the next generation of writers with his
shift from raw sensibility to an exploration of stoicism.
　　Perhaps more women than men were prepared to contemplate the
dark side of the colonial experience because women tended to dominate the
next generation of writers to tackle these themes. Henry Handel Richardson
explored questions of belonging and not belonging in Australia while
Katharine Susannah Prichard and Mary Gilmore tried to locate the silenced
voices of the indigenous people. Eleanor Dark reinterpreted the story of

settlement in the Sydney region in her historical novel *A Timeless Land*, published in 1941. In the meantime, May Gibbs, who used her imagination and empathy to bring alive an enchanted forest for Australian children, had pioneered a genre of children's literature that would continue to grow; ensuring that more Australians grew up with a heightened awareness of the bush.

The dark side of colonisation and empathy with the non-human world were prominent themes in the early work of poet Judith Wright, who sought out stories embedded in the New England landscapes that had nurtured her as a child. She also joined the search for the silenced voices of the indigenous people and this search eventually led her into a close friendship and collaboration with Aboriginal poet Kath Walker (later known as Oodgeroo Noonuccal) and a lifetime commitment to the causes of the Aboriginal people. As both poet and conservation activist (see chapter 6), Wright displayed an ability to make white Australians feel simultaneously more at home and more unsettled in Australian landscapes that 'spoke in the language of the leaves'.

Feeling from childhood that she was an empathetic outsider in landscapes that she loved, Wright was easily drawn into a deeper engagement with the landscapes around her and, from there, into the onerous work of conservation activism. In the early 1970s she was both surprised and delighted to find herself sharing some public platforms with a distant relative of patrician origins, Patrick White, who, as the son of a transplanted rural aristocracy, followed the well-trodden path back to 'mother England' before realising that the real inspirations for his creative work as a novelist lay in his real homeland, Australia. White was not the first Australian to discover that life in England was the life of an exile, but he probably made more of his return home by exploring the psycho-cultural roots of belonging and not belonging. A string of novels – from *The Tree of Man* (1955) to *A Fringe of Leaves* (1976) – took him towards the platforms that he shared with Wright. His Nobel Prize of 1973 was partly a recognition of an emerging and distinct Australian literature that was drawing inspirations from Australian landscapes.

Between them, Wright and White were able to inspire the next generation of writers who were then able to bring a higher level of landscape literacy into their creative work. Among them was David Malouf, who has said that the task of writers is to mythologise the places in which people live in order to stimulate curiosity and a desire for engagement. Influential Australian writers have certainly been able to stimulate a stronger desire to explore the interrelationships of people and places; at the same time they have had the task of re-examining the fundamental mythologies of our cultural identity; an identity forged through an uneasy relationship between people and the land.

THE *BULLETIN* WRITERS: BANJO PATERSON AND HENRY LAWSON
On July 9, 1892 Henry Lawson opened his now-celebrated debate (in verse) with Banjo Paterson in the pages of the *Bulletin* with a poem called 'Borderland', which began:

> I'm back from up the country – very sorry that I went
> Seeking out the Southern poets' land whereon to pitch my tent
> I have lost a lot of idols, which were broken on the track
> Burnt a lot of fancy verses, and I am glad that I am back.
> (in Semmler, 1974, p. 81)

He had not mentioned Paterson by name, but his jibe about 'Southern poets' and their 'fancy verses' was aimed at those who, like Paterson, lived in the city and wrote romantically about the experience of 'bush life'. Paterson opened in the July 23 edition with a lengthy rejoinder that began:

> So you're back from up the country, Mister Lawson, where you went
> And you're cursing all the business in a bitter discontent . . .
> (pp. 81–82)

and concluded with:

> Did you hear the silver chiming of the bell-birds on the range?
> But, perchance, the wild birds' music by your senses was despised . . .
>
> You had better stick to Sydney and make merry with the 'push',
> For the bush will never suit you, and you'll never suit the bush.

After this exchange the stoush became more willing with Lawson addressing 'Mr Banjo' by name, accusing him of travelling the country 'like a gent'. In a verse published on August 6, he asked, rather accusingly:

> Did you ever guard the cattle when the night was inky black
> And it rained, and icy water trickled gently down your back,
> Till your saddle-weary backbone started aching at the roots
> And you almost heard the croaking of the bullfrog in your boots?
> Did you shiver in the saddle, curse the restless stock and cough
> Till a squatter's blanky dummy cantered up to warn you off?

The contest raged on for a couple more months, with a couple of other bards weighing in on Lawson's side, before Paterson tried to end it with a conciliatory poem suggesting that 'we must agree to differ in all friendship, you and I'.

Lawson did not respond and, although both men made references to the debate in much later poems, their bout had finished inconclusively after a flurry of punches. Most commentators feel that Lawson probably landed the heavier blows, a point that Paterson virtually conceded when he wrote, many years later:

> ... I think that Lawson put his case better than I did, but I had the better case, so that honours (or dishonours) were fairly equal. An undignified affair, but it was a case of 'root hog or die'. (Semmler, 1974, p. 85)

This is a pretty dramatic appraisal by Paterson because, in truth, the debate was fairly contrived. Paterson also later acknowledged that the idea of the debate was put to him by Lawson one day when he suggested it would give them both a chance to attract attention to their work. According to Paterson's recollection, Lawson had said: 'We ought to do pretty well out of it, we ought to be able to get in three or four sets of verses before they stop us.' (Semmler, 1974, p. 81). By all accounts, the sharp exchange did not damage the relationship between the two Sydney-based writers. We know for sure that Paterson later visited Lawson at his home and that, as a solicitor, he represented Lawson in his dealings with the publishers Angus and Robertson. It seems that the two men had a respect for each other's work that was not apparent in the exchange. But there was substance to the debate because the two writers represented different traditions that were current in Australian literature. Paterson was from the romantic tradition, pioneered by Adam Lindsay Gordon, which suggested that the authentic Australian experience was to be had in the 'bush' not in the 'squalid' cities. Lawson, by contrast, was pioneering a tradition that suggested rural life was a stern test of character that could, at best, make people more aware of their dependence on other people. This might be called the 'stoic' tradition in bush writing and the support Lawson got in his debate with Paterson suggested it was gaining ground. Of course, what they had in common was the view that the 'bush experience' was pivotal in shaping an Australian identity.

By character and circumstance, Lawson and Paterson were as different as canvas and silk. Lawson had come from a very poor rural family and struggled financially all his life. He was warm and gregarious but highly vulnerable. Paterson was born into a more prosperous rural family. His family also experienced financial problems when the fortunes of farming turned against them; forcing them to retreat to Sydney. But Paterson was able to build a successful career as solicitor, journalist and newspaper editor. Among his recreational pursuits he included playing polo and frequenting Sydney's rather exclusive Australia Club. By personality he was quite aloof and formal. Over the years, the literary critics have been much kinder to Lawson than Paterson. They point out that Paterson's literary output mainly came out of an intense burst between 1889 and 1899 and that, although he lived until 1941, he produced nothing of note after that (see Roderick, 1993). It has often been said that he is really only remembered for three works – 'The Man from Snowy River', 'Clancy of the Overflow' and 'Waltzing Matilda'. It is also acknowledged that Lawson's literary career peaked around a self-proclaimed 'high point' in 1901, when he completed his most complex set of stories, published under the title *Joe Wilson and His Mates*. After that his output also declined in both quantity and quality before he died a lonely death in 1922. But the critics insist, with good reason, that Lawson dipped into a much deeper well of human experiences and emotions and tackled much more difficult questions about the nature of Australian identity. His influence on the writers who came after him was much more profound.

However, those who dismiss Paterson's literary legacy fail to explain the enduring popularity of his best work. Even if he is only remembered for a narrow range of poems, written in a narrow span of his life, those poems seem to have touched the soul of the nation, to be integrated into its psyche. Many writers would cut off their writing hand to have that sort of success. Again, it is argued that these poems have such popularity because they were rammed down our throats when we were in primary school (Roderick, 1993). But we are exposed to a number of 'classics' at a tender age and few lines stick in the memory as much as the opening line of *The Man from Snowy River* ('There was movement at the station, for the word had passed around . . . '). More Australians can remember the words to 'Waltzing Matilda' than the national anthem. Most of us retain a visual image of 'the Banjo' sitting in his 'dingy little' city office exchanging correspondence with a man on horseback, following a herd of cattle out on the 'sunlit plains extended'. Of course, Paterson's poems were designed to make us feel good about being Australian. That was the aim of the romantic bush poets. Lawson is less remembered for individual poems, stories, or lines. People may remember funny incidents, tender moments, or particular characters. They are likely to know that Lawson lived a troubled life every bit as challenging as any presented in his fiction. Memories of Lawson are not likely to make us feel good about ourselves, but they may make us think more deeply about the experience of living in Australia.

Despite their differences, Paterson and Lawson were both adopted as '*Bulletin* writers'[1] because both had work regularly published in the magazine and both received support and encouragement from its editor J. F. Archibald. Archibald migrated to Australia in 1885 and used the magazine he established to argue that Australia needed to come of age and establish its own identity. He probably had views similar to those of the Brisbane-based journalist, publisher and trade union activist William Lane.

But, whereas Lane's utopian vision for Australia soured in the wake of big industrial battles in the early 1890s (and he took a band of followers off to Paraguay to start a 'New Australia'), Archibald kept up the good fight. He enticed A. G. Stephens, who had worked on Lane's paper *The Worker*, to Sydney and together they assembled a stable of writers who could continue to build the magazine's reputation (see Palmer, 1954). Archibald was not particularly fond of poetry and he had a prejudice against anything that seemed bookish or lacking in humour, but he wanted to make the magazine lively and diverse so that it might appeal to 'shearer or selector, factory-hand or man of business' (Palmer's phrase, p. 107). For this reason, both Paterson and Lawson appealed to Archibald; for him their contrasting tones were probably like the flip sides of the coin of national identity. Paterson's work had more immediate public appeal, but Archibald knew that Lawson was well worth indulging, despite his weakness for the bottle. As a group, the Sydney-based *Bulletin* writers can be likened to the Melbourne-based Heidelberg painters, with Paterson having the easy appeal of Streeton and Lawson the intensity of Roberts (see chapter 3).

Over the years, cultural historians have made a similar critique of the *Bulletin* writers as that made of the Heidelberg painters – that they presented outdated and misleading interpretations of the emerging Australian identity.[2] They built a rural legend, it is argued, at a time when two-thirds of Australians were already living in the cities or major towns; when the future of the country was being forged in the urban centres. But this ignores the fact that coming to terms with the 'bush experience' was an important part of overcoming a cultural cringe that had suggested that harsh Australian landscapes could not inspire artists and writers. To the extent that culture was thought to exist in Australia it might be found in the cities where cultural centres (theatres, galleries etc.) had been constructed according to English models. In his 1957 novel *Voss*, Patrick White suggested that nineteenth century Australians 'huddled' into towns on the coast because they were intimidated by the challenge of the wide-open spaces of the vast, dry continent and the 'mysterious' Aboriginal people who lived there. Even in the 1890s, most Australians were living on the coast looking out towards the sea, casting nervous glances over their shoulders at the landscapes behind them. In this context, it was timely and courageous for the *Bulletin* writers to turn the attention inland. Lawson, in particular, was prepared to confront the 'dark side' of the Australian experience. Ecologically, it was more important to consider the experience of coping with 'foreign' and challenging landscapes than to dwell on the impact of the built environment (even if this had developed a character of its own[3]). The strong public response to the work of the 'bush poets' suggests there was an unconscious awareness of this imperative.

Critics have also pointed out that both Paterson and Lawson wrote about 'bush life' from the relative comfort of the city (although Lawson's life was rarely comfortable). However, both men had plenty of direct experience of rural life and maintained contact with it. Lawson, of course, also turned his attention as a writer to the struggles of city life. But, somehow, his sensitive portrayals of the injustices to be found in the city lacked both the conviction and impact of his 'bush stories'. He is remembered for work that captured the attention of his readers and inspired the next generation of writers.

Henry Lawson was born in 1867 in a 'tent'[4] on the Grenfell goldfields in New South Wales.[5] His father, Niels Hertzberg Larsen, was a Norwegian sailor who jumped ship in Melbourne, lured by the prospect of digging up a quick fortune in gold. Travelling north through the goldfields of Victoria and New South Wales he washed up at a place called New Pipeclay (near Mudgee) where he married 18-year-old local girl Louisa Albury after a very brief courtship. The couple had gone to Grenfell to try their luck when Henry arrived as their oldest child. Failing to make their fortune, they returned to the Mudgee area where Louisa convinced her husband to take up a selection on the road between Mudgee and Gulgong. Predictably, the farm failed and when, in 1870, hopeful prospectors began streaming past his house on their way to a new find at Gulgong, Niels – now known as Peter Larsen – decided to try his luck once more. This time he had a little success and while he was picking at the dirt, Louisa and one of her sisters were picking at seams in a little dressmaking business and general store that they opened on the goldfields. Between them they salted away enough money for Louisa and her two sons to enjoy a holiday with relatives in Sydney. However, by 1873, luck had abandoned Larsen once again and the family returned to the property at New Pipeclay where he pieced together a living out of farming, prospecting and building. It was as a house-builder that he finally began to prosper.

By all accounts Henry was fond of his father. His first published story, *His Father's Helper*, certainly suggested this was so. But he must have also felt his father's frustrations when his dreams of a good life in the new country consistently evaporated. Larsen tried to fit in – changing his surname to Lawson – but his life was never easy and when his wife looked at the prospects facing the family in 'the bush' she left the marriage and took her children to live in Sydney. Although much younger than her husband, Louisa was a strong woman and she was largely responsible for awakening Henry's interest in literature. With the family moving so often, Henry only ever had about four years of formal education in his life, but his mother introduced him to the fiction of writers like Charles Dickens and Marcus Clarke and, in Mudgee, he struck a teacher who introduced him to the short-story writing of Bret Harte.

Henry Lawson. (National Library of Australia)

Louisa left her husband, and rural life, in 1883 and started a modest boarding house in the Sydney suburb of Marrickville. Once she was established in Sydney she sent for Henry, who had been working with his father as a house-painter, and found him a job as a coach-painter at railway workshops at Clyde. Not long after Henry arrived the family moved into a house in Phillip Street near the centre of the city. From there he had to walk three kilometres through city streets to Redfern every day to catch the 'workmen's train' to Clyde and it was on these walks that he came across homeless and destitute people begging for money in the grey light of dawn. If his own life in the country had been tough, he thought these people had it worse and he began to write Dickensian stories about the slum-dwellers.

Louisa's interests broadened and, by taking herself to public meetings and discussion groups, she came across people who seemed to be able to articulate some of her frustrations. One such person was a Scottish migrant, George Black, who was both a spiritualist and a staunch socialist. Louisa's house was centrally located and so people like Black began calling in there to meet with like-minded folk. It became a republican stronghold and, under such influences, Henry wrote a republican anthem called 'Sons of the South' which he submitted to the *Bulletin* in 1887. It was published as 'A Song of the Republic'. In 1888 Louisa began to publish her own journal, *Dawn*, which campaigned for a better deal for women.

Henry kept in contact with his father and was devastated when he died in the Blue Mountains in 1888. Remorse over the death of his father seemed to worsen his problem with alcohol and his mother suggested a trip for him and his younger brother Charles to Western Australia to take their mind off their loss. Lawson returned from this trip to take up a position on the newly established *Republican* newspaper. From there he was offered a position on a new Brisbane journal called *The Boomerang*. The new venture failed but Lawson was in Brisbane long enough to meet William Lane and his future sparring partner as literary editor at the *Bulletin*, A. G. Stephens. After the failure of *The Boomerang*, Lawson returned to Sydney where J. F. Archibald took him more directly under his wing. In 1892 he sent him on a trip 'up the country' to Bourke and Hungerford (on the Queensland border) – a trip that resulted in a number of famous stories. In 1894 Louisa Lawson published the first collection of her son's work under the title *Short Stories in Prose and Verse*. The following year he signed a contract with Angus and Robertson.

As mentioned, Lawson is best remembered for his 'bush stories' published in collections with titles like *In the Days When the World was Wide* and *While the Billy Boils* (both published in 1896). While his poetry was limited, he mastered the art of short-story writing, showing greatness in his ability to capture poignant moments using an Australian vernacular. Many critics say he reached the peak of his craft in a series of stories published under the title *Joe Wilson and His Mates* in 1901, and this series is the closest he ever came to tackling the complexity of a novel. His stories of city life are also well-crafted, even if they did not achieve the same resonance. However, it is interesting to note that in writing about the city he often adopted the perspective of an outsider. This is evident in the poem that is arguably his best and probably his best-known commentary on city life – 'Faces in the Street', published in 1888. It begins:

> They lie, the men who tell us, for reasons of their own
> That want is here a stranger, and that misery's unknown;
> For where the nearest suburb and the city proper meet
> My window-sill is level with the faces in the street –
>> Drifting past, drifting past,
>> To the beat of weary feet –
> While I sorrow for the owners of those faces in the street . . .

While his writing career may have peaked around 1901–02, his personal life was already in turmoil by then. On a special trip to Melbourne in 1885 he had learnt that doctors could do nothing about his deafness and, as mentioned, his battles with the demon drink began in earnest around 1888. In 1890 he thought he had found a perfect match in a young school-teacher who had come to Sydney after years of experience in the country and lived for a short time at his mother's house. Her name was Mary Cameron, but she later became known to the world as the poet Mary Gilmore. It is not entirely clear if Cameron was put off by Henry's drinking or by the experience of living with his mother – she later recalled that Louisa was 'without exception the dirtiest housekeeper I ever knew' (Dutton, 1991, p. 11). But either way she spurned Henry's advances. In 1896 he proposed to the daughter of an owner of a well-known Sydney bookshop (McNamara's), Bertha Bredt. Although she was advised against the marriage by the publisher George Robertson – who told her Lawson was a 'temperamental genius' and 'a drunkard' (Roderick, 1982) – she accepted. The marriage was already rocky when the couple, now with two children, went to London for a two-year stint in 1900. Lawson's work was better received in London than he could have imagined, but, away from friends and family, Bertha couldn't cope with her husband. She had a 'breakdown' and was admitted to Bethlem Royal Hospital, and when Mary Gilmore and her husband, Colin, visited in 1902 they offered to take Bertha and the children home leaving Henry to complete the proof-reading for his latest publication, *Children of the Bush*.

Soon after Lawson arrived back in Australia, late in 1902, he tried to commit suicide by jumping from a 25-metre cliff at Manly. By April 1903 he was in Prince Alfred Hospital being treated for alcoholism and Bertha filed for a divorce. Over the next 10 years he was in and out of hospitals, mental hospitals, and jails (for failure to pay maintenance for Bertha and the children). On September 2, 1922, he died, alone, in the backyard of his latest dwelling, in Abbotsford.

His death seemed to set off a cathartic response for those who had known and loved him and his work. He was given a huge public funeral with his estranged wife as chief mourner. A wide range of friends and associates contributed to the book *Henry Lawson By His Mates*. A statue of the 'people's poet' was erected in the Sydney Domain. His work was as popular as ever and his life began to achieve legendary status. It all seemed like a story he might have written himself. As the central character he would have been known for his habit of firing questions at people he had just met and, offering those he liked a firm handshake and the phrase 'You'll do me.'[6] The dark side of Lawson's life is often downplayed in commentaries about his life and work. Yet, as David Tacey (1995) has pointed out, he has resonated with Australian readers largely because he was prepared to confront the 'abyss' – the awful fear that we might become spiritually impoverished people in a land that offers little or no comfort. Only by admitting such fears, Tacey argues, can we begin to achieve a peace of mind in an accommodation with the land that holds us. But if you stand too close to an abyss you can fall into the darkness, as Lawson did in his own life.

Of course, there is a delightful, light side to Lawson's best work – a sense of humour that is as dry as the country; a humour that encourages us to laugh our way out of despair. Very few of his stories mention the natural environment directly, yet it is always present as the essential context for the story. It takes human solidarity to survive in harsh conditions and the brief flowerings of human kindness in the desert of difficult lives is what Lawson liked to describe most (as in 'The Union Buries Its Dead'). It is a message about interdependence and coexistence – important ecological themes.

The *Bulletin* writers have been heavily criticised for presenting racist and sexist notions of an Australian identity. Certainly it is true that many radical nationalists of the 1890s, including some who contributed to the *Bulletin*, were openly racist – particularly towards the Chinese. They certainly didn't have any progressive views regarding indigenous Australians. But they preached a radical message of egalitarianism that would actually challenge discrimination on the basis of class, race and sex. Lawson, in particular, stood unequivocally on the side of 'outsiders' and – with a mother like Louisa – he was quite sensitive to the plight of women. In one of his most famous stories, 'The Drover's Wife', Lawson presents a sympathetic portrayal of a strong woman character. In this and other stories he was interested in the stoic resistance of women who faced even greater hardship than the men (a theme that Russell Drysdale later returned to with paintings like 'The Drover's Wife' – see chapter 3). Certainly, the concept of an egalitarianism born of 'mateship' is based on male activities and most of Lawson's characters were men. He was, of course, a man of his times, but he had a sensitivity that has consistently impressed women commentators, including the poet Judith Wright.[7]

WOMEN'S VOICES: MILES FRANKLIN TO ELEANOR DARK

Lawson's direct dealings with women were problematic. The breakdown of his marriage has been mentioned above and, although he clearly admired his mother's achievements, he dubbed her the 'Chieftainess' because he found her intimidating. Still he managed to maintain a friendship with Mary Cameron/Gilmore even after she had rejected him and he provided warm encouragement to the emerging woman writer Miles Franklin. When Franklin sent the manuscript for *My Brilliant Career* to the publishers in 1901 they sent it to Lawson to ask if he would write a preface. He responded by praising the book for its freshness and insight but he said he first wanted to meet the author to find out if she was 'a mate or a mere miss'.[8] Once he was convinced Franklin was the genuine article, he offered her support for which she was eternally grateful. In an address delivered beside the statue of Lawson in the Sydney Domain in 1942, Franklin praised his achievements by saying that he was:

> ... one of the most powerful of that band which in the 'nineties helped Australians to a realisation of their country. He quickened their instinctive reactions towards it. To remain unrooted in the soil of one's permanent residence is to be forever a prey to nostalgia – a drug so potent, that uncontrolled it can enervate purpose and defer destiny ...
>
> Henry Lawson lighted lamps for us in a vast and lonely habitat ...[9]

When *My Brilliant Career* was published (in 1901) it had an immediate impact, with the *Bulletin*'s A. G. Stephens describing it, in rather exaggerated terms, as 'the very first Australian novel to be published' (Barnes, in Dutton, 1985, p. 183). A largely autobiographical story of a young woman growing up in a remote area, it was quite an immature work – not surprising in view of the fact that Franklin wrote it while she was still at school (Roderick, 1982). Joseph Furphy had written his novel *Such is Life* in 1896, but couldn't get in published until 1903. There is no doubt that Furphy's novel is superior to Franklin's, but *My Brilliant Career* had more early impact because it was written from the perspective of a young woman, quite a novelty at that time. Franklin was so encouraged by the response that she moved to Sydney to become a full-time writer. This was not so easy and she wrote a sequel called *My Career Goes Bung*, which was not published until 1946. Feeling frustrated, she left Australia in 1906 and did not return for almost 30 years. However, her early success encouraged other women to take their writing more seriously and women tended to dominate the next generation of Australian writers.

One of those women writers, May Gibbs, who tends to be neglected in most reviews of Australian literature, was the creator of a whole genre of Australian children's literature. Gibbs, who migrated from England to Perth as a young woman, started her professional life as an illustrator rather than a writer but she discovered that the Australian bush offered up a unique cast of characters for stories written for children. It was in 1916 that May Gibbs managed to convince the publishers Angus and Robertson to publish two stories she had written and illustrated about mythical babies living in the fruit and flower of gum trees and they were an instant success (Walsh, 1985). She had an inkling that they would work because she had earlier (in 1914) struck an enthusiastic response when she included drawings of the 'gumnut babies' in the jacket design for a book she had been asked to illustrate. The first two stories, *Gum-Nut Babies* and *Gum-Blossom Babies*, were followed in 1918 by the even more popular *Tales of Snugglepot and Cuddlepie* and by this time the artist Norman Lindsay had joined the fun with the release of his immortal, illustrated tale – *The Magic Pudding: Being the Adventures of Bunyip Bluegum and his friends Bill Barnacle and Sam Sawnoff*. After the horrors of the war in Europe, magical stories set in the Australian bush were probably welcomed by parents as much as children, but they did encourage children to see the bush as a place of enchantment, not just danger. They enabled parents and children to explore the bush together, with a greater sense of fun and imagination.

In 1924 May Gibbs began a regular comic strip *Bib and Bub* (which she managed to continue until 1967, when she turned 90). More of her books appeared in the 1920s. In the 1930s she was joined by Dorothy Wall, the creator of the adventurous koala Blinky Bill. When Gibbs began, writers like Beatrix Potter and Kenneth Grahame were also locating their stories of animals in the farms and woodlands of England, but Gibbs had the more difficult task because Anglo-Australians had generally felt that Australian landscapes lacked the charm of the English countryside. Furthermore, Gibbs' vulnerable gumnut babies had to confront both the real and imaginary threats that people associated with the Australian bush – snakes, spiders and 'Banksia men'. But the naming of such fears probably had a cathartic effect because only then could people begin to see the other, more gentle, side of bush life.

However, if the bush stories written for children in the period following World War I helped both children and their parents confront some of their fears of the bush, it was more difficult for the writers of adult fiction to confront the dark side of colonisation. Another woman writer who took on this task in a fairly explicit way was Katharine Susannah Prichard, who travelled extensively in the rural districts of Western Australia in the 1920s to gather material for the novels that would follow. A single mother, she even went to work on a remote cattle station in the state's north-west, and out of this experience came the novel *Coonardoo* (1928), which was very courageous for its time in dealing with a love affair between a white cattle station owner and an Aboriginal woman. The station owner in *Coonardoo*, Hugh Watt, is characteristically paternalistic towards both the land that he 'owns' and the Aboriginal people who were there before him when he meets the young Coonardoo, whose name translates as 'wells in the shadows' (see Haynes, 1998). However, he detects in the young woman a 'force ... much closer to the source of things' and his affair with her awakens a recognition that:

> ... here in a country of endless horizons, limitless sky shells, to live within yourself was to decompose internally. You had to keep in the flow of the country to survive ... How could a man stand still, sterilize himself in a land where drought and sterility were hell. (in Haynes, 1998, p. 187)

As long as Coonardoo is around, the station flourishes. But when the affair collapses, through Watt's inability to overcome his paternalism, she leaves and the drought returns. The station goes into steady decline. Although the novel, which was awarded the prestigious Bulletin Prize in 1928, was innovative in identifying Aboriginal people as a source of life and hope, it probably went too far in suggesting that Coonardoo's lust for life would take the form of a virtual addiction to sex. However, it boldly challenged some prevailing stereotypes and challenged its readers to believe that white Australians had much to learn from the Aborigines about how to live with the land and its cycles.

While Prichard's novel really did not get inside the head of Coonardoo, it was the first to cast an Aboriginal woman as a central character. Another woman writer who had a strong interest in Aboriginal people was Mary Gilmore who, as already noted, was associated with the *Bulletin* writers in the early 1890s in Sydney before joining William Lane's ill-fated attempt to set up a New Australia community in Paraguay. Gilmore was out of Australia from 1893 until 1902 and when she returned she built a reputation as a poet and, perhaps more importantly, as a generous mentor for younger women writers. Both Gilmore's maternal grandfather, Hugh Beattie, and father, Doug Cameron, had been unusual members of the rural settler society in that they had an empathy for, and strong interest in, Aboriginal society and culture. In 1934, at the age of 70, Gilmore published a retrospective volume of stories, titled *Old Days: Old Ways*, which included several very interesting essays on what she had learnt about Aboriginal nature conservation practices from her father.

Gilmore's essays had some impact on a reading public that was probably beginning to feel more uneasy about what had been done to the indigenous people in the process of colonisation. However, a book that had far greater public impact was Eleanor Dark's reinterpretation of the settlement of Sydney in *A Timeless Land*, published in 1941. This was Dark's most thoroughly researched book and is undoubtedly her best-known work, but her work in general deserves some consideration in a book about pioneering ecological thought.

Eleanor Dark was born in Sydney in 1901 – the year in which Australia became a nation and *My Brilliant Career* was published.[10] Her mother, after whom she was named, died when she was just 13 and her father, Dowell O'Reilly, remarried a cousin in England after a courtship by mail. Dowell O'Reilly's father had come to Australia from Ireland and worked at a range of physical jobs before becoming a fire and brimstone Christian minister. Dowell was one of eight children and his family's needs made it very difficult for him to pursue his desire to become a writer but, after his father's death, he took some studies in English at Sydney University, where he made the acquaintance of Professor John Le Gay Brereton, who was also something of a mentor for Henry Lawson. He also entered into correspondence with the emerging poet Christopher Brennan. In 1894, at the age of 29, he was elected to the New South Wales Parliament as the Labor Member for Parramatta. He served one term and tried, unsuccessfully, to re-enter the Parliament in 1905, after his daughter was born. His political leanings were to the left and he had many friends in the Labor Party.

O'Reilly cultivated his friendships with influential people but he never had much money himself. The family moved houses quite often and, after her mother's death, Eleanor lived for a period with her maternal grandmother in Mosman. While at Mosman she started at the expensive Redlands school for girls at Neutral Bay. She thrived academically but felt isolated without her father and brother. The family seemed to be at its happiest when they were reunited, together with Dowell's second wife Molly, and living in a house in Vaucluse that had a splendid view of Sydney Harbour. Eleanor later recalled that she enjoyed sitting on the verandah watching the ferries to Watson's Bay go by. They featured in one of her novels – *Waterway* (1938).

Eleanor was only 20 when she married Dr Eric Dark, a former pupil at a school where her father had taught. Dark had returned from the battlefields of France as a decorated war hero and took his young wife Kathleen to Bungendore in the New South Wales southern highlands to establish himself in a medical practice. In 1920 the couple had a son, John, but, tragically, Kathleen died soon afterwards with septic peritonitis. Dark returned to Sydney with his son and began a habit of visiting his former teacher,

O'Reilly, often taking Eleanor for a ride in his car. Eric was as spontaneous as he was courageous and, near the end of his life, he told Eleanor's biographers (Brooks and Clark, 1998) that he proposed to Eleanor in a rather impulsive and dramatic way. He made his decision in an instant when she had turned to smile at him after he had dropped her outside her house. He got out of the car, put his arms around her and caressed her breasts. 'You're very sudden, Eric Dark', she had said to him. Shortly afterwards, in 1922, they were married. The same year they moved to Katoomba, in the Blue Mountains, where Eric would make his second attempt to become a small town doctor.

By any measure, the Darks had an extraordinary partnership. It lasted 63 years and when Eleanor died first, in 1985, Eric said he could not get her out of his mind. According to Brooks and Clark: 'Eric was an optimist, and Eleanor was a pessimist. Eric had grown transparent; Eleanor had known blackness, dark and dreamy introspection' (p. 437). But in partnership they both prospered. Eric not only succeeded in establishing a thriving medical practice; he also became a well-known Blue Mountains identity. Physically strong as well as strong-willed, he built a reputation as a daring rock climber and adventurer, particularly in partnership with the barefooted Dot English/Butler (see chapter 6). He got involved in politics as a prominent member of the Katoomba branch of the Labor Party and wrote about medical and social issues. Eric introduced his wife to the joys of bushwalking and they had similar views on social issues. But she was fiercely independent in the way she pursued her literary career.

As they were settling into Katoomba, Eleanor found herself busy caring for Eric's son, John, and helping establish Eric's practice. Then her own son, Michael, arrived in 1929. In 1965 she told her biographer A. Grove Day (1976) that before going to Katoomba she had harboured a desire to become an actor but when this dream sagged under the weight of her household chores and other obligations she decided to return to writing. Before her marriage, she had dabbled in a bit of poetry and short story writing and had some pieces published in magazines such as *Triad*, *Bulletin*, and *Australian Woman's Mirror*. However, she knew she had not found her medium so she bravely decided to tackle a novel. For this project her relative isolation in Katoomba had some advantages and she started to enjoy the house they had bought, 'Varuna', and its two-acre block near the bush at the bottom of Cascade Street.[11]

Her first two novels, *Slow Dawning* (1932) and *Prelude to Christopher* (1934), explored the emotional lives of people who were in situations similar to her own (both featured doctors and strong women characters). *Prelude to Christopher* was a complex novel with interactions between a wide range of characters and many see it as one of her best (see reviews in Day, 1976). It was published in England to considerable acclaim before being

Eric and Eleanor Dark at their 'secret cave' in the Blue Mountains in 1946. (Michael Dark)

awarded a gold medal by the Australian Literature Society. At the time when Dark was building her reputation, Australian publishers were still reluctant to take a gamble on Australian writers. After the success of *Prelude to Christopher*, Dark had all of her subsequent work published in England, with some of her books also being published in the United States. This led A. Grove Day to suggest, in 1975, that she was probably better known as a writer outside Australia. This may have been the case for many of her novels but it was not the case for the trilogy of historical novels that opened with *A Timeless Land* in 1941. These works firmly established her reputation in Australia.

Dark was also a pioneer writer. She concentrated on contemporary themes and was very innovative in her style and structure. Although *Prelude to Christopher* had established her reputation she followed that work with three quickly written novels that were less ambitious but distinctive in the way they focused on events taking place over a short space of time but involving a diversity of characters. In each of these novels the settings also became very important. They were *Return to Coolami* (1936), *Sun Across the Sky* (1937) and *Waterway* (1938).

Sun Across the Sky clearly follows an environmental theme with its focus on the clash between local residents and a property developer in a small seaside resort. There is something of Dark herself in the character of Lois Marshall, a widowed painter who falls in love with the town's married doctor. Although her dead husband had left her a comfortable legacy by writing 'potboiler' novels, she has decided to follow her creative impulses and this leads her into a sharp clash with the values of the developer Sir

Frederick Gormley. In *Return to Coolami*, most of the 'action' takes place in a car carrying two couples on a journey from Sydney to a cattle station in central western New South Wales. In this case, the journey out of Sydney and across the 'barrier' of the Blue Mountains (where they almost come to grief in a car accident) also becomes a journey of self-discovery for the travellers. It's as if the journey out into wide-open spaces allows them to look at their lives in a different way. The third novel, *Waterway*, is very much a Sydney novel, focusing on people who live near the sea and the harbour and travel to the city on the ferry (that Pixie O'Reilly used to watch). Along with Christina Stead's *Seven Poor Men of Sydney* and Dymphna Cusack's *Jungfrau*, it has been described as one of the first novels to feature its Sydney setting (Brooks and Clark, 1998, p. 186). But it is the place and its history, more than the contemporary city, which matters in *Waterway*. Early in the novel, we find Oliver Denning, the doctor in *Sun Across the Sky*, gazing at the harbour and wondering what it might have been like before the arrival of the whites. This allows him to feel a deeper sense of place, captured in the passage:

> What you see now, spreading itself over the foreshores, reaching back far out of sight, and still back into the very heart of the land, is something in whose ultimate good you must believe or perish. The red roofs and the quiet grey city become intimate and precious – part of the story of which you yourself are another part, and whose ending neither you nor they will see. (in Brooks and Clark, 1998, pp. 186–187)

Living in the Blue Mountains probably reminded Dark of how the early white settlers had felt hemmed in by this rugged landscape. This may explain why the travellers on a journey of self-discovery in *Return to Coolami* nearly came to grief in the Blue Mountains before they were able to reach the open spaces beyond. When she was conducting her research for her historical novels about the Sydney settlement, Eleanor and Eric set out to trace the steps of William Dawes, who had been sent by Governor Phillip in 1789 to find out what lay beyond the hills behind the Nepean River. However, they failed in their attempt to emulate Dawes' experience because Eric's superior knowledge of the topography meant that they found their way to the top of the range.

Eleanor Dark got interested in the idea of an historical novel when, in 1937, she wrote an article about Caroline Chisholm, to be published on the 150th anniversary of the founding of Sydney (1938). As she was conducting research for this in Sydney's Mitchell Library, she was surprised to see how much material the library had about the early contact between the white settlers and the Aborigines in the Sydney region. She decided, out of personal interest, to make a study of the life of Bennelong (she used the spelling Bennilong) – a man who befriended the settlers and was sent to

England as a kind of curio, before ending up a broken figure not respected by either his own people or the settlers. In setting out to write a novel – a trilogy, as it turned out – that would present a radical reinterpretation of the experience of colonisation, Dark committed herself to years of bone-rattling travel on the train down to the Mitchell Library and countless hours of copying out information by hand. She ended up with a case of bursitis of the elbow, but she also had a wealth of material and a whole new understanding of Aboriginal society.[12]

The first of the trilogy, *A Timeless Land* (1941), is woven around historical figures like Bennelong, Governor Arthur Phillip, and Captain Watkin Tench. Dark also added fictional characters in the convict Andrew Prentice and his family and wealthy settler Stephen Mannion and his family in order to discuss a range of colonial experiences, as reflected in the lives of particular people. We discover that Bennelong was not surprised at the arrival of the great ships because his father had frequently taken him to a point overlooking the sea where he described what he saw when Captain Cook's *Endeavour* had sailed by in 1770. He had even made a rock drawing of the winged ship. So Bennelong was prepared and offered a hand of friendship to the newcomers. The novel presents Phillip as a noble-minded and sensitive man, but others take advantage of Bennelong's naïve trust and he is eventually destroyed. By contrast, Andrew Prentice escapes from the colony and manages to survive through the assistance of Aborigines living outside the settlement. It is almost as if Bennelong and Prentice cross paths with the latter having a happier experience. The second book, *Storm of Time* (1948), focuses on the turbulent times surrounding the governorships of Hunter, King, and Bligh and the third volume, *No Barrier* (1953), covers the more successful governorship of Macquarie and the successful crossing of the Blue Mountains in 1813 – 25 years after the founding of the colony.

The trilogy was not only an epic in its subject matter, it was also an epic to write, taking up 15 years of Dark's creative life. After completing the third volume in 1953 she only managed to complete one more novel – *Lantana Lane* (1959). When she started *A Timeless Land* she was very unsure as to how it might be received, but she need not have worried because, even in the midst of World War II, it found publishers in both England and the United States. In London the *Saturday Review of Literature* put Dark on its cover and rated her book 'among the best of the year' and in the US the Book of the Month Club made it their choice for September 1941 (Brooks and Clark, 1998, p. 357). It took a little longer for the impact to be felt at home, but steadily the novel was taken to heart by many Australians. It remained in print all through the 1940s and 50s. It was broadcast on ABC radio and when television came to Australia talk began of making it into a series. Eventually the book was used for an eight-part television series first screened by the ABC in 1980 and repeated in 1981 and 1984. Even those who have not read the book are likely to have heard the title, which has entered Australian folklore.

Apart from its achievement in presenting a view of history 'from the other side', *A Timeless Land* encouraged its readers to see the folly in the efforts of the white settlers to 'tame' the 'foreign' landscapes and recreate outposts of English society. A phrase uttered by Governor Phillip to the ambitious Stephen Mannion at the end of the book may well have been the theme of the entire trilogy: 'You intend to exploit this land; have a care, Sir, that it does not end by exploiting you.' Writing the book probably made Dark more aware of the ecological impact of colonisation. In an essay about 'Australia and the Australians' written in 1946, she said that considerable achievements had been made in a land that had once seemed so hostile, but:

> Against them must be set the ignorance and greed that used the land too recklessly, overstocking it till pastures become deserts; denuding the earth of its vegetation till the precious soil is eroded, and the still more precious rivers silted up; felling trees irresponsibly, without knowledge or forethought, using valuable timber for posts and rails, or even firewood; building barbarously with no thought of beauty. And, darkest of all blunders, heaviest upon our conscience, the blunder of our dealings with the black Australians whose land we stole. (in Day, 1976, p. 134)

Another Australian writer who became interested in an Aboriginal perspective on colonisation was Xavier Herbert – born in the same year as Eleanor Dark (1901) but on the other side of the continent (at Geraldton, Western Australia). After training as a pharmacist in Perth, Herbert went to Melbourne to study medicine, but after a year of that he decided he really wanted to become a writer. So he took off on a long journey around the country to gather material, trying his hand at everything from deep-sea diving to flying aeroplanes. He ended up basing himself in Darwin where he served as Superintendent of Aborigines in 1935–36. While Dark had got her understanding of Aboriginal society from a study of the literature, Herbert got his by working with Aboriginal communities right across the Top End. His first novel, *Capricornia* (1938), was based on his Top End experiences, thinly disguised by fictional names. In the following years he had a number of novels and essays published, but he is chiefly remembered for *Capricornia* and a massive volume that stood as a bookend at the other end of his career, *Poor Fellow My Country* (1975), which begins in Darwin but takes its story as far as Sydney and Melbourne.

THE CONSERVATIONIST POET: JUDITH WRIGHT

In the same year that Xavier Herbert had his first novel published, 1938, a group of writers based in Adelaide decided it was time to more fully embrace Aboriginal influences in their work and they called themselves the Jindyworobaks. A key leader of this group was the poet Rex Ingamells, who was one of the first white Australians to acknowledge the Aboriginal name for Ayers Rock when he wrote a poem titled *Uluru: An Apostrophe to Ayers Rock*. However, despite taking some trips together to central Australia, the city-based Jindyworobaks failed to get beyond a fairly token acknowledgement of Aboriginal influences. A much more serious attempt to locate the hidden voices of the indigenous people was made by the poet Judith Wright who had been haunted by the thought of what had been done to the indigenous people in the New England district where she grew up.

As Wright's biographer Veronica Brady has put it, Wright was born 'under the shadow of war on 31 May 1915, just a few weeks before the landing at Gallipoli' (Brady, 1998). She announced her arrival as a poet with a collection of verse published under the title *The Moving Image* in 1946, just as the shadows of the next global conflict were beginning to fade. This was a time when many Australians were preoccupied with thoughts of the country's place in the global landscape left at the end of years of devastating warfare. But Wright, who felt a need to return from Sydney to New England during the height of the war, wanted to focus the attention of her readers back on the landscapes into which we are born. For her the New England tablelands would always be her 'blood's country'[13] and she had been reminded of the importance of having a sense of belonging during the uncertainties of the war.

Wright was born into a family that, on her father's side, could trace its lineage back to the Speaker in the English Long Parliament at the time of the English civil war, and to the MacGregor clan of Scotland (Brady, 1998, p. 5). Her paternal great-grandfather, George Wyndham, arrived in Australia in 1828 and established the grand 'Dalwood' estate in the Hunter Valley. His son Albert struck out further afield, spending years trying to develop a cattle station on the Dawson River in Queensland before coming back to a fairly rough and mountainous stretch on the eastern fall of the New England plateau. Albert and his young wife May were determined to sink their roots into this land and they chose the name 'Wongwibinda' from the local Aboriginal language – meaning 'Stay here always' (Brady, 1998, p. 9). Albert Wright died not long after the birth of Judith's father Phillip and everyone expected May to sell Wongwibinda and return to the womb of the family in the Hunter Valley. However, with the help of her eldest son, Arthur, she made a success of the farm and gradually accumulated a portfolio of properties in New England, Queensland and western New South Wales. By the time Phillip married local girl Ethel Bigg he had become used to a life of luxury. He inherited a property called 'Wallamumbi' from his mother and this is where Judith was born.

In her autobiography,[14] Wright has said that there was confusion in her family about the meaning and origin of the property's Aboriginal name, which was similar to the name of a small township nearby. Although colonisation of the area was less than 100 years old when Wright was young, local Aborigines were only present as 'a few dark shadows . . . on the fringes of our lives'. Noting that the landscapes had largely become 'bare of any life but our own' she commented:

> This bare landscape had once been an important gathering place for the Aborigines of the eastern side of the tableland. The ground we walked on as children, once had been wet with Aboriginal blood. But such matters were never referred to. When the Anglican churchmen came to the little township of Wollomombi to conduct their services and afterwards to Sunday dinner at Wallamumbi, the injunction against murder in the Ten Commandments was never applied to the past. (pp. 32–34)

Tragically, Ethel Wright contracted Spanish influenza just a few years after Judith was born and she remained in poor health until she died in 1927 (when Judith was 11). Phillip had to be the backbone of the family and Judith became very attached to him. She preferred to be outside the house because her mother's chronic illness created a rather sombre atmosphere inside; so she spent most of her time in the male world of stockmen and farm-workers. She liked to sit and listen to their stories, including those of Ted Chalker, who became the voice in the poem 'South of My Days'. She also liked to explore the property on her own and caused quite a panic one day when she took herself on a long walk, across a dangerous creek, to visit the family of one of the stockmen.

Wright did her early schooling by correspondence and was introduced to poetry and Australian literature by her mother – she later recalled that she identified quite strongly with Sybylla in *My Brilliant Career*. When she was in the house she tended to retreat into the world of books, where her mother's illness and her own growing deafness could be ignored. She later wrote (*Half A Lifetime*, p. 38) that she felt some guilt about not being able to enter more fully into her mother's world, saying that she developed a 'heedless dislike of sewing, crochet and most of the indoor female occupations'. Her preoccupation with books was not thought to be very healthy and, in order to avoid directions to 'do something more useful', she often took herself to a hidden corner of a tankstand that was adjacent to the kitchen wall. When her mother died she was sent off to boarding school in Armidale. Her isolated life, partial deafness and short-sightedness made her feel awkward and out of place with the other students. But she did well academically; well enough to qualify for entry to Sydney University.

While living in Sydney, Wright often bought a copy of the Communist Party's *Tribune* newspaper, attended lectures by the left-wing

Labor parliamentarian Eddie Ward, and went to the left-wing New Theatre, where she once heard a talk by her childhood hero Miles Franklin (see *Half A Lifetime*). When she discovered that Mary Gilmore lived a short distance from where she had lodgings in Kings Cross, she knocked on her door with a bunch of flowers to pay respect to someone she admired for what she had done as a person more than for her writing. Walking through suburbs like Redfern and Surry Hills (as Henry Lawson had done many decades earlier) she saw signs of poverty that disturbed her and she became more acutely aware of her relatively privileged background. She became committed to the cause of social justice but felt that the communists she met were as one-eyed in their view of the world as the people they opposed (*Half A Lifetime*).

When she finished at Sydney University (although not with a degree because she did not have the prerequisites to enrol in a required subject) she left for Europe with a student colleague, Cecily Nixon. Travelling through Europe in 1937 (including Germany) she felt the threatening presence of fascism and became aware of the gathering clouds of war. She felt a sense of relief when she returned to London and an even greater sense of relief when she set foot back in Australia in 1938. By now she was determined to make it as a writer. She felt fortunate to land a clerical job at the university where she could keep in contact with other aspiring writers and she started getting poems published in magazines like *Southerly* and *Meanjin* (then known as *The Meanjin Papers*). When Australia got drawn into the war she volunteered to help organise the Air Raid Precautions Centre, which developed plans to pack people into the underground railway system in the event of a Japanese attack. For her literary work she was beginning to think she would be better off in Brisbane, where Clem Christesen was producing *Meanjin*. But Brisbane was an even more likely target for Japanese bombs, given that US General Douglas MacArthur was based there, and she felt a responsibility to return to New England to help her father, who had taken on the additional responsibility of co-ordinating a civil defence plan for the region. When she returned to Wallamumbi she felt this was the right place to be at a time of uncertainty. It gave her the incentive to write many of the poems that appeared in her first published volume, *The Moving Image* (1946).

She was well established in Brisbane by the time *The Moving Image* was published but the poems in that volume established her as a New England poet. With sad memories of the war still hanging in the air, readers were captivated by the love of the land that emerges in these poems. Although poems like 'South of My Days' and 'Bullocky' hinted at a nostalgia for the past, they provided entirely fresh interpretations of familiar landscapes and quickly became popular. Outside poems by Paterson and, to a lesser extent, Lawson, 'South of My Days' may be one of the best-known Australian poems, remembered for its startling opening:

A youthful Judith Wright in the 1930s; Judith Wright in 1996.
(Meredith McKinney)

> South of my days' circle, part of my blood's country
> rises that tableland, high delicate outline
> of bony slopes wincing under the winter,
> low trees blue-leaved and olive, outcropping granite –
> clean, lean, hungry country . . .

Through the voice of an old stockman (modelled on Ted Chalker), the poem goes on to tell the story of past events that transpired in this country; stories now embedded in the landscape.

However, in telling the stories of the pioneers, Wright lay herself open to the charge that she was 'celebrating' the colonisation of the land. Those who make this criticism often cite 'Bullocky' and its reference to a 'promised land' fertilised by the blood and bone of the pioneers (see, for example, Tacey, 1995, pp. 73–74). Wright has said that those who criticised 'Bullocky' missed the fact that the 'hero' is actually mad (Wright 1992). However, because of the way the poem was interpreted she decided to withdraw it from all future anthologies not long before the celebration of the nation's bicentennial in 1988. But if 'Bullocky', and other poems in the first volume, were open to this kind of interpretation, Wright made it apparent in other work that she certainly did not 'celebrate' the achievements of those who had come to conquer the land. For example, *The Moving Image* also includes a poem called 'Country Town' which decries attempts by people to impose themselves on the land:

> This is a landscape that the town creeps over;
> a landscape safe with bitumen and banks.
> The hostile hills are netted in with fences
> and the roads lead to houses and the pictures.

And, in 1953, she made the point even more strongly in a poem titled 'Eroded Hills':

> These hills my father's father stripped;
> and, beggars to the winter wind,
> they crouch like shoulders naked and whipped –
> humble, abandoned, out of mind.
> (*A Human Pattern: Selected Poems*, pp. 49–50)

Soon afterwards, in a poem called 'At Cooloolah', she noted that, compared to a blue crane fishing in the water of a coastal lake, 'I am a stranger, come of a conquering people'. Contemplating both the crane and the memory of the 'dark-skinned people who once named Cooloolah she was 'made uneasy, for an old murder's sake'. In subsequent work she went on to develop the theme that white Australians will remain spiritually impoverished as long as we fail to overcome the mentality of conquerors and learn to enjoy the land for what it is. This is what shaped her views on Australia's identity as a nation and, in 1961, she wrote:

> Australia is still, for us, not a country but a state – or states – of mind. We do not yet speak from within her, but from outside: from the state of mind that describes, rather than expresses, its surroundings, or from the state of mind that imposes itself upon, rather than lives through, landscape and event . . . We are caught up in the nineteenth-century split of consciousness, the stunned shock of those who cross the seas and find themselves, as the Australian ballad puts it, in a 'hut that's upside down'.[15]

According to Strauss (1995), 'Wright's initial notion of nationalism relevant to poetry had little to do with political structures or with issues of class or social justice; it had everything to do with being at one with the land.' However, identity with the land is not possible at some abstracted national level; it is something that must be experienced locally and a new concept of Australia, of being Australian, must be built through an amalgamation of regional struggles for identity. According to Strauss, Wright saw a role for poetry in 'living meaning into the landscape . . . enabling Australians to "feel" both their alienation from, and their possibilities of identification with, that landscape as it was experienced in its local particularities' (Strauss, 1995, p. 50). Wright first announced herself as a 'regional poet' with the New England poems published in *The Moving*

Image (1946). However, when she later moved to south-east Queensland and, subsequently, to Braidwood near Canberra, her poetry moved with her, reflecting her attention to the stories in the landscapes wherever she chose to live.

If there is ambiguity or tension in the New England poems about celebrating a sense of place while also recognising that we sit uncomfortably in this land, this is not something that Wright ever sought to resolve or explain. Indeed it was the tension – the uneasy relationship between people and nature – that she wanted to highlight because we are not likely to treat nature with greater respect if we continue to see it as passive, as taken-for-granted. As David Tacey put it, she wanted the settlers to feel 'unsettled' so that we might begin to rethink our attitudes and past behaviours (1995, pp. 70–76). In an essay titled 'Conservation as a Concept', she said we need to approach nature with 'renewed humility' so that we might develop more 'imaginative participation in a life-process which includes us' (in *Quadrant*, 12.1, 1968).

As mentioned, Wright was living in Brisbane by the time *The Moving Image* was published in 1946. She had a paid job at the University of Queensland and was doing voluntary work to help Clem Christesen in the production of *Meanjin*. She was disappointed to find that it was not easy to work with Christesen because she felt he treated her as a secretary rather than a collaborator, but, through him, she came into contact with a lot of other writers. She started attending gatherings of writers who were younger than her and, through these, formed lasting friendships with Barrie Reid and Barbara Patterson (who became Barbara Blackman after her marriage to painter Charles Blackman). Reid, Patterson, Thea Astley and others were part of a 'Barjai Group' of writers who had some contact with the Angry Penguins in Melbourne (see chapter 3) and it was this group that encouraged Wright to publish her first volume of poems. It was also through Christesen that she met Jack McKinney – the man who would become her cherished partner and husband.

McKinney was 24 years older than Wright and a little world-weary by the time they met. During the First World War he had disguised the fact that he had an Achilles heel to join the army, returning from the killing fields of France with a rather severe case of war trauma (he was still prone to nightmares during his life with Wright). He had married in 1920 and had four children, living for a period on a farm at Kingaroy and later at Surfers Paradise where he managed to eke out a living for the family by writing the script for a long-running radio serial called 'The Noonans'. He also managed to get some short stories published and, in 1935, completed a novel about his war experience, *The Crucible*, which won first prize in a competition sponsored by the Returned Servicemen's League (RSL). By the time he met Wright (at Christesen's house) his wife had left him, taking their two youngest children with her, and he had turned his attention to a study of philosophy to try to make sense of a world that he thought was becoming increasingly insane. He came to Christesen with a manuscript and to get help in locating relevant books for his research in philosophy. He had a relaxed manner and most people, including Christesen, took a liking to him. Wright immediately found him comfortable and stimulating to be with. She found that many of his interests were the same as hers and she thought his mind was 'like none I had come across in its quickness of apprehension, its surprising and, to me, often illuminating comments on books and theories I thought I knew about' (Brady, 1998, p. 116). They began to seek out each other's company.

McKinney was a fan of Wright's poetry and he introduced her to the work of thinkers such as Freud, Whitehead and Bergson. He was attracted to philosophy because he wanted to understand the 'allure of violence' and conversations with McKinney triggered many of the poems published in Wright's second volume of poetry – *Woman to Man* published in 1949. After the war was over McKinney's wife returned with the children to reclaim the cottage at Surfers Paradise and for a while McKinney was homeless, taking to the road with his 'bicycle and swag'. Around the same time Wright was becoming concerned that her growing deafness might put an end to her paid job at the university so she started to contemplate the idea of buying a house in the country where she could concentrate on her writing. When she noticed an advertisement in a newspaper for a little cottage at Mt Tamborine, in the hinterland of the Gold Coast, she suggested to McKinney that they check it out together. When they got to Mt Tamborine the cottage they had come to inspect had already been sold but another one, about three kilometres from the village, was up for sale and Wright had enough money saved to make the deposit. For a period Wright kept her job at the university and a small apartment in New Farm, travelling at weekends to join McKinney at Mt Tamborine. However, when she felt more confident in her relationship with McKinney, and in her ability to earn enough money as a writer to live on, she quit Brisbane and moved into the tiny two-roomed forester's cottage that she had dubbed Quantum (because McKinney had told her that the word meant a 'small package of energy').

Although they had little income, Wright has described the period living at 'Quantum' as the most intensely happy period of her life (in *Half A Lifetime*). Wright wanted to know more about the surrounding rainforest but could not find any books to help her in this quest. She did learn, however, that one of the locals, Hilda Curtis, had conducted her own research on the local birds and orchids and was happy to share her impressive knowledge with the newcomers. Through Curtis, Wright and McKinney also learnt more about the community they had joined and, slowly, they began to feel as if they belonged.

While they were living at Quantum Wright fell pregnant. She was 35 and McKinney was 57. His age may have been a matter of some concern but of much greater concern was the fact a doctor who attended Wright after she had suffered a bad riding accident as a child, resulting in a smashed pelvis, had told her that childbearing may threaten her life. As a precaution, Wright returned to Brisbane for much of the time she was pregnant and her father gave her the money to cover the cost of regular visits to a doctor. Fortunately, the pregnancy proceeded smoothly and a baby daughter, Meredith, arrived, without complications, in 1950. Wright has written that, like most men, McKinney showed little interest in the newborn baby but once Meredith was able to walk and talk he became her 'willing slave' (*Half A Lifetime*). Now the problem was that Quantum was too small for three people but, in a happy coincidence, Wright unexpectedly inherited enough money from a distant aunt in England to pay for a larger house closer to the village that had become available. So the family moved into 'Calanthe' – named after a local white orchid that grew in the garden – where they remained for the next 16 years.

For obvious reasons, Mt Tamborine became the first place to jostle with New England for a place in Wright's heart. It was when she became aware that much of her treasured local rainforest might be felled to make way for new housing developments that she joined forces with her friend Kathleen McArthur to form the Queensland branch of the Wildlife Preservation Society in 1954 (this story is covered in chapter 6). Through her work as a conservationist – especially in the 1960s – she became strongly committed to many special places in Queensland. But Mt Tamborine was the place she always came home to (even after night meetings in Brisbane) and it was the place that constantly renewed her commitment to conservation. For both her art and her work as a conservationist she found that an attentive appreciation of the living world around her could renew her inspiration; by making her think more deeply about herself. A good example of this is found in the poem 'The Flame-tree', which begins:

> How to live, I said, as the flame-tree lives?
> – to know what the flame-tree knows; to be
> prodigal of my life as that wild tree
> and wear my passions so? . . .
> (*Collected Works*, 1994, p. 95)

It seems that Wright was struck by flame-trees soon after moving to sub-tropical Queensland. In her second volume of poems, *Woman to Man*, published in 1949, she wrote about a rather surprising encounter with a flame-tree in flower in the poem *Flame-tree in a Quarry*:

From the broken bone of the hill
stripped and left for dead,
like a wrecked skull,
leaps out this bush of blood . . .

Out of the very wound
springs up this scarlet breath –
this fountain of hot joy
this living ghost of death.

Maybe it was a shared fascination because just after they moved into 'Calanthe' McKinney planted a new flame-tree in their garden. After 16 years, Wright had just about given up hope that it would ever come into blossom. But it did, just weeks after McKinney died!

McKinney's death in 1966, after a prolonged illness, closed the door on the Mt Tamborine experience for Wright. Soon afterwards she rented out 'Calanthe' (not yet willing to sell it) and moved with Meredith into a small flat in Brisbane. This was much more convenient for her work in the conservation movement (the campaign to protect the Great Barrier Reef was reaching its peak – see chapter 6) and it was better for Meredith's education. Meredith duly progressed to university and mother and daughter both got involved in protests against Australia's involvement in the Vietnam war. Wright publicly welcomed the election of the Whitlam government in 1972, although she was disappointed with its response to the threat of sand-mining on Fraser Island (see chapter 6). The new government invited her to serve on an Enquiry into the National Estate. She was also a member of the national council of the Australian Conservation Foundation and soon found herself spending as much time in Canberra as in Brisbane. By the time she was sure that Meredith could fend for herself she decided to move much closer to Canberra and eventually found a suitable property surrounded by bush near Braidwood, where she felt able to continue her writing as well as her political work.

It's hard to know if Wright became better known as a poet or conservationist. Her legion of fans will argue that the two go together, but there have been critics who have argued that her 'partisan' political activity in pursuit of particular causes robbed her poetry of a necessary creative independence.[16] This was a debate that Wright herself was drawn into and, in 1965, she fired her own salvo at poets and critics who favoured an academic approach to the art. In disdaining the fashion for long, discursive poems, she said that she preferred to think of her own poems as

small things that flower and waft their seed
and die to make a soil again

Judith Wright and Jack McKinney at 'Calanthe' circa 1964.
(Meredith McKinney)

on sand laid bare by human greed –
a soil to hold and breathe the rain,
and clothe the dunes, and change their air.
(Brady, 1998, p. 221)

It was certainly not easy for Wright to combine her taxing work as a conservationist with her more contemplative work as a poet and many times she tired of the effort. In a number of interviews during the 1970s she said she looked forward to a time when 'the end of my conservation work is in sight' and she once advised Patrick White to stick to writing novels in order to avoid becoming 'too harassed and driven by the battle to concentrate on writing' (Bonyhady, 2000). However, late in her life she looked back on her experiences as an activist with fondness and when interviewer Richard Glover asked her in 1993 if she regretted her decision to allocate so much of her time to activist work on the grounds that there are 'so few great poets and so many activists', she replied 'But there aren't. I get poetry flooding my desk every day, most of it no good, but a really good activist . . .' (Bonyhady, 2000). There can be little doubt that Wright's poetry did sometimes suffer because her political work took so much of her time and energy. A volume of her poems, *Fourth Quartet*, published in 1977 was not well received by the critics with one reviewer saying that many of the poems had a 'flat, complaining, nagging tone' (Bonyhady, 2000a). Yet, as Bonyhady has pointed out, this was a time when she was producing some high quality prose about her own experiences and family background in works like *The Coral Battleground* and *The Cry for the Dead*. And, in retrospect, it is easy to see that her conservation work helped to nurture her great love of nature, which remained the wellspring of her prodigious poetic output. As early as 1975, the poet and academic A. D. Hope – who could have been one to take umbrage with Wright's 1965 broadside aimed at academic poets – wrote that it was impossible to separate Wright the poet from Wright the conservationist because it was precisely the connection between the two that gave her work a special vitality (Pope, 1975).

Although Wright had become a little unfashionable in academic literary circles by the 1980s she remained one of the best-known poets the country has produced and had accumulated a string of national and international awards – including the Encyclopaedia Britannica Award for Literature in 1964, the World Prize for Poetry in 1984 and honorary doctorates from the Australian National University (Canberra), Monash University (Melbourne), the University of New England, the University of Melbourne, the University of Sydney, and Griffith University (Brisbane). She produced six volumes of poetry, three children's books, a range of published essays, an account of the lives of her ancestors in Australia, several works on relationships between white and indigenous Australians, and some highly praised literary criticism. When she died in July 2000, other writers and academics alike, acclaimed her for her capacity to produce art that was driven by a passion for life (both human and non-human). For example, the academic and poet Kevin Hart wrote that he had been captivated by Wright's poetry from his schooldays to the extent that:

> It was hard to tell if I was reading a poem by Judith Wright or living deep inside one. Her poems taught me how to see the country. Or, better, they gathered me inside the country and its people. (Stephens, 2000)

The writer Thomas Keneally also paid a tribute to Wright that touched on her undoubted influence on other writers:

> At a time when it was every writer's sacred duty to be alienated by Australia – to be a European soul descended into this terrible place – she was unaffected. Instead she made her myths out of this place. The spaciousness of her spirit has always been so grand. (Stephens, 2000)

Strauss (1995, p. 93) has said that Wright's immersion in nature led her to a 'vaguely Buddhist' view of the interconnections between all living things (p. 94) and a sense of responsibility for the fate of other forms of life. A sense of guilt about what humans have done to nature in the past may be a useful starting point but it is not enough to achieve genuine compassion. It requires both compassion and humility to realise that our own window of perception is a small one. We cannot hope to decipher the language of the scribbly gum she wrote in a semi-humorous poem of 1955, published in a collection titled *Alive* (1973), but we can learn to

> Listen, listen
> latecomers to my country
> eat of wild manna
>   There is
>   there was
>   a country
> that spoke in the language of the leaves
>   (Wright 1994, p. 329)

In our desire to place ourselves above the 'tyranny' of nature, we dream of immortality and lose sight of the unity constituted by cycles of life and death. As she put it in a poem titled 'Rainforest' in *Phantom Dwelling* (1985), we might try to imagine what the humble frog is telling us in his distinctive way, but

> We cannot understand that call
> unless we move into his dream
> where all is one and one is all
> and frog and python are the same
>
> We with our quick dividing eyes
> measure, distinguish and are gone.
> The forest burns, the tree-frog dies,
> yet one is all and all are one.

In an earlier, more angry poem, 'Australia, 1970', she described human beings as 'self-poisoners' and praised the resistance of those we would kill:

> Die, wild country, like the eaglehawk,
> dangerous till the last breath's gone,
> clawing and striking. Die
> cursing your captor through a raging eye . . .
>
> I praise the scoring drought, the flying dust,
> the drying creek, the furious animal,
> that they oppose us still;
> that we are ruined by the thing we kill.

Wright herself has said that she owed much of her inspiration in the second half of her career as a writer to her friendship with Aboriginal poet and activist Kath Walker. Walker was the person who put a face to the silent voices that still dwelt in the land; the 'dark-skinned people' who had treated the land so differently to the white 'conquerors'. Wright has recalled that she first met Walker through her poetry (Brady, 1998, p. 218) when the publishers Jacaranda Press sent her the manuscript for a collection of poems called *We Are Going*, seeking her opinion as to whether or not they should be published. Wright has said that the poems clearly stood out from the 'general run of largely boring and cliché-ridden verse that thudded on to publishers' desks' (Brady, 1998, p. 218) because they were full of passion and she not only recommended, but insisted, that they be published. She was particularly moved by the poem that gave the collection its title, which included the lines:

> The bora ring is one
> The corroboree is gone
> We are going
> > (in Brady, 1998, p. 219)

because she had once discovered an old bora ring on her father's property and wondered what had become of the people who had made it. She had written a poem about it in *The Moving Image*. She had also included a haunting and disturbing poem in that volume about a place called 'Niggers' Leap, New England' where white settlers had once driven a group of Aboriginal men, women and children over a cliff in retaliation for the killing of some cattle. Her father had told her this story when they passed the place one day during the war, and she wondered if he wanted to get a dark secret off his chest – although he made no direct comment to her after reading the published poem, she gained the 'impression' that he was relieved the story had been told (*Half a Lifetime*). So Wright was well aware of the tragedies that had been inflicted on the Aborigines. However, hearing it from Walker directly made it more real and personal. Knowing Walker intensified Wright's anger about what had been done in the past but the friendship also gave her hope for a reconciliation in the future. In the late 1970s she became convinced that the process of reconciliation needed to be formalised at a political and legal level and she worked together with her good friend H. C. 'Nugget' Coombs to form an Australian Treaty Committee to promote this idea. In the coming years, Wright and Coombs managed to spark a major public debate about the idea of a treaty before they decided the time had come to wind up their committee. However, while working on the committee, Wright managed to complete two books – *The Cry of the Dead* (1981), which she had been researching for many years, and *We Call for a Treaty* (1985). In 1992, the federal High Court accomplished much of what Wright believed a treaty would accomplish by simply ruling that the concept of *terra nullius* (empty land) was a legal fiction and that a form of native title did exist (see chapter 9). However, the need for a treaty as an act of reconciliation was still being discussed at the time of Wright's death in 2000.

THE NOBEL NOVELIST: PATRICK WHITE

As mentioned, Wright publicly welcomed the election of the Whitlam government in 1972 and she was surprised to find that the Nobel-Prize winning novelist Patrick White had done the same. White had come from a family that made its wealth from properties in the Hunter Valley and New England. When Judith Wright was at school in Armidale she was well aware of the 'New England Whites and their "grand houses"' (*Going On Talking*, 1992, p. 50). She knew that Patrick had lived with his mother in Sydney and that he had been sent to school in England, 'ignoring the public schools of the local aristocracy' (p. 49). Then she heard nothing more of him until one day in Sydney in 1942 when she came across a copy of his first novel *Happy Valley* on a second-hand bookstall. Although he was living in England when he wrote it, Wright was very impressed with the way he had evoked the atmosphere of an Australian country town. She later read *The Tree of Man* (1955) 'with deep recognition' and greatly admired *Voss* (1957), becoming a committed fan of White's work. But she had not expected him to publicly associate himself with a Labor government. Then when the Whitlam government failed to act to stop sand-mining on Fraser Island, she wrote to White (more out of hope than expectation) asking him to let Whitlam know that he was disappointed in this failure. She was delighted to find that he went further in airing his disappointment publicly and in nominating the leader of the campaign to protect Fraser Island, John Sinclair, as Australian of the Year (see chapter 6). Clearly Wright had more in common with her distant relative that she would have imagined.

All through the 1970s and 1980s, Wright and White found themselves on the same public platforms, particularly in protests against uranium mining and nuclear weapons. In fact, Wright eventually complained that the only time they ever seemed to meet was at some kind of protest. But she also admitted that she had been inspired to see a man of his background take such strong stands in favour of the natural environment. In a tribute to White, she wrote:

> Both of us stem from days when the Australian tale was just beginning, with all its uneasy European legacy, its own contribution of blood and ugliness, its struggle towards understanding. That so great a writer has emerged from that background can't help but give me hope that we're after all redeemable.[17]

Certainly there were times in White's early life when it would not have seemed possible that he would find himself as a major contributor in the telling of the Australian story. He was born in England (in 1912) and, although he spent his childhood in Sydney and the Blue Mountains, he was sent to preparatory school in Moss Vale so that he could return 'Home' to attend Cheltenham College and then Cambridge University.[18] Between school and university he spent two years in Australia, working as a jackaroo on family properties at Adaminaby and Walgett. But after finishing at Cambridge he settled down in England with hopes of making it as a novelist and playwright. His first successful novel (after a couple of unpublished manuscripts) – *Happy Valley* (1939) – was set in Australia and reflected his time at Adaminaby.[19] The next one, *The Living and the Dead* (1941), made no reference to Australia or Australians and might have signalled the start of his career as a European writer. However, he soon found himself drawn back to Australian characters and Australian settings.

During World War II, White signed up for the Royal Air Force and did some uneventful service in the Middle East; uneventful, except for the fact that, in Egypt, he met a young poet serving in the Greek army, Manoly Lascaris, who became his lifelong companion. When the war was over, White came to Australia for a visit and decided he wanted to return permanently. He went back to England to make arrangements and visited Lascaris in Alexandria to convince him to come too. They arrived together in 1948 and settled on a small farm at Castle Hill, on the outskirts of Sydney, where they grew flowers and vegetables, kept goats, and bred Schnauzer dogs. White had finished writing *The Aunt's Story*, which was something of a summary of his life until that point, on the long journey by boat back to Australia. Now he was ready to start new novels set in Australia.

The experience of plunging his hands into the soil at Castle Hill probably gave him the inspiration for the first novel written on his return – *The Tree of Man* (1955). It opens with the central character Stan Parker camped in the bush, clearing some land to build a house for himself and his young bride, Amy. White later said that he wanted to write a novel about 'the lives of an ordinary man and woman . . . to discover the extraordinary behind the ordinary, the mystery and poetry which alone would make bearable the lives of such people, and, incidentally, my own since my return' (Kiernan, 1985). For a novel about an ordinary man and woman, it became quite an epic with a range of characters trying to make meaning out of their existence. At the end of his life, Stan Parker seems to find a sense of fulfilment in the garden he and his wife have created but he can't communicate this to her. It is left to his grandson to dream of writing a novel about his life.

The Tree of Man was followed quickly by the even more ambitious *Voss* (1957) – an historical novel based loosely on the story of German explorer Ludwig Leichhardt who disappeared (reputedly at the hands of Aborigines) on a long trek in northern Australia in the 1840s. Although the death of Leichhardt did not attract as much attention at the time as the death, 13 years later, of Burke and Wills (see chapter 3), he was generally seen as a 'fallen hero', lauded for his bravery and service to the goal of pushing back the frontier. However, as with Burke and Wills later, the admiration was diluted by a degree of cynicism about the real motivations of men who seemed arrogant and driven, and the hero myths were steadily overtaken by more critical appraisals. In the case of Leichhardt, the cynicism was probably intensified by the fact that he was not English but German, and White's interest in him, in the wake of World War II, probably stemmed from the fact that many Australians wanted to blame the rise of the Nazi movement on Germanic cultural traits. White was interested in the ambivalent public perception of such a man, but he was also interested in exploring the fine line between courage and madness; between being driven by a strong desire to push the boundaries of social achievement and being consumed by a personal desire for immortality that can appear, at times, to take the form of a death wish.

We know that White was influenced by a critical assessment of Leichhardt written by the acclaimed Victorian nature writer Alex Chisholm in 1941 (Haynes, 1998, p. 239). We also know that he was influenced by Sidney Nolan's interpretation of the mythology surrounding Burke and Wills, because he asked Nolan to design a dustcover for *Voss* (see chapter 3). But even more importantly, perhaps, the experience of World War II had led many Australians to question their assumptions about the achievements of European 'civilisation' and this may have seemed a good time to re-examine some of the foundational myths about our own white settler society. As a German, Leichhardt may have represented the Europe that was capable of giving birth to Nazism, and, at the same time, he could be seen as a strong figure defeated in his attempt to conquer the unknown. Whereas Chisholm, and others, tried to turn the hero myths about Leichhardt precisely on their head, White began his novel by introducing a hero who was driven by a mixture of idealism and egotism – a flawed character who, nonetheless, seemed more admirable than the rather weak and shallow 'leaders' of the colonial outpost (i.e. Sydney) which clung to the coast. A sense of anticipation builds up as this hero single-mindedly brings together the men and equipment he needs for a long voyage but even as his focus is fixed on the journey he brashly calls at the home of a young beauty, Laura Trevelyan, for whom he has taken a fancy. The explorers set out with nervous excitement and we follow the thoughts of various members of the party as they travel deeper and deeper into the unknown. Increasingly, it becomes obvious that there will be no return and that the journey has become a kind of ritualised self-sacrifice and withdrawal from reality, ending with Voss's violent death at an Aboriginal sacred site. Back in Sydney, Laura's thoughts are constantly with the man who had the courage to leave the colony and its tedious people behind and Voss finds he can draw her into his dreamworld, so that she becomes part of an erotic fantasy that leads to his death. Laura knows, intuitively, that Voss has perished, but her last statement in the novel is defiant: 'He is still there, it is said, and will always be. His legend will be written down, eventually, by those who have been troubled by it.' We are troubled by powerful landscapes and the stories that are embedded in them; we know that in seeking to understand the land we might have to accept considerable sacrifice. But still we are drawn to it, just as White was drawn to return to live and work in Australia.

Neither hero nor villain, White's Voss is a complex, ambiguous character who is driven by both an apparent desire for martyrdom and for some kind of spiritual enlightenment in the mysterious and frightening interior. He tells a potential recruit:

> You will be burnt most likely, you will have the flesh torn from your bones, you will be tortured probably in many horrible and primitive ways, but you will realize that genius of which you sometimes suspect you are possessed and of which you will not tell me you are afraid.

In White's treatment the desert is neither 'the enemy' that cannot be tamed nor a *tabula rasa* on which heroic tales will be transcribed. Rather, it is complex, multi-faceted and actively involved in the shaping of the human story. It demands both understanding and respect. For both Nolan and White the changing mythology about the early explorers provided an opportunity to explore deep psychological aspects of belonging and not belonging – a persistent theme in Australian art and literature. They were both able to play a role in deepening the mythology, particularly as it touched on the inter-relationships between people and their environments. *Voss* was White's only direct venture into this field, yet it is widely considered to be his finest work. As one critic, Andrew Taylor, wrote:

> [*Voss*] has become a powerful influence in the way many Australians . . . see their country's past and, therefore, on the way they inhabit its present. (Haynes, 1998, p. 239)

By 1964 White and Lascaris had had enough of their farming experience and moved from Castle Hill to Centennial Park, near the centre of the city. Thereafter, most of his novels were city-based, although *The Eye of the Storm* (1973) and *A Fringe of Leaves* (1976) were both inspired by a visit to Fraser Island. White first heard the story of Eliza Fraser from Nolan (who had earlier completed a series of paintings based on the story, see chapter 3) when he visited him in Florida in 1958 (Marr, 1991, p. 378). Nolan told him that after the Aborigines had killed other survivors of the shipwreck they also took Fraser's clothes from her, so that all she had to maintain her modesty was 'a fringe of leaves' in which she kept her wedding ring. White visited the island for himself in 1961 and found that it more than lived up to his expectations. Ten years later he was able to remember what he had seen in great detail:

> ... the columns so moss-upholstered or lichen-encrusted, the vines suspended from them so intricately rigged, the light barely slithered down, and then a dark watery green, though in rare gaps where sassafras had been thinned out ... (Marr, 1991, p. 380)

No doubt White had excellent qualities as a writer. When he was awarded the Nobel Prize for Literature in 1973, the Swedish Academy (which picks the winner) commended him for 'wrestling with the language in order to extract all its power and all its nuances, to the verge of the unattainable' (Bliss, 1991). But it also said he had been chosen because he had 'introduced a new continent into literature'. This last comment may say more about the cultural myopia of Europeans than it does about the evolution of Australian literature, but it does show that White's appeal rests heavily on the fact that his writing was rooted in a distinctly Australian experience. His novels are not easy to read, yet the author with the patrician family which thought of England as home created a new sense of home for many Australians.

White's great achievement in winning the Nobel Prize provided great encouragement for the next generation of Australian writers to write about Australian experiences. Some of them came to be strongly associated with particular places, just as White became known as a Sydney writer. It is not easy to think of a single Melbourne equivalent of White but, from the 1970s onwards, a wide range of Melbourne writers began to explore the 'Melbourne experience' in novels, plays and filmscripts. Perhaps it is typical of the Brisbane experience that a largely exiled writer, David Malouf, came to be most closely associated with that city and with south-east Queensland more generally. Malouf has proudly acknowledged that growing up in Brisbane shaped his sense of identity and, consequently, the character of his work. In the autobiographical *12 Edmondstone Street*, for example, he has explained that the experience of living in a classic, stilted Queensland weatherboard house gave him a strong sense of connection with the surrounding natural environment, because:

> They have about them the improvised air of tree-houses. Airy, open, often with no doors between the rooms, they are on such easy terms with breezes, with the thick foliage they break into at window level, with the lives of possums and flying-foxes, that living in them, barefoot for the most part, is like living in a reorganised forest. The creak of timber as the day's heat seeps

away, the gradual adjustment in all its parts, like a giant instrument being tuned, of the house-frame on its stumps, is a condition of life that goes deep into consciousness. It makes the timber-house dweller, among the domesticated, a distinct subspecies.

To some extent, Tim Winton has emerged as a Perth-based writer, but his sense of 'home' might be more broadly defined as the coast. Whereas earlier writers had played a pioneering role in turning our gaze inland, away from the sea, Winton has suggested that the time has come to return that gaze to the coast because that is where so many of us live and recreate. 'We love the sea, but treat it carelessly' ('Time and Tide' in *Good Weekend*, November 8, 1997), he has said. Exploring the paradox that many Australians can recall magical experiences by the sea when they were young, yet as a nation we treat it as either an exploitable resource or dumping ground, Winton wrote:

> The Australian process thinker Charles Birch talks of our need to treat nature as 'subject' instead of 'object'. Only the rigorously averted gaze can let the old attitudes, the thoughtless exploitation, continue. If we come from the sea, as scientists suggest, then the sea is our ancestral home, our source, and somehow we belong to it. Australians have long been preoccupied with the idea of belonging. Non-indigenous Australians are trying to learn what it means to be part of our landscape, to learn what it takes to belong . . .

> In coming to terms with the sea, in particular, Australians have to will themselves to turn their heads and look hard into the glare and see the reality of our relationship with it. And half of that reality involves our relative ignorance of it . . .' ('Time and Tide')

Understanding Australian landscapes also involves understanding the relationships between the land and the sea and the transition zone of the coast.

David Malouf has suggested that creative writers have an obligation to bring to their readers a sense of magic about the places where their stories are set. In an interview about writing he once said:

> . . . you have to find some way – I call it mythological – to find some real spiritual link between us and the landscape, us and the cities, us and the lives we live *here*. And to do that you have to give people – in books – something like a mythology that they can have . . .

> You don't just describe a place, you mythologise it, turn it into a symbolised place that your work can exist in. (Neilsen, 1996)

At one level, this mythologising of places and landscapes has helped white Australians to overcome a sense of fear and alienation, as reflected in the work of such diverse writers as May Gibbs and Judith Wright. On another level, generations of writers have depicted a shift of mood from a melancholic response to the land, through forms of stoicism, to a more overt appreciation of the delicate and intricate beauty found in places that were once feared. In this transition, for example, the 'dead heart' has become the red and vibrant centre.

While writers undoubtedly have a role to play in mythologising the places in which we might otherwise live rather uninspired lives – from the city to the bush – it is an exaggeration to say that they create the mythologies that get passed from one generation to the next. Stories that become myths (sometimes passing through the status of 'legends') grow out of broad social and cultural processes. Individual writers can enhance such processes, as Patrick White did with his exploration of explorer myths in *Voss*. In doing so they can help to take us beyond the constraints of rational thinking (which dominates the world of science) into more emotive and enchanted relationships with our environments. In this sense, contemporary writers like Malouf and Winton follow in the traditions established by people like Lawson, Dark, Wright and White. Along with May Gibbs, they are the writers who can be seen as ecological pioneers of Australian literature.

CHAPTER 5

TAKING NATURE TO THE
PUBLIC:
*Journalists, Broadcasters and
Writers as Educators*

INTRODUCTION

Tim Bonyhady (2000b) has suggested that nature conservation work in Australia probably gained its first serious impulse in Victoria in the 1860s. Certainly there were individuals before then who campaigned for the protection of endangered species and the first official action came as early as 1790 when the commandant of Norfolk Island restricted the killing of endangered birds on the island (Bonyhady, 2000b, pp. 13–39). However, there was no serious *movement* for nature conservation until the 1860s and 1870s. During those two decades the Melbourne-based *Argus* newspaper took up the cause, arguing for the preservation of areas of natural beauty – especially forests being targeted by loggers. In 1880, the Victorian Field Naturalists Club was formed – the first in Australia – and during the 1880s *Argus* journalist Donald Macdonald replaced that paper's art critic James Smith as the foremost advocate of landscape conservation. Through his regular columns in the *Argus*, Macdonald worked hard to educate the public on matters of natural history and struck a surprisingly good response. He created a legacy of public education work that was expanded in other ways through the creation of specialist magazines and the production of natural history books. Eventually this kind of work would expand into radio and television.

Philip Crosbie Morrison succeeded Macdonald as a nature writer on the *Argus* and he was the one who took this work into radio with his regular Sunday evening broadcasts starting in 1938. By the 1940s, Morrison's work on radio had become so popular nationally that he was widely known as the 'Voice of Nature', once being introduced to an audience of nature enthusiasts in Perth as simply 'The Voice'. Morrison's premature death in 1958 cut short his promising career in television and it was left to a younger man who had been inspired by him – Perth-based Vincent Serventy – to pioneer Australian nature programs on TV. Morrison and Serventy shared a grounding in biological sciences and a profound respect for the work of amateur naturalists. They were able to demonstrate that the growing rift between the amateur movement and the 'professional' movement of scientists (to be discussed in chapter 7) was neither inevitable nor desirable. Many scientists have been quick to disparage the work of popular nature writers and broadcasters, accusing them of engaging in 'pop science'. Yet they laid the foundations for the professional movement and they have had more success in influencing public opinion because they worked hard to make their work accessible to a wide range of people and because they never lost the passion of amateurs.

Those who followed in the footsteps of people like Macdonald and Morrison were able to win broad respect for their craft and their influence. They pioneered a genre of nature writing and broadcasting that attained an enduring popularity, particularly after it recognised the importance of acknowledging the ecological knowledge of the indigenous Australians.

'THE WOODLANDERS': CHARLES BARRETT AND DONALD MACDONALD

In the first years of the twentieth century, three young nature lovers from Melbourne established a regular 'bush camp' in an old hut set in an abandoned orchard at Olinda Creek, just out of Lilydale at the foot of the Dandenong Ranges. Like the Heidelberg painters, who had set up their bush camp a little closer to Melbourne in the same general direction less than two decades earlier, the three wanted to spend some serious time in the bush to get to know it better. Like 'Bulldog' Roberts and his 'gang', they also adopted nicknames for each other: Charles Barrett was dubbed 'The Scribe' because he recorded their experiences for articles that were subsequently published in the magazine *New Idea*; Claude Kinnane was 'The Artist' even though he used a camera rather than paint and canvas; and Brooke Nicholls, a city dentist, was 'The Doctor' (Griffiths, 1996, pp. 127–129). Collectively they called themselves The Woodlanders in reference to Thomas Hardy's work on rural English life and they called their hut Walden after the dwelling place of their main hero, Henry David Thoreau. Unlike Thoreau, they could not attract the external sponsorship required to abandon their urban life and live at their Walden full-time; 'unlike our Master, we paid taxes, having no Emerson to pay them for us', wrote Barrett (Griffiths, 1996, p. 127). However, they did head for their 'retreat' at every available opportunity in order to

> . . . experience that return to Nature of which so much has been written in recent years; to leave the din and dust of the great city, and dwell awhile in the forest among birds and flowers and trees. (Griffiths, 1996, p. 128)

The intention of the three adventurers was to make a detailed study of the natural history of one specific area, in the tradition established by the English naturalist Gilbert White (see chapter 1).

Barrett was undoubtedly the intellectual leader of the group. The son of a wealthy mayor of the boom suburb of Hawthorn who lost most of his wealth in the depressions of the 1890s, Barrett was forced, by dint of his family's financial situation, to abandon his hopes of spending time under the 'dreaming towers of Oxford'; turning instead to the meditative habit of 'rambling in the paddocks near home' (Griffiths, 1996, p. 129). As The Scribe for The Woodlanders he was able to both discover and demonstrate his talents as a writer, landing a job on the *Herald* in 1906 where he continued to write a nature column for the next 30 years. In 1910 he also became assistant editor of *Emu*, the journal of the Royal Australasian Ornithologists Union, and from 1925 to 1940 he was editor of the influential *Victorian Naturalist*. By the end of his writing career he had published over 60 books on Australian natural history and Aboriginal 'folklore'. He was one of the finest flowerings of the 'back to nature movement' that had been inspired in Australia by northern hemisphere writers like Thoreau and White.

According to Tom Griffiths, Barrett represented a departure from the earlier generation of field naturalists who had formed the Victorian Field Naturalists Club in 1880, because those pioneers had been driven by a long-established mentality of hunting and collecting exotic things to place in museums. By contrast, Barrett was much more interested in learning how nature worked in an ecological sense –that is *in situ*. Amateur naturalists like Barrett were, in turn, challenged by professional biologists working mainly in universities – people whom Barrett disparagingly referred to as 'armchair theorists' (Griffiths, 1996, p. 129). However, the amateur movement did not disappear; it continued to play a crucial role in educating a broader public about Australian nature.

Barrett's professed hero may have been Thoreau but he had a much more significant mentor a lot closer to home in the acclaimed nature writer on the *Argus* newspaper, Donald Macdonald, who was a regular visitor to Barrett's Walden during the years when Barrett and his colleagues were spending time there. Long before Barrett convinced his friends to begin their adventures at Walden, he had been a fan of Macdonald's work in the *Argus* and he had met him in Melbourne some time in the 1890s. Presumably Macdonald was quick to recognise Barrett's potential because he seems to have nurtured the relationship with the younger man. Not only did he visit Barrett and his friends at Olinda Creek, he also invited Barrett to his own retreat at Black Rock and Barrett's daughter has written that her father loved visits to Macdonald's retreat where 'in the peace and warmth of a book-lined study, the two men would talk and reminisce' (Griffiths,

1996, p. 128). Barrett dedicated his first book, *From Range to Sea: A Bird-Lover's Way* (1907), to Macdonald.

Donald Macdonald was born in 1859 in western Victoria and grew up in the then-rural township of Keilor on the Maribyrnong River, north-west of Melbourne city (Griffiths, 1996, pp. 118–127). His forebears had been pioneer settlers in the state's western districts where they lived alongside substantial Aboriginal communities. As a child, Macdonald was influenced by a great-aunt who had spent so much of her early childhood with Aboriginal children that she learnt their language at the same time that she learnt to speak in English (Griffiths, 1996, p. 119). By the mid-1880s Macdonald wrote that those same Aboriginal communities had vanished almost without trace because 'instead of studying [them] we shot them' (Griffiths, 1996, p. 119). From his home at Keilor, Macdonald landed a job on the *Argus* as a sports reporter, where he established a reputation for innovative work in reporting cricket test matches between Australia and England. However, it was an essay about Keilor, written in the early 1880s, which marked his entry into nature writing. Describing Keilor as a village 'set in a deep hollow in the plain' he gave a lively account of the plants and animals that were sharing the area with the human residents. According to Griffiths, Macdonald's article was clearly inspired by the work of Gilbert White. He presented a largely Arcadian image of the town where he grew up; making note of indigenous species of plants and animals that were jostling for space with the introduced species, but writing fondly of the blackberries, thrushes and starlings that reminded the settlers of the

Donald Macdonald (left) with 'The Woodlanders' at 'Walden'.
(Oxford University Press)

mother country. He welcomed 'suggestions of the old world' but also noted the delights of the native species that were increasingly being confined to remnant communities. Pointing out that you had to go to the steepest slopes of the Grampians to see beautiful Australian heath, he wrote that 'Nature's better part is often hidden in remote corners, her gems in almost inaccessible places.' Meanwhile, he saw Keilor as a 'peaceful place', but one that 'possesses a tragedy' in the suppressed story of the annihilation of the indigenous people of the area.

Macdonald's 'sketches' of rural life became popular among *Argus* readers, but his focus changed radically when he was invited to begin a column called 'Nature Notes and Queries'. Through this he became much more interested in the native species of plants and animals and discovered that many readers shared this fascination. His column generated a large number of letters, many of them from young people, prompting the editors of the *Argus* to give Macdonald a second column called 'Notes for Boys'. Macdonald was not the pioneer who gave the *Argus* a reputation for being in favour of nature conservation. As mentioned in chapter 3, that honour probably goes to the art critic James Smith, who wrote front-page editorials in 1861 condemning plans to log forests at Ferntree Gully in the Dandenongs, made famous by an 1857 painting by Eugene von Guérard. Stemming from this specific interest, Smith carried out some research on temperate forests throughout the world and he wrote in 1880 that the clearing of such forests on a world scale had become a 'peril of unparalleled magnitude' (Bonyhady, 1994, p. 174). However, Macdonald was the writer who developed the expertise to be able to respond to a widespread desire to know more about Australian plants and animals. Not only did he receive a large volume of mail, he was also sent specimens of things, in various states of decomposition, that people wanted identified. He was able to enter into a dialogue with his readers that encouraged him to develop his expertise even further and to encourage younger people like Barrett to follow suit.

'THE VOICE OF NATURE': PHILIP CROSBIE MORRISON

Macdonald kept up both his nature columns until his death in 1932. At this point both columns were temporarily assigned to a young journalist still learning the ropes in his chosen profession – Philip Crosbie Morrison. Morrison had to give way to the very experienced nature writer Alex Chisholm, who first made his mark with a 1922 book called *Mateship with Birds*, when he joined the *Argus* soon afterwards. However, Morrison had enjoyed his experience of this kind of writing and he had demonstrated an aptitude for it. When Chisholm became the editor of the *Argus* in 1937, Morrison was given a more permanent custodianship of the two treasured columns (although 'Notes for Boys' had by now been renamed 'Notes for

Boys and Girls'!). It could not have been easy to follow in the footsteps of Macdonald and Chisholm, but Morrison tackled the task with enthusiasm, maintaining the level of reader interest generated by his illustrious predecessors. When he got the opportunity to branch out into radio the following year he came to overshadow them both.

Morrison's career in radio began indirectly and inauspiciously (Pizzey, 1992). First he was asked by his employers at the *Herald and Weekly Times* organisation to become founding editor, in 1938, of a specialist magazine called *Wild Life* and then, because the *Herald and Weekly Times* also owned radio station 3DB-3LK, he was asked to go on radio to promote the new magazine. The station managers responded to the request to give Morrison some airtime by allocating him a 'dead spot' that followed the news on Sunday evenings and they told him he could have it for six weeks in order to build some interest. Morrison had never been on radio before, but his rich warm voice made him a natural and he maintained an informal chatty style. After the first broadcast some letters arrived welcoming the new program and asking Morrison to provide information on particular topics. Increasingly, he constructed his broadcasts around responses to listeners' inquiries and this triggered an even stronger flow of letters. The station management was staggered and delighted by the strong public response to the broadcasts and asked Morrison to extend the series beyond the six weeks that had been allocated. The program continued, without interruption, for the next 20 years, although it was shifted to 3UZ in 1954 when the *Herald and Weekly Times*, then under new management, closed *Wild Life* magazine. In 1943, a radio listening survey in Victoria found that an incredible 74 per cent of Victorian radios turned on at this time on a Sunday were tuned in to Morrison (Pizzey, 1992, p. 158). The following year the program went national through a network of radio stations in all states and was even relayed into New Zealand. In 1952 another survey in Victoria had Morrison listed as the most popular personality on radio, followed in the top ten by nine full-time, professional announcers (Pizzey, 1992, p. 167). By now he was widely known as 'the Voice of Nature'.

Morrison was not the first person to make regular broadcasts on radio about nature. In 1937 Professor William Dakin from the University of Sydney began a regular broadcast on ABC radio called 'Science in the News' as part of his successful campaign to have Biology included on the curriculum for the New South Wales High School Leaving Certificate.[1] Dakin was unusual for scientists of his time, in that he saw a need to campaign for public support in order to promote the status of his field of study. However, Dakin's foray into radio was relatively brief (lasting less than a year). By contrast, Morrison made broadcasting his main career, supplemented by print journalism, and he displayed great skill in establishing a dialogue with his listeners. He became a very effective public educator.

The success of his radio broadcasts may not have taken Morrison completely by surprise. The long-standing success of the columns he inherited at the *Argus* had demonstrated that there was a 'market' for well-presented information about nature. Morrison probably knew that he simply had to find a way of building a relationship with his listeners (as distinct from his readers). However, he could not have been prepared for the personal praise he would attract or for the notoriety that went with it. His wife Lucy has said that his life became extremely busy after the radio broadcasts began and he started to be inundated with requests to give talks or make public appearances (Pizzey, 1992, p. 160). He tried to meet as many requests as possible but he already had a stressful job in trying to keep a specialist magazine afloat. Increasingly, he found that he had little time for himself or his family. Even though he worked on radio he found himself being constantly recognised in public. The *Herald* had printed photos of him in publicising the new magazine and, of course, there was *that* voice. Lucy told his biographer that one day during the war they had stopped to give a lift to a soldier. Before they had travelled two miles the young man had worked out where he had heard that voice before and then there was 'no stopping him'.

Although it was the radio broadcasts that made Morrison famous, his wife has confirmed that he personally got more enjoyment out of the production of the magazine because it was something tangible that could be perused and kept rather than something ephemeral like a broadcast. He continued to see the broadcasts as a way of encouraging people to buy the magazine, although most people were happy with what they got on the radio. It is never easy to make a specialist magazine commercially viable and Morrison felt the burden of this quite keenly. However, he saw it as his major legacy. It was his baby, although the idea had actually originated with another nature lover – the Melbourne industrialist and philanthropist Russell Grimwade.

Since he first joined a 1916 committee that gave birth to the Council for Scientific and Industrial Research (subsequently the Commonwealth Scientific and Industrial Research Organisation – CSIRO), Grimwade had maintained a strong interest in scientific research. He was also an active member of the Victorian Forest League and financially supported its quarterly journal *The Gum Leaf*. In the early 1930s he had suggested to his friend Keith Murdoch, owner of the *Herald and Weekly Times*, that he should take over *The Gum Leaf* and build it into a popular magazine on natural history and science. Murdoch was not convinced so Grimwade put the file back in his cabinet and waited for another opportunity. Morrison's work at the *Argus* – especially in the column 'Nature Notes and Queries' – began to catch Grimwade's attention and he decided that he had found the right person to make the magazine a success. This time he took the respected director of the National Museum (as the Museum of Victoria was then called), D. J. Mahony, with him to see Murdoch and they convinced him to talk to Morrison. Murdoch was impressed by Morrison and decided to run with the project. He probably suspected that it would never turn a profit but could also see that it would bring credit to the organisation. He continued to subsidise it until his death in 1952.

Morrison began working for Murdoch in July 1938. Before the magazine could be launched he had to find the writers and photographers who could make it work. This was sometimes tricky because some of the people he wanted were contracted to other organisations. However, almost everyone he approached was keen on the project and was willing to contribute, even if they had to use their own time to do so. He was particularly pleased to get the support of Robert Croll, a public servant who loved to get out into the bush and wrote perceptively about it. It was a feather in his cap to get David Fleay, a biologist who had built his reputation by designing the Australian section of the Melbourne Zoo and by being a high-profile director of a wildlife sanctuary at Healesville. Ray Littlejohns, a skilled nature photographer and acknowledged expert on lyrebirds, became one of the most reliable contributors and a personal friend.

When the first edition of *Wild Life* appeared in October 1938, it featured an article on a family of boobook owls, which opened with a startling flashlight photo of an adult owl arriving back to a nest full of fledglings with the headless body of a honeyeater in its beak. Such photography had never been seen in print before and it made quite an impact. Articles were informative but readable and often a little quirky. It was pitched as a family magazine, yet did not pull any punches in pushing the conservation message, as the following excerpt from one of Morrison's early editorials shows:

> . . . the whole tradition of Australia, more perhaps than of any other country except America, is a tradition of pioneering, in which man appears fighting a lifelong battle against the virgin bush and the noxious creatures in it.
>
> Inevitably, in the pioneer's view, a tree was a thing to be hewn down, its roots to be grubbed out with infinite labor, to make room to grow grain. The kangaroo was an enemy, eating the crops that were the product of such toil; the emu, the wombat, the possum, the majestic wedge-tailed eagle, all appeared as enemies to be subdued.
>
> So the appreciation of nature in Australia had a bad start. The second factor is the failure of the schools to alter the earlier tradition. In recent years some attempt has been made to remedy this, but in most States Nature Study is still relegated to a position far inferior to that which it should occupy. An enlightened central administration, and a qualified and enthusiastic teaching staff, are required to put the study of nature in its due place in the curriculum. (April 1940; in Pizzey, 1992, pp. 132–133)

In keeping with the tradition established by the nature columns at the *Argus,* Morrison also started an interactive column in *Wild Life,* called 'Along the Track with the Editor', which became one of the most popular features of the magazine. Again, the success seemed to come from the relationship established between writer and reader. Long after he had become famous, Morrison said in an interview that he believed his success was based on a habit of trying to address each one of his readers or listeners one at a time, remaining conscious that others were also listening in (Pizzey, 1992, p. 148). He might have in mind, he said, that he was talking to a small boy who had a question about worms and remember at the same time that a fisherman had come by a little earlier and had other questions about worms. While he was addressing the young boy he could imagine the fisherman catching the conversation and wandering back into the room to listen in. He took a similar approach in his written articles.

There was nothing in Morrison's early life to suggest that he would build a high-profile career in journalism and broadcasting. He was born, in 1900, in quite humble circumstances in the Melbourne suburb of Hawthorn, where his father had established a small business as a draper and blind-maker. James Crosbie Morrison's own father had been a bricklayer who had migrated from Ireland, but James had married a girl from Auckland whose father had been a successful draper. From his father Philip probably got the gift of the gab; from his mother he got a love of sacred church music and when he joined the choir in his local Methodist church he was said to possess 'a lovely voice' (Pizzey, 1992, p. 23). It seems that he developed an early interest in collecting 'bugs' because when his Sunday School wanted to honour him they chose to give him *The Wonders of the Insect World,* by the renowned British naturalist Edmund Selous, as a prize.

At school Morrison was a diligent, but not outstanding, student. However, he did well enough to be offered a place at the University High School in Carlton, which had a proud record of getting its students into university courses at nearby Melbourne University. He spent four influential years at the school, where his academic progress was steady rather than spectacular, but where he also discovered a love of theatre, playing prominent roles in a number of school plays. When the time came to matriculate at university, however, he did not even bother to fulfil all the requirements because World War I was at its height and he assumed he would soon be required by the army. Fortunately for him, he remained too young to go off to the killing fields in France. Instead he managed to wangle a position as assistant teacher at the Wesley Preparatory School, where his duties involved chemistry demonstrations. At this time he took another important step by joining the Field Naturalists Club of Victoria, which included in its ranks such luminaries as Melbourne University biology professor Sir Baldwin Spencer. He enjoyed field trips to various locations in and around

Philip Crosbie Morrison in the field with a small specimen of the giant Gippsland
earthworm, *Megascolides australis* in the 1940s. (State Library of Victoria)

Melbourne and must have been impressed at the quality of guest presentations the club could organise. He was certainly impressed with a presentation by the geologist Griffith Taylor who had survived the ill-fated 1911–12 expedition to Antarctica led by Captain Robert Scott (Pizzey, 1992, p. 27). Taylor illustrated his talk with 'lantern slides' from photographs taken by the famous Antarctic photographer Herbert Ponting; a presentation style that Morrison subsequently used himself.

While Morrison was conducting chemistry lessons at the Wesley school he decided that his future lay as an industrial chemist. This meant he had to complete the requirements for matriculation and then complete a science degree at university. Even before beginning his university study, he managed to find a job with a company that was harvesting resin from grass-trees (*Xanthorrhoea tatiana*) on Kangaroo Island in South Australia. On a visit to the island he had qualms about the practice of burning the ancient plants in order to extract the resin and could see that many of them did not survive the ordeal. He was probably feeling the gap between his interest in natural history and his chosen career as a chemist, but he could not imagine who would employ a person who had specialised in biology. He continued to major in chemistry but added some studies in zoology, in which he excelled. At the end of his first year he was awarded the recently created Baldwin Spencer Prize in Zoology and was being urged by his teachers to make this his major. In a sense the choice was made for him when he was offered a scholarship at the beginning of 1924 to do postgraduate studies in zoology. He began studying under the tutelage of Baldwin Spencer's successor; an English zoologist, Professor W. E. Agar, who was very popular with research students (Pizzey, 1992, p. 30).

During his period of postgraduate study, Morrison had the opportunity to accompany a high-powered research team on a six-month study of the Great Barrier Reef. He had his own independent research project to complete as part of his studies, but he also had a rare opportunity to meet a wide range of people with an interest in the future of the reef. He found these opportunities to talk to people even more interesting than his research on the reef and he began to see an opportunity for himself as a journalist with a strong grounding in science.[2] Back in Melbourne, his teachers were rather perplexed by his decision, but one of his mentors, Professor Alfred Ewart (a botanist), wrote a strong recommendation on his behalf to the editor of the *Argus*, and this was enough to get him a start on the paper.

With his formal background in science, Morrison was clearly different to other pioneers of nature writing in Victoria, such as Macdonald and Barrett. However, he would not be given an armchair ride in the tough world of journalism and he was expected to learn his craft in the traditional way, starting with the mundane task of compiling 'Shipping Notes'. Only on rare occasions did he have an opportunity to write about developments in science and only then could he include his qualifications in his by-line. He got a taste of what he wanted when he was asked to take over Macdonald's columns in 1932, but had to move aside for Chisholm. His apprenticeship in journalism was probably longer than he had expected, but at least he had the opportunity to learn from two significant mentors in Macdonald and Chisholm. When he got his chance to build on the legacy created by these two men at the *Argus* he was able to draw strength from his scientific training as well as his long apprenticeship in journalism. He had prepared well to make the most of the opportunity when it came, but his wife Lucy has also said that the secret of his rapid success was that he was so interested in his field that he enjoyed doing the research that would make him sound like an expert on so many topics. For example, when he first took up birdwatching in a serious way after the launch of *Wild Life* he had to rely on lessons from Ray Littlejohns and others (Pizzey, 1992, p. 189). But he learnt enough to become an active and respected member of the Royal Australasian Ornithologists Union (now called Birds Australia) and counted birdwatching as his favourite pastime.

When World War II broke out in 1939, Morrison was just settling into a career for which he was ideally suited. Whereas he had been too young to join the army in World War I, this time he was clearly too old. He tried to make a contribution to the war effort by firstly accepting an invitation to act as state censor for coverage of the war in Victorian newspapers and subsequently as broadcaster in the Department of Information Broadcast Division. However, both of these endeavours turned sour for him and he decided that the best contribution he could make would be to give people something other than the war to think about. So he put his efforts back into the magazine and radio broadcasts, with the latter climbing rapidly in its public appeal. His guess that people wanted to be distracted from the war proved correct, even for the soldiers. In 1944 he made a highly acclaimed lecture tour of army camps in the Northern Territory.

During the war he also stepped up his work as a community educator. As already mentioned, the popularity of his radio program meant that he got many invitations to give public talks and he tried to accept as many as possible. In early 1939 he also took on a position as part-time lecturer in natural history for the Workers' Education Association and Melbourne University Extension Board, conducting evening classes in the School of Botany at the university. He did as well in live presentations as he did in taped radio broadcasts and after he took over these classes the attendance soared from around 20 to over 200 (Pizzey, 1992, p. 172). He enjoyed the direct contact with audiences and continued a busy round of lecturing commitments after the war was over. In 1945, with the war reaching its climax in the Pacific, he also accepted an invitation to serve as deputy chairman to his friend and mentor Russell Grimwade on the board of the Victorian Museum (then called the National Museum) and took over as chairman when Grimwade died in 1955. In 1946, he also took on the task of preparing a visitors' guide for the world-famous Royal Botanic Gardens.

By the end of the war Morrison had become a respected member of the Melbourne 'establishment'.[3] Yet he risked this new-found respectability by throwing himself into a protest campaign over the state of Victoria's national parks. He was particularly alarmed at what he heard about the degraded state of Wilson's Promontory, which had been used as a military training ground during the war, and he ran an article about it in *Wild Life*. On his prompting, the Field Naturalists Club of Victoria called a special conference to consider the state of the parks in June 1946 and this led to the establishment of an Investigating Committee with Morrison as its chairman. The committee released a report in 1948 that attracted considerable public interest. It included a recommendation that the state government set up a new National Parks Authority responsible for the maintenance of all existing parks and reserves and with the task of recommending the acquisition or proclamation of new areas that it believed warranted protection. Morrison was asked to take the report's recommendations to government. Although he managed to win the support of Liberal Premier T. T. Holloway, the minister responsible for national parks in the coalition government was from the Country Party and he blocked the full implementation of the recommendations. The coalition government soon lost office and the Labor government led by John Cain had other priorities. It was not until a new coalition government, headed by Henry Bolte, was elected in 1955 that Morrison was able to get the report's recommendations back on the negotiating table. Finally, in 1957 – eleven years after the conference that began the process – a National Parks Authority was established with Morrison as its inaugural chairman.

In 1952, Morrison and his wife were participating in a Royal Australasian Ornithological Union Camp-Out near Alice Springs when he received some news he had been dreading – his great beneficiary Sir Keith Murdoch had died. Morrison was sure that his successors at the *Herald and Weekly Times* would decide that *Wild Life* was a luxury they could not afford to sustain. As it turned out, the demise of the magazine was not swift, because an effort was first made to move it more in the direction of outdoor recreation, changing the title to *Wild Life and Outdoors*. Morrison did not enjoy the change of direction and he was probably relieved when the new venture was deemed to be a failure. He prepared his last edition of the magazine in January 1954. By this time he had been invited to contribute a column to the *Sun News-Pictorial* and before long he was back at the *Argus*. The *Argus* collapsed in 1957 but the blow was softened for Morrison by the fact that he had been offered a position to work on national radio and television with the ABC.

Morrison was excited by the challenge of moving into television. During World War II he had bought a bulky 16 mm cine camera to take on field trips and subsequently had convinced Prime Minister Robert Menzies, a personal friend, to donate a more portable one. He had this camera with him on the 1952 bird-watching Camp-Out near Alice Springs and shot some footage later used for a documentary called *Beyond the Alice* (1955). After this film had been screened at a special showing at the Geelong Town Hall, an admirer wrote a letter to Morrison to give his account of the response:

> Bob Menzies himself could not have dragged half of Geelong to listen to him give a free show, while your show was a three bob [shillings] touch. My wife said 'it's a pity Crosbie struck such a wet night, there will be no one there'. We . . . reached the Town Hall at 7.30, and found [the] . . . street packed with cars. The crowd extended 100 yards . . . while the front doors were packed with a struggling mass of people fighting to get in . . . (Pizzey, 1992, p. 194)

Morrison's new job at the ABC seemed to be the perfect move for him and in February 1958 he began work on a documentary about marine life. That film was never finished and, tragically, his new career was cut short when he collapsed and died from a brain haemorrhage at his home on March 1, 1958. The nation was shocked by the news and messages of condolence to his family came not only from his friend Prime Minister Menzies, but also from British Prime Minister Harold Macmillan who had been taken on a tour of the Dandenong Ranges by Morrison during a state visit to Australia. Even more telling than the messages from two prime ministers was the flood of mail from his fans. Typical of the sentiment expressed was the following message to Morrison's wife from a Mrs J. E. Betheras of Clematis:

> May I, a stranger to you, ask you and your family to accept my sincere sympathy . . . we regard his loss as a calamity for Australia. (Pizzey, 1992, p. 257)

THE CONSERVATION CAMPAIGNER: VINCENT SERVENTY

One of Morrison's admirers who was shocked to hear of his death was a young man living in Perth, Vincent Serventy. Less than a decade later, Serventy would get an unexpected opportunity to make his own series of nature broadcasts for television, a pioneering effort called *Nature Walkabout*. However, at the time he heard about Morrison's death, Serventy had no idea that his own career would follow the path that Morrison had blazed. When the authors interviewed Serventy at his home at Pearl Beach, just north of Sydney, in 1996, he insisted that we should include Morrison in our selection of ecological pioneers. According to Serventy:

People have forgotten the influence Crosbie Morrison had, but I'll never forget the time he came to speak at a wildlife show organised by the Western Australian Naturalists' Club. No one knew he was coming yet the announcer only had to say that our special guest was 'The Voice' for the whole crowd to go wild. They knew it was him even before they saw him. He was really admired all over the country.

By the end of the 1960s, Serventy himself was renowned nationally as a newspaper columnist and documentary maker. He was national president of the Wildlife Preservation Society of Australia and, by the mid-1990s, he had over 60 published books to his credit.[4] He had been awarded the Dutch Golden Ark Award in 1985 for his international contribution to nature conservation and was made an emeritus fellow of the Australia Council's Literature Board in 1993. At the peak of his career he was Australia's best-known naturalist and was frequently interviewed by the media on contemporary issues. Yet, as we will see, he came from a conspicuously humble background, even less propitious than that of Morrison.

In the early part of his career, Serventy's mentor was his own brother Dominic, who built a highly successful career, against the odds, as a research biologist. Vincent told the authors that he was inspired by his older brother's ability to move from a small country school to becoming a university lecturer. Dominic found lecturing 'a torment' because he suffered a stammer when he spoke, so he moved from Perth to Sydney to work in the CSIRO's newly established fisheries division. From there he was then able to work his way up to becoming head of the bird section in the Wildlife Division led by Francis Ratcliffe (who features in chapter 7). However, when Dominic Serventy was teaching at the university in Perth, he introduced Vincent to the WA Naturalists' Club, which Vincent described as a very lively affair:

> Of course, Perth was a pretty small place in those days. It was easy to get to know a lot of people and in the club we had everyone from the mayor and university lecturers to wharf-lumpers. We all worked happily together. It was like joining a big family, so that's how I became a naturalist. My brother was very well known. He got an offer from the Museum to conduct evening classes in natural history and he got me involved. I had to learn a lot. But it was easy to get things done then. The club got support from all sorts of people. Even the government departments were friendly to us. The director of the Museum offered the club facilities to use, like lecture theatres.[5]

Serventy had his first taste of campaigning for conservation in 1946 when plans were announced to build a swimming pool in Perth's Kings Park – renowned for its display of native plants. A manifestation of public anger at the idea stopped the development and Serventy was impressed with the efficacy of 'people's power'. Around this time he also joined the Wildlife Preservation Society, after being introduced to it by his sister. Twenty years later he became the organisation's national president.

The Serventy boys both did very well to build high-profile public careers after starting life in rather humble circumstances. Their parents had migrated separately from Croatia, then part of Austria-Hungary, and met on the Kalgoorlie goldfields. Their father, Victor Vincent Serventy, was an adventurous character who had served in the Austrian army before coming to Australia. But once he made the decision to settle in Australia he wanted to forget about Europe. 'He was opposed to any sort of enclaves; not really a supporter of multiculturalism.' So Serventy's parents refused to speak their native Croatian, even at home. When Vincent was born, in 1926, the family was living on a small farm – 'we had an orchard and a vineyard' – near the Darling Escarpment east of Perth. He was the youngest of eight children.

> I didn't wear shoes until I was about 11. We used to run through the bush like brumbies. To the east of us was all bush and to the west just a few small settlements. We weren't all that far from Perth but it felt pretty remote. It was a great place for children and that's where I got my love of the bush.

Vincent was able to follow Dominic's success in getting through school and into university. He chose to do a Bachelor of Science, majoring in geology and then took a second degree in education in order to become a teacher, because 'there was no future in natural history'. But his interest shifted from geology to psychology because he had become fascinated by the study of animal behaviour. Near the end of his time at university, he heard that the world-renowned animal behaviourist Nico Tinbergen was coming to Australia to carry out a study of gulls. He applied to be his assistant and was accepted but, unfortunately, the project fell through and Tinbergen did not come. Serventy fell back to his safety net as a teacher. However, he was always an ideas man, not content with a career of straight teaching. So he started to get involved in curriculum development and the preparation of new textbooks and teaching materials. Meanwhile, he maintained his involvement in the Naturalists' Club, becoming the driving force behind its annual wildlife shows, which became so popular that the club started taking them out of Perth to various country towns.

In the mid-1950s, Serventy took up a position as senior lecturer in science and mathematics at the Claremont Teachers' College in Perth, where he had the opportunity to show trainee teachers how they could introduce nature conservation into their pedagogy. While he was in this position, he married fellow naturalist Carol Derbyshire who became an important partner in his future career. She wrote several books with Vincent, including one called *Australian Mother and Baby Animals*, published in 1981. She also built a separate career as a supporter of museums, rising to the position of national president of the Friends of Australian Museums and, subsequently, president of the World Federation of Friends of Museums.

In 1962 Serventy had the opportunity to visit Britain to attend various conferences on science education and conservation. While there he gave a number of public talks about the situation in Australia that surprised and alarmed his audiences. On one occasion a listener suggested that he should publish a book about the endangered Noisy Scrub Bird. He replied that the whole continent was in danger and this gave him the idea for one of his most influential and enduring books – *A Continent in Danger* (1966). When he returned from this first trip overseas he decided the time had come to return to study and complete a PhD. He had earlier been invited to undertake a doctorate at the University of Texas and he now decided he would pursue that invitation. But first he had an idea he wanted to try. What about a whole series of nature documentaries filmed in the then underrated Northern Territory to be screened nationally on television? He decided that the legendary Sydney media proprietor Frank Packer was the man to put this idea to, so he wrote to him saying he would be prepared to delay his trip to Texas if this project was accepted.

It was probably a long shot, but the reply came back: Come to Sydney and we'll discuss the idea. In fact, Packer went on to suggest that if

Vincent Serventy with wife Carol on their 'bush camp honeymoon' in 1955. (Vincent Serventy)

Serventy wanted to become a player in the future of the nation he should move to Sydney or Melbourne permanently. So, in 1965, Vincent and Carol wound up their multitudinous affairs in the west and headed for Sydney, with three young children in tow. An agreement was reached to make 13 half-hour programs under the title *Nature Walkabout* with filming to begin immediately. The dust had barely settled from the trip across the Nullarbor when the family had to relocate to a caravan that would be their home for the next six months as 13 episodes grew into 26. 'I had a great time', Vincent told us, 'but, with three young children, Carol hated it. She has hated caravans ever since.' *Nature Walkabout* became the first nature series made in Australia and was a popular success. It made Serventy a household name.

By the time Serventy had finished making *Nature Walkabout* he had gone cold on the idea of going to Texas. New opportunities would now open up as a result of the success of the television series and Vincent and Carol had a number of book projects in the pipeline. If they left now it might break the flow. So Serventy asked Packer if he could find him a position on his Sydney newspapers. He was offered a column, first in the *Daily Telegraph* and later in the *Sunday Telegraph*, which had a bigger, statewide, circulation.

Serventy's first book for a popular audience was *The Australian Nature Trail*, published in 1965. This was followed by *A Continent in Danger* (1966) and, over the following decades, his books appeared at the rate of around two a year. *Landforms of Australia* (1967) began a series of popular, but scientifically sound, books on the character of the continent and its wildlife, culminating in a huge collaborative project in 1973–74 to produce the 105-part *Australia's Wildlife Heritage*, for which Serventy acted as editor-in-chief. He consistently returned to conservation themes and *Australia's World Heritage Sites* (1985) was the first book to document these sites of world significance. In 1970 he published one of his most acclaimed works, *Dryandra – the story of an Australian forest*, which was selected as one of 100 great books in Australian literature by the renowned literary critic Geoffrey Dutton. The book drew public attention to the threat posed to the forest by a mining lease that had been granted to a company called Alwest. Serventy sent a copy of the book to Rupert Murdoch, managing director of the company, with a letter saying that mining the forest would amount to 'sacrilege'. He was pleasantly surprised to find that Murdoch was convinced by his argument and ordered the company to relinquish the lease.

In 1966, Serventy also agreed to a request from Judith Wright to take over the editorship of a magazine she had established some years earlier called *Wildlife in Australia* (see chapter 6). After the death of her husband and with her own health failing, Wright decided, in 1965, that she could not continue as editor even though she had succeeded in getting funding from the Myer Foundation. A successor had not worked out and she remembered that Serventy had contacted her from a caravan park in Brisbane (while working on *Nature Walkabout*) to express his admiration for the magazine. Having Serventy as editor posed some problems because he would not be living in Queensland and the magazine was 'owned' by the Queensland branch of the Wildlife Preservation Society. So it was a bit of a gamble. But he made such a success of the job that he remained as editor for 16 years, building the magazine's reputation nationally.

Serventy probably reached the peak of his public career in the 1970s. He had established his profile with *Nature Walkabout* and through his newspaper columns (moving from the *Sunday Telegraph* to the *Sydney Morning Herald*). His books were reaching a wide audience and were being used in schools. He reached a more specialist audience with *Wildlife* magazine and had strong standing with conservationists through his work with the Wildlife Preservation Society. He was frequently contacted by the media for comment on issues and contributed to the Melbourne *Age*, the *Bulletin* and numerous other magazines in Australia and overseas.

Throughout his career he was involved in some major conservation battles. As mentioned, he cut his teeth in the campaign to prevent the building of a swimming pool in Perth's Kings Park in 1946 and in 1970 he was living in Sydney's Hunters Hill when three local women formed a committee to try to stop a new housing development that would destroy a three-hectare patch of native bush on the harbour foreshore, known as Kellys Bush. Serventy was a member of the committee that decided to ask the Builders Labourers Federation, led by Jack Mundey, to ban the construction of the houses – a request that led to the imposition of that union's first Green Ban (see chapter 10). In 1971, he responded to calls from Tasmanian conservationists for national protests against plans by the state's Hydro-Electricity Commission to build a dam in the south-west wilderness that would flood the extraordinarily beautiful Lake Pedder (see chapter 10). As national president of the AWLPS, he mobilised the organisation to raise money for the campaign and to raise the issue with federal parliamentarians. He went to Tasmania to give public talks in Hobart, Launceston and Devonport on reasons for saving Lake Pedder. The conservationists lost this battle and the lake was drowned, but the campaign paved the way for later success in preventing the construction of another dam that would have flooded the Franklin River in the same region (see chapter 10). Serventy had become the first conservationist to operate on a truly national scale.

Between his work as writer, publicist and conservation advocate Serventy seemed to have his hands full. Yet he always sought to maintain some kind of involvement in scientific work in order to test his knowledge and understanding. In 1951 he had been the zoologist on the Australian Geographic Society's research expedition to the Recherche Archipelago and wrote reports on his findings. He also worked with his brother Dominic whenever the opportunity arose. Together with Dominic and John Warham, he prepared the first definitive *Handbook on Australian Seabirds*, published in 1971. Together with Bob Raymond, he produced television documentaries and books on Australian rivers and rainforests that involved extensive reviews of the latest research. When we interviewed him in 1996, he was particularly proud of a recently released book titled *Flight of the Shearwater*, based on research done primarily by Dominic over a 40-year period. Dominic did not find the time to write the book himself, so Vincent put it together after his death, as a kind of memorial to his brother and mentor.

Serventy was critical of conservationists who did not make the effort to keep up with research developments:

> In the WLPS we have maintained a council of scientific advisers. I mean people like Bill Williams in Adelaide who is one of the top men on wetlands, or Harold Cogger on reptiles and amphibians. So we've got these people we can ring up for information and advice before we stick our neck out on anything. I mean, if you don't get your facts right you can end up looking pretty silly. Of course, some scientists will talk on any side, like lawyers. So you have to make judgements about their integrity. But you always try to get the best information available.

Serventy had his first real taste of politics at university when he served for a while as president of the Labor Club, with a younger Bob Hawke on his committee. His later friendship with Frank Packer might have opened other political doors for him but he did not hesitate to accept when asked to help shape Labor Party environment policy in the lead-up to the 1972 elections. This came at the height of the campaign to save Lake Pedder and Serventy was hopeful that a Labor government would be more willing to override the states to protect important areas. By the time the Whitlam government was elected, plans to flood Lake Pedder were already well advanced, but Serventy was pleased to hear the new prime minister offer the Tasmanian government $5 million to build a pipeline that might enable the Hydro-Electricity Commission to have its dam without drowning the lake. Unfortunately, the Tasmanian Labor government rejected the offer. Of course, the common complaint made by politicians in Canberra was that conservation was really a state issue and that they could do little if the state governments were not willing to act. However, this view was challenged in an article written by a minister in the Whitlam government, Tom Uren, who argued that the federal government had the power to intervene when it came to a matter of protecting the 'national heritage', and the government decided to establish an Australian Heritage Commission that would have the power to override state governments in the protection of the 'national estate'. The Labor government lost power before such a commission could be established and most people thought that the incoming Fraser government, strong on rhetoric about 'states' rights', would abandon the idea. But just as Fraser surprised conservationists by overruling the Queensland government on sand-mining on Fraser Island (see chapter 6), he also confounded his critics by going ahead with the commission. Even more surprising to Serventy was the fact that he was among the commissioners selected by the Prime Minister. Impressed by Fraser's stand, Serventy subsequently asked him to support the idea of establishing an Australian branch of the Worldwide Fund for Nature. A few years later the organisation was established with Serventy and Fraser as co-founders and Serventy had learnt how to work with politicians on either side of the two-party divide.

The Heritage Commission that Fraser established, with the Victorian lawyer David Yencken as its chairman, worked hard and fast to get parts of Australia on the list of World Heritage Sites, so that they would have stronger legal and moral protection against the vagaries of the political process. Serventy was very proud of the achievements:

> We got five of them – the Great Barrier Reef, Lord Howe Island, Lake Mungo (in western NSW), Kakadu, and South-West Tasmania. The last one was one of the most important and we got it by sheer accident. The Labor government in Tasmania had earlier rejected an offer of $5 million from Whitlam to save Lake Pedder so we didn't get very far with them. But there were changes in the Labor leadership in Tasmania and when a young bloke called [Doug] Lowe became premier we thought we had a chance. There was no certainty that he would last very long as premier so we knew we had to act fast.[6] So we rushed the nomination forms down to Lowe, he signed them and Fraser shot them off to Paris. And that's really what saved the Franklin River. Of course the public campaign led by Bob Brown put the pressure on the federal government to act but it could only do so if it had the legal power to intervene and the World Heritage listing gave them that power. You know, I warned Fraser that he would lose the 1983 elections if he didn't act to save the Franklin. He didn't and he lost and Bob Hawke [by now the Labor leader] promised me that his government would take action. The government used its power and the Franklin was saved.

In discussing politicians past and present, Serventy mentioned a few that had a genuine interest in the environment – such as Malcolm Fraser and NSW premiers Neville Wran and Bob Carr. But he also pointed out that one of the most effective was Graham Richardson, environment minister in the Hawke government, who had no interest in the environment at all. What Richardson had was political influence within the government and a determination to be a successful minister. He was also smart enough to acknowledge his ignorance and consult experts. Ideally, said Serventy, you would like a minister who had both interest and influence but Richardson showed you can do pretty well if you are prepared to take advice.

To mark his fiftieth anniversary as a conservationist in 1996 (dating from his involvement in the Kings Park campaign), Serventy was honoured by a special luncheon at Parliament House in Sydney. Following in the footsteps of Philip Crosbie Morrison he had built a diverse career and had opened some new pathways for the next generation of advocates and educators. One younger man who was directly inspired by Serventy came to outshine his mentor for a period of time when he was awarded a Golden Sammy in 1977 as best male personality on Australian television and was voted Australian of the Year for 1979. It was Harry Butler, star of a nature program *In the Wild* that screened in a prime-time slot on ABC TV for a total of 26 episodes.

THE TELEVISION NATURALISTS: HARRY BUTLER AND LES HIDDINS

Serventy first met Butler when the latter was a ten-year-old schoolboy at a primary school in the hills near Perth.[7] Serventy and other members of the WA Naturalists Club were invited by the headmaster to come to the school to take the children on a nature walk and flaxen-haired Butler stood out as the most enthusiastic learner of the children. Wanting to foster his enthusiasm, Serventy promised to send him some wildlife magazines and he recalled how the boy ran after the bus carrying the visitors crying out 'You won't forget, sir?' He didn't forget and he next met Butler as a pupil at Northam High School, where Serventy was a science teacher. Although keen on nature studies Butler didn't take to formal education and left school at 15 to become a mechanic. However, he joined the Naturalists Club and the Gould League of Birdlovers and liked to help Serventy organise the Wildlife Shows that the naturalists started taking out to various country towns (see earlier). Through his work on the Wildlife Shows, Butler came to the attention of the Western Australian Director of Education who offered him a special job as adviser to the lecturer in nature work at Claremont Teachers' College. In this position he was able to give advice on nature studies to teachers at primary, secondary and even tertiary level. He also joined the Nature Advisory Service that was headed by Serventy and the two men often went together on field trips to collect specimens for museums, sometimes staying with Aboriginal communities. In 1963 Butler set himself up as an environmental consultant, working for government and resource development companies. He was in this line of work when he put the idea of a new nature series to the ABC in the late 1960s and he had established enough credibility for the ABC to take the proposal seriously. 'He certainly became better known for his television work than I did', Serventy told the authors, 'but then he had a lot more production assistance. *Nature Walkabout* was low budget and self-directed. But you've got to admire Butler for what he achieved as a lad who left school at 15.'

Critics of Butler's programs have said that they were excessively stage-managed; there was something unusual under every rock he turned over. He carefully cultivated the tough 'bushie' image and enraged many conservationists by coming out in support of drilling for oil on the Great Barrier Reef and mining uranium in Kakadu National Park. But he confirmed that Australian nature can rate highly on television. His laconic style and enthusiasm for his subject made him popular and encouraged many more people to take an interest in the unusual natural heritage of the country.

A 'bushman' who seemed even more authentic had a similar success in the late 1980s and early 90s with the series called *Bush Tucker Man*. Major Les Hiddins was still a serving member of the Australian army when he started making the programs and part of the appeal was to show his audience that a deep knowledge of the bush could enable a person to survive, even in country that seemed hostile and frightening (Mellor, 1990). Again the image of a laconic man totally at home in the bush captured the public imagination, but this time the appeal was extended by the fact that he freely admitted that most of what he knew he had learnt from Aborigines in northern Australia. Hiddins first developed an interest in bush tucker by pondering the fate of early explorers in Cape York who had died of starvation when surrounded by edible food. With some difficulty, he convinced the Army brass that learning something from the Aborigines about how to survive in the Australian bush could be quite useful for all soldiers and they gave him leave to begin a research project through James Cook University in Townsville in 1980. While doing this work he appeared in two episodes of a series called *Big Country* and the response was so strong that the ABC offered him a series of his own. This proved to be so popular that he was offered a second series and became a well-known identity on Australian television.

By the time Hiddins enjoyed his success, nature programs were commonplace on Australian television. English presenter David Attenborough had probably been the most successful with expensive productions filmed in all parts of the world, including a lot of footage from Australia. Stunning photography brought exotic creatures and places into the living room, with various series and one-off programs shown in prime time on most networks. However, people like Serventy, Butler and Hiddins had ensured that Australians realised their own land was as interesting as any on the planet. In doing so, they confirmed the potential that Philip Crosbie Morrison had seen in film.

From Donald Macdonald right through to Les Hiddins a spirit of amateurism has illuminated the work of successful nature writers and broadcasters. Of course, many of them have developed a great deal of expertise and some have had scientific training, but the sort of passion for plants and animals that motivated the early amateur naturalists has shone through the work of those who could communicate well with a broad public audience. In the early years of the twentieth century, amateurism came under fire from the 'more professional' biological scientists and Charles Barrett shot back that the scientists were nothing but 'armchair theorists'. The rift became a bitter one, yet the amateurs – operating in the tradition established by Gilbert White in England – pushed on with effective public education work. In recent times, more scientists have become convinced of the need to develop skills in communicating with a broad audience, acknowledging a role for passion and subjectivity in their work.[8]

Because of his family background, Macdonald was aware that many landscapes contained dark secrets about what the white settlers had done to the indigenous people, yet conservationists who campaigned for the preservation of 'pristine' nature subsequently turned a blind eye to this sad story. They also turned their back on the wealth of ecological knowledge that Aboriginal societies had accumulated through countless generations. By making *Nature Walkabout* in northern Australia, Serventy was one of the first to suggest it was not too late to learn from the ecological wisdom of the indigenous people. It's rather ironic that many white Australians got their first inkling of the survival skills of Aboriginal people by watching television programs about 'bush tucker' made by a white Army man. Les Hiddins may have been little more than a conveyor of knowledge that had long been neglected by the settler society, but, once again, it was his passion for his subject matter that made him an effective communicator.

CHAPTER 6

TOWARDS A CONSERVATION ETHIC:
Birth of the Conservation Movement

INTRODUCTION

As earlier chapters have explained, Australians began to rethink their attitudes towards Australian landscapes and ecosystems in the last decades of the nineteenth century: partly in response to the search for a new national identity but also because of the work of passionate nature lovers who were beginning to find ways to influence public opinion. Birdwatching was a popular pastime brought to Australia from England and the early birdwatchers were probably astounded by the diversity of species found in the 'new land'. Yet, while their chosen 'hobby' was seen as a passive recreational activity, it seems that the birdwatchers also became an effective political lobby because a Bird Protection Act was adopted in New South Wales in 1881, long before an Animal Protection Act was proclaimed in 1903 (Hutton and Connors, 1999). As Tim Bonyhady (2000) has pointed out, specific legislation to protect specific species of birds and animals had been enacted from 1790 (on Norfolk Island) through until the 1860s, when more systematic legislation was adopted in Victoria. In general, birds were the first to get the benefit of protection, followed by animals, which were seen as being in more direct competition with farmers. Legislation to protect native plants was not considered until much later, reflecting a hierarchy of values placed on the different elements of the indigenous ecosystems.

The early, patchy efforts to halt the degradation of Australia's unique and fragile biodiversity suggest that the early conservation movement was not strong, but, perhaps more importantly, they also indicate the size of the task involved in turning around the mentality wanting to 'conquer' the 'untamed' land (see chapter 2). Faced with such a challenge, it is probably encouraging to think about what pioneering conservationists did manage to achieve. In hindsight, we can say that, after a long period of falling on rocky ground, the idea of nature conservation took root in a range of isolated spots in the last part of the nineteenth century before it could revegetate barren areas of our national consciousness in the early part of the twentieth century. Rather than simply bemoaning the fact that it took a long time for the nature conservation movement to develop enough clout to tackle the legacy of nature colonisation, it is also important to note that the movement somehow took root in different parts of Australia at a similar period of time; suggesting, perhaps, that it was an idea whose time had come.

It is often noted that Australia was the second country in the world – after the United States – to declare a national park when the Royal National Park (although the word 'Royal' was not included until much later), just south of Sydney, was proclaimed in 1879. However, this bushland reserve was clearly modelled on recreation reserves that were in vogue in England and the clear intent was to preserve a tract of 'scenic' nature as a refuge for city-dwellers seeking a break from urban landscapes. The proclamation of the park did not herald the emergence of an indigenous Australian conservation movement. As discussed in chapter 5, the movement of amateur naturalists in Victoria added its weight to conservation battles begun as early as the 1860s to make Melbourne the early hub of conservation work. Tim Bonyhady has suggested that a conservation group formed in Bendigo in 1888, the Northern District Forest Conservation League was probably the first conservation organisation in the world (2000b, p. 184). However, the idea of preserving tracts of bushland in order to preserve nature 'for its own sake' probably started first in Queensland, where R.M. Collins, and then Romeo Lahey, campaigned for the declaration of national parks based on the model pioneered in California (Yellowstone and Yosemite). In the year 1908 the photographer John Watt Beattie began a campaign for a substantial nature reserve in the region of the Gordon River in Tasmania's south-west and in 1909 David Stead and others founded the Wildlife Preservation Society in Sydney – probably the first conservation organisation to develop political clout and a national framework for its work. In the 1920s, Myles Dunphy, a pioneer of recreational bushwalking in Australia, turned his attention to active conservation work and, with the support of his bushwalking colleagues, he was able to launch a string of successful campaigns. By this time a 'preservationist' conservation movement, quite distinct from recreational or resource-oriented conservation movements, had been established with some striking similarities to the preservationist movement established by John Muir in the United States.

In pushing for the preservation of significant tracts of 'pristine' nature, the preservationist movements in both the US and Australia were driven by an ecological awareness of functioning natural systems. However, not surprisingly, both these movements continued to reflect the legacy of frontier societies in which 'undisturbed' nature was seen as being on the other side of the frontier from human settlement. In recent times the cultural blindness of this movement in failing to recognise that 'pristine' nature in Australia was actually the product of many thousands of years of interaction between 'natural' systems and the indigenous people has been pointed out, as has the paralysing effect of suggesting that nature 'worth keeping' must be kept separate from people. Nevertheless, the movement forged by people like Romeo Lahey and Myles Dunphy has been primarily responsible for the creation of an extensive system of nature reserves in areas that might otherwise have been substantially changed by human settlement or resource exploitation. It has also raised public awareness of issues related to biodiversity and the relative 'fragility' of many Australian ecosystems.

Lahey and Dunphy perfected the art of lobbying politicians and public servants to achieve their aims. They relied mostly on preparing well-researched submissions followed by quiet persuasion. However, a boom in resource development in the 1960s made this approach ineffective and a new generation of conservationists, including Myles' son Milo Dunphy, pioneered new tactics that focused heavily on building public support for a preservation proposal. Watershed battles took place over the Great Barrier Reef, the Colong Caves in New South Wales and Fraser Island off the Queensland coast, and a 'second-wave' conservation movement was formed under the leadership of people like Milo Dunphy, Judith Wright and John Sinclair. This more radical conservation movement subsequently found itself working alongside global environmental organisations that took root in Australia, such as Friends of the Earth and Greenpeace, and the conservation movement became part of a broader environmental movement.

THE MOUNTAIN TRAILS CLUB AND MYLES DUNPHY

An intriguing legend surrounds a visit by a party of Sydney-based bushwalkers to the Blue Gum Forest in the Blue Mountains west of Sydney at Easter in 1931. At the time bushwalking was still emerging as a recreational activity and the delightful forest of mature bluegums at the confluence of the Grose River and Govetts Creek, now a Blue Mountains heritage icon, was fairly remote and little known. What is certain is that the party of walkers was led by Alan Rigby of the Mountain Trails Club and that while they were camped in the forest they had a disturbing encounter with a Clarrie Hungerford, who informed them that he had acquired a lease to the area that enabled him to harvest its resources, including the magnificent bluegums. But that is where certainty ends and legend takes over.

According to the most popular account of the event,[1] the bushwalkers were peacefully enjoying their time in the forest when they were shocked to hear the unmistakeable sound of an axe biting into a live tree and when they rushed to investigate the sound they found Hungerford engaged in the process of ring-barking one of the forest giants. According to this account, the bushwalkers engaged Hungerford in conversation and when they heard about his lease someone came up with the idea of raising the money to purchase the lease from him. By the end of the conversation the two parties had reached an agreement on the amount Hungerford would want in order to sell the lease and the bushwalkers returned to Sydney to start raising that amount. It is true that the bushwalking fraternity of Sydney, organised mainly through the Mountain Trails Club and the Sydney Bushwalkers, did raise money to pay Clarrie Hungerford for the Conditional Purchase Lease covering the bluegum forest and surrounding bushland, however, the popular account of how this came to pass conflates a much more complicated series of events. Having conducted some detailed research on the topic for his book *Back from the Brink: Blue Gum Forest and the Grose Wilderness,* Andy Macqueen believes that the encounter between Hungerford and the bushwalkers began when Hungerford and a friend, Bert Pierce, rode their horses into the forest after descending a new track they had cut into the valley from the Bells Line of Road (a trail unknown to the bushwalkers) and found the bushwalkers having lunch. Accepting an invitation to share a cup of billy tea, Hungerford then stunned the campers by telling them that he intended to clear the forest to plant a grove of walnut trees (Macqueen, 1997, p. 242). Descendants of Hungerford told Macqueen that this announcement about planting walnuts might have been 'a ruse' hatched by an alert mind sensing the opportunity to make a tidy little profit by on-selling the lease to the bushwalkers (Macqueen, 1997). What adds credence to this suggestion is that the forest encounter took place at the time of the Great Depression when Hungerford and Pierce were struggling to survive on marginal farms along the Bells Line of Road, boosting their income by joining road construction gangs whenever they could. Apparently Hungerford had obtained the lease to the 'unimproved' land in the Grose Valley for next to nothing and a quick sale would be safer than any attempt to 'work the land'.

Whatever the precise nature of the encounter, the records reveal that Alan Rigby raised the matter at a meeting of the Mountain Trails Club (MTC) on April 17, 1931, asking the club to do everything possible to prevent the logging of the forest (Macqueen, 1997). The meeting agreed to follow up the matter by writing to the Blackheath Progress Association alerting them to the threat posed to a particularly scenic part of the mountain wilderness and by finding out more about the nature of Hungerford's lease. The secretary of the MTC, Myles Dunphy, became a kind of secretary to a *de facto* sub-group of the club that followed up this decision and, having established the details of the lease, he wrote to Hungerford saying they wanted to discuss alternatives for saving the forest. It may have been this letter that first raised the possibility of the bushwalkers raising the money to buy out the lease; at any rate Hungerford replied by saying he wanted £150 to settle his interest. By the time this letter arrived, a special Blue Gum Forest Committee had been formed, with Rigby and Dunphy from the MTC, representatives of the Sydney Bushwalkers, and Roy Bennett from the Wildlife Preservation Society as chairman. On behalf of this committee, Dunphy told Hungerford his price was too high but they were willing to continue negotiating while they lobbied the Department of Lands to have the area declared a public reserve.

Exchanging letters with Hungerford proved to be a slow and cumbersome process and so the bushwalkers suggested a face-to-face meeting in the forest itself, open to all those interested in the matter. A date for such a meeting was set for November 1931 and the bushwalkers then made plans to go to the area a day earlier so that some of them, Dunphy included, could actually see it for the first time. Interest was so great that three different parties of bushwalkers converged on the forest from three different directions, undoubtedly giving Hungerford an uneasy feeling of being 'surrounded'. However, Hungerford also decided to 'play hard' by arriving even earlier and felling one of the bluegums as a kind of threat. So, by the time the negotiations began, emotions were already running high. Hungerford dropped his price to £130, provided the conservationists could pay a deposit of £25 and the rest by the end of the year. The conservationists agreed, knowing that the Wildlife Preservation Society could pay the deposit as a loan. Following the agreement, two individuals who were in the bushwalking clubs put up the balance owed to Hungerford as two-year loans and that gave the committee two years to raise the full amount. They began by organising a successful Blue Gum Ball and within a year had cleared the loans and had convinced the Department of Lands to then change the title of the land from leasehold to public reserve.

Despite the price paid, Alan Rigby, Myles Dunphy and their colleagues had won an important symbolic battle in preserving remaining areas of Blue Mountains wilderness. The campaign had also provided an important focus for an emerging conservation movement and had introduced the idea of nature conservation to a broader audience. For Dunphy, it was a resting place on the trail of a much longer campaign – the creation of a Greater Blue Mountains National Park. The Blue Gum Forest campaign demonstrated that the Sydney bushwalkers had established themselves as an effective force for conservation.

The Blue Gum Forest campaign achieved success much more quickly than the first conservation campaign launched by members of the MTC in 1924. This campaign focused on a proposal to extend the Royal National Park south of Sydney to incorporate an area of land along the coast from Garie Beach down to Stanwell Park. The area was dubbed Garawarra – a composite of Garie and Illawarra – and so this was known as the Garawarra campaign. The Sydney Bushwalkers joined the MTC in this campaign after the club was formed in 1927 and members of the two clubs managed to collect 5,000 signatures on a petition calling on the state government to accept the proposal. Eventually, in 1934, the Minister for Lands, E. A. Buttershaw, agreed to the establishment of a Garawarra Park but, initially, it only included 1,300 acres of the proposed 2,500. The bushwalkers kept up their campaign and over a number of years, as some private leases expired, further areas were added to the park which, in turn, was eventually integrated into the Royal National Park.[2]

The two campaigns taken up by the bushwalkers overlapped and they absorbed a lot of time and effort. Initially, Myles Dunphy had struck considerable resistance when he asked his colleagues in the MTC to take up the Garawarra campaign. However, they were buoyed by their hard-won successes and by the middle of the 1930s conservation had become an integral part of the activities of the bushwalkers, both in the MTC and the Sydney Bushwalkers, and many of the bushwalking pioneers became lifelong conservation activists. The first and most prominent of them was Dunphy, who built a life around bushwalking, campaigning for conservation, and teaching architecture. He remained an active conservationist until his death in 1985.

Myles Dunphy and his mates were the pioneers of recreational bushwalking in Australia. The MTC was formed at a meeting held in Dunphy's Annandale home in 1914 organised by Dunphy, his school friend Bert Gallop and his work colleague Roy Rudder. Six members were inducted into the club at its first meeting and a decision was taken to admit only those who had clearly demonstrated their love of the bush and a desire to venture into remote areas. It would remain a rather exclusive club. From 1914 until 1970 the total number of members admitted reached only 55 – all of them men. The MTC was not Sydney's first walking club. That honour belongs to the Warragamba Walking Club, formed in 1895 by William Hamlet. But members of this club were primarily interested in walking for exercise and their journeys rarely strayed away from established paths. At night they would stay in country inns. By contrast, the members of the MTC sought out journeys where no paths or detailed maps existed. They always camped out and for a long time had to make their own swags, tents and billies because lightweight versions of such things were not otherwise available. The term bushwalking was not invented until the MTC agreed to sponsor the formation of another, more inclusive club to operate in parallel. Discussion about the name of the new club settled on Sydney Bushwalkers and the term was then picked up by people in a range of other clubs that came into existence during the 1920s.

The MTC was certainly not Australia's first conservation organisation. Tim Bonyhady (2000b) has suggested that this honour goes to the conservation group formed in Bendigo in 1888 referred to earlier. Others point to the formation of the Field Naturalists Club of Victoria, which was formed in 1880 with Frederick McCoy, a professor in natural history, as its founding president (Hutton and Connors, 1999, p. 36). A Natural History Association of New South Wales was discussed in 1880, but not established until 1887, around the same time that the Royal Society of Queensland established its Field Naturalists Section (Hutton and Connors, 1999, p. 37). In 1904 the Tasmanian Field Naturalists Club was formed (Hutton and Connors, 1999, p. 36). While the organisations of field naturalists engaged in some lobbying work to achieve legislative protection for endangered species, this was a consequence of their primary interest, which was to study Australian plants and animals. Arguably, the first organisation specifically set up to influence government policies was the Wildlife Protection Society formed in Sydney in 1909 (Hutton and Connors, 1999, p. 35). This organisation was formed by naturalists who felt that existing legislation aimed at protecting flora and fauna had been ineffective. The key founder was David Stead (father of the novelist Christina Stead) who turned a private interest in nature into a career and established a considerable profile as a nature advocate (Hutton and Connors, 1999, pp. 49–50. In the same year that the WPS was formed in Sydney, the Gould League of Bird Lovers was formed, first in Victoria then spreading to New South Wales, Queensland and Western Australia (Hutton and Connors, 1999, p. 42). Dedicated to the memory of the English naturalist John Gould and his artist-wife Elizabeth Gould, who visited Australia in the 1830s and produced a beautiful book on *Birds of Australia* in 1848, the Gould League set out to build a public base, particularly among young people, and, indirectly, this built support for conservation policies.

Contrary to popular belief, Dunphy was also not the first Australian to campaign for national parks along the line of those being established in the United States. In 1878 Queensland nature lover R. M. Collins, who had been born into one of that colony's oldest pastoralist families, visited California and heard about Yellowstone National Park (the world's first).[3] On returning to Australia he kept in touch with people in California campaigning for the establishment of further national parks and he began lobbying for the establishment of such a park in the McPherson Ranges, near his family's property. In 1896 Collins was elected to the Queensland parliament and used his influence to campaign for legislation allowing the establishment of national parks. The legislation was finally adopted in 1906 and two years later the first, small, park was established at Mt Tamborine. After two more years small national parks were also established at Cunningham's Gap and Bunya Mountains, but Collins had been frustrated in his attempts to get landholders in the region of the McPherson Ranges to agree to the establishment of a park there. However, when he tired of this campaign it was taken over by a younger member of the Royal

Geographical Society, Romeo Lahey. Lahey was the son of an Irish-born timber mill manager in southern Queensland and he had spent much of his childhood in the shadow of the McPherson Ranges.[4] A walking trip into the ranges at Easter in 1908 convinced him to take up where Collins had left off in lobbying for the establishment of a park. In 1911 he moved to Sydney to study engineering at Sydney University, but continued his campaign of lobbying and, at every available opportunity, returned to the area to visit landholders to convince them to support the idea. Lahey put seven years of intensive work into the campaign and finally, after the election of a Labor government in June 1915, the Lamington National Park was created.[5] Unfortunately, Collins, who died in 1913, did not live to see this outcome.

Although Collins and Lahey might be called the 'fathers' of national parks in Australia, it was Dunphy, after 1924, who became the best-known advocate of the idea. Whereas the Royal National Park had been established in New South Wales as early as 1879 as a reserve for human recreation, Dunphy campaigned for national parks that would be primarily aimed at preserving wilderness and, in 1932, he formed the National Parks and Primitive Areas Council, which has been called Australia's first 'wilderness society' (Thomson (ed.), 1989, *Myles Dunphy: Selected Writings*). The big advantage Dunphy enjoyed over people like Lahey was that he had a dedicated band of like-minded colleagues to work with.

The thing that probably motivated Dunphy the most was a concern that forest clearance had become an ingrained national habit, even a test of Australian 'manhood'. In a letter published in the *Katoomba Daily* in 1934, he wrote:

> The task of subjugating wilderness in the past rightly was reckoned to be a manful job. Sturdy men and trusty axes, confronted with primeval bushland, steadily hewed a wide and wasteful way through it and out the other side. Later on, tree destruction became a kind of national complexus, it went altogether too far; it became spiteful. For some settlers the very zenith of land 'improvement' was a holding absolutely short of trees – a grassy desert. Rain-drags never were considered; wind-breaks rarely. Sometimes a settler – after much mind travail – might plant a couple of pine trees for a little shade which he thought his beasts might need. (Thomson 1989, p. 27)

The letter went on to argue that the time had come to replace the 'Age of Wastefulness' with the 'Age of Conservation'.

Like John Muir in the United States (Stephen, 1981), Dunphy campaigned for a shift in public perception regarding the concept of wilderness. The term had already undergone major transformations since it emerged in the English language around the thirteenth century as a word meaning 'home for wild animals' (Robertson, Vang and Brown, 1992). It was subsequently used in translations of the Bible to refer to 'barren' arid regions largely devoid of life. When English explorers travelled the world to find vast forests that seemed unknowable and beyond human control they reached into their vocabulary and retrieved the world 'wilderness' to refer to landscapes that were anything but barren. Through its various transformations the word had retained an eclectic mix of meanings, ranging from the dangerous and unknown to barren and unproductive to unknowable and uncontrollable. It was used to refer to landscapes that contrasted sharply with the pastoral scenes of the English countryside. For Dunphy, wilderness was a source of infinite variety and personal inspiration and he sought to share his excitement about wild country in his writings. He argued that significant tracts of wilderness must be preserved in order to preserve the rich diversity of plant and animal species. By the end of the 1980s the public perception of wilderness had gone through another major transformation.[6] By this time the term would more often be linked with biodiversity than with barrenness; fear of wild country had been matched by reverence. The term had almost turned into its opposite. Clearly, this perceptual shift was not confined to Australia but Dunphy's efforts had been vindicated and his ideas can be seen as having been visionary.

As Tom Griffiths (1996) has pointed out, the first wave of Australian conservationists, operating in organisations like the Victorian Field Naturalists Club, were still embedded in the European tradition of collecting exotic things to store, as prized possessions, in a multitude of museums. Griffiths sees this fascination for collecting 'specimens' as an extension of a hunting culture which, for Europeans, was less a matter of survival and more an 'ultimate test' of a man's knowledge, skill and courage. He points to an 1891 photograph of a party from the Victorian Field Naturalists Club posing for the camera before going on an excursion, with most of the men (and there were only men) proudly nursing their rifles. Even the early photographs of Myles Dunphy in the bush show him with a rifle over his shoulder and, although it has been suggested this was needed to collect food, one suspects that it was also the actualisation of an 'appropriate' image. However, if Dunphy and his colleagues were still influenced by the mentality of the hunter, they found that time spent in the bush began to erode their sense of separation and mastery.

Peter Meredith, author of a book (1999) about Myles Dunphy and his famous conservationist son Milo, has said:

> Myles Dunphy's preservationist ideas, his entire attitude to the outdoors and outdoor recreation, were a product of his time . . . [H]e had grown up in an era of imperial adventure when explorers and scientists were pushing back the boundaries of the known world. Their feats were recounted in an adventure literature, fictional and non-fictional, for which the Australian public, Myles included, had an insatiable appetite.

Myles Dunphy, circa 1915, with the swag he designed for recreational bushwalkers. (Enviro Books)

Yet, as we have seen, Dunphy became critical of the idea that conquering wilderness was an appropriate test of Australian 'manhood' and he urged policy makers to have the courage to defy a culture that fostered this perception. Because his ideas were radical for their time he had to put his arguments carefully and patiently. He became famous for meticulous research. As Meredith put it:

> As fluent in speech as in writing, he articulated his ideas with awesome thoroughness, using what commentators termed the 'salami technique', delivering slice after slice of assiduously researched material and backing this up with gentlemanly contact with bureaucrats.

He may not have been the 'father of the conservation movement', as some have claimed, but he was an effective pioneer and, as Meredith put it:

> Myles's dreams ultimately became the dreams of the Australian public and nearly all of his schemes succeeded . . . (p. 2)

Myles Dunphy was born in Melbourne in 1891. His mother came from Tasmania and his father from County Kilkenny in Ireland (Meredith, 1999). When Myles was young the family lived for a while in the small Victorian town of Outram before moving to Sydney. Myles was about 11 when the family moved again – this time to Kiama on the coast south of Sydney. Six years later they were back in Sydney. The years spent in Kiama were important ones for Myles. He later wrote fondly of days spent with younger brother Bryan and friends trekking up into the hills behind the town or wandering and fossicking along the coast. He also recalled that while the family was living in Kiama his father had given him a copy of *The Boys' Own Book of Natural History* by Rev. J. G. Wood, which awakened an interest in understanding how nature works. The gregarious Myles had little difficulty making friends and, soon after the family returned to Sydney in 1907, he gathered some of his mates into the Orizaba Cricket Club (named after a mountain in distant Mexico where cricket is totally unknown). In the cricket off-season of 1908 Dunphy formed the associated Orizaba Tourist Club and the destination for their first excursion was Kiama. On this trip Dunphy began a lifelong task of keeping a detailed record of his adventures.[7]

In 1912 he renewed his friendship with an old schoolmate from Kiama, Bert Gallop, who had also moved to Sydney. With Gallop, Dunphy made his first visit to the Blue Mountains and the pair was so excited by what they found that they returned for a series of long journeys through some largely uncharted country. On one particular trip from the Boyd Plateau (near Jenolan Caves) down into the valley of the Kowmung River and out to the old gold-mining town of Yeranderie (west of Camden) they took a wrong turn and got hopelessly lost. They had run out of food before they found their bearings again and eventually they staggered into Yeranderie, days late and very hungry. But the experience had not dimmed their enthusiasm. After they formed the MTC in 1914 the pair were involved in many excursions into the country surrounding the Kowmung River. It became the club's favourite area.

In 1919 Dunphy teamed up with another friend, Roy Davies, for an epic journey by foot and canoe, starting at Nowra on the New South Wales south coast and ending up at the mouth of the Murray River in South Australia. The first stage was a walk from Nowra to Harrietville in the Victorian alps and the second a canoe trip from Albury to Port Elliott. The journey took almost a year to plan. With the attention to detail that he was already becoming famous for, Dunphy made a special study of ideas and trends in canoe construction internationally, and he had his own design turned into a sturdy vessel by a boatbuilder in Sydney's Neutral Bay. In between his adventures Dunphy built a very successful career as an architect and teacher, first at Sydney Technical College and later at the University of New South Wales. In 1925 he married Margaret Peel and they had two children: Milo, who followed in his father's footsteps as bushwalker, conservationist and architect, and Dexter, who went on to become a professor in the Graduate School of Management at the University of New South Wales. After their marriage in 1925, Myles and Margaret spent their honeymoon in a canoe – the same one that Myles had built for the voyage with Roy Davies.

The declaration of the Garawarra National Park in 1934 was a personal milestone for Dunphy because he had initiated the campaign ten years earlier. In the same year (1934) his newly formed National Parks and Primitive Areas Council was able to celebrate another major victory with the declaration of the Tallowa Primitive Reserve (later incorporated into the Moreton National Park). At this time the distinction between a national park and a primitive reserve was important because the parks were still seen as being essentially for human recreation. The reserve at Tallowa was only the second area in the world to be set aside specifically for the preservation of wilderness. The first such reserve was the Gila Wilderness established in the US in 1924. The success at Tallowa showed that Dunphy was looking beyond his beloved Blue Mountains. From 1932 onwards he campaigned for a system of national parks in New South Wales, stretching from the Snowy Mountains to the Hastings River. But the area surrounding the Kanangra-Boyd Plateau remained his first love. When his first son, Milo, was born he was given the middle name of Kanangra. Eventually Dunphy's conservation work focused largely on the idea of a Greater Blue Mountains National Park that would stretch from near Mittagong north to the Hunter Valley. While a single park of that size has not been created, the creation of three separate parks, Kanangra-Boyd, Blue Mountains, and Wollemi, has meant that this dream has been substantially realised.

In 1939 Dunphy first became alarmed at the possibility of a mining lease being granted for a belt of limestone in the area of the Colong Caves just to the south of the Kanangra-Boyd plateau. From then until 1957 he wrote regularly to relevant ministers in the state government appealing for the area to be permanently exempted from any threat of mining, but his requests fell on deaf ears. Then, in 1968, his fears were realised when the Askin government granted a mining lease for the area to an international resource company. Dunphy's colleague Alan Rigby, who had led the 1931 trip to the Blue Gum Forest that uncovered the logging threat there, took the lead in the formation of a Save the Colong Committee, with Dunphy as its patron. The committee waged a vigorous public protest campaign about the granting of the lease and prepared a detailed submission outlining the case for the preservation of the area as wilderness. By 1973, the battle had been won and the following year the Colong leases were added to the Kanangra-Boyd National Park. The committee had shown that a direct appeal for public support could win the day on conservation issues; this proved to be a significant turning point for the conservation movement nationally.

Dunphy's legacy also included the preparation of detailed maps of many of the areas he visited. He had a habit of taking whatever maps were available with him on trips and adding to them as he went. His colleagues have said that he also loved to talk to people who lived in the places where he travelled so that he could collect information and stories about such places. This work gave him the opportunity to name quite a few landmarks that had not previously been mapped. He used this opportunity to honour some of his bushwalking colleagues, but he also liked to think up colourful and memorable names. For example, the place where he and Gallop took a wrong turn on their trip to Yeranderie in 1913 was called Squatting Rock Gap. As early as 1932, Dunphy's mapping skills were recognised by the Lands Department and he was asked to be a chief adviser on a project to create a Blue Mountains Tourist Map. Later he was appointed as a councillor on the Geographical Names Board of New South Wales. When the Board announced in 1969 that it was going to undertake a review of all names in the Blue Mountains, Dunphy worked to ensure that colourful names like Wild Dog Mountain and the peaks called Rip, Rack, Roar and Rumble were not replaced by something more 'respectable' and mundane. Dunphy argued that place names often reflected stories associated with those places and the names could encourage people to find out more about those stories.

Dunphy's journals reveal much about the character of the man. His many adventures are recounted in great detail and places are described with loving care. The close bond forged between Dunphy and his MTC colleagues is obvious, especially when some of them had to walk into the

Kowmung valley and carry him out on a stretcher after he had suffered a severe case of heat stroke on a trip with Norman Colton in 1934. Tributes to Dunphy, especially at the time of his death in 1985, repeatedly stress that he was a good raconteur and very good company for a long walk. His journals also show that he was a good listener – to other people and to the land. His journals support the point that has often been made about him; that he was meticulous in every task he undertook. After Alan Rigby died in the Colong area during the campaign to stop the mining threat, his son, Byron, wrote to Dunphy from his home in London asking for an account of the time when his father had helped carry Dunphy out of the Kowmung valley. What Dunphy sent back was a 105-page account of the event taken from his journal! His attention to detail helped him make convincing submissions to government bodies. By the late 1960s, 14 proposals for parks or reserves that he had worked on had come to fruition (Meredith, 1999, p. 6).

Dunphy treasured the bond he forged with other members of the MTC. He wrote in his journals that journeys into the wilderness tend to have an 'equalising' effect because social status means very little when you have to rely on each other for survival. Among his colleagues were successful lawyers and people high up in the public service. But when they went on trips together they became part of what Dunphy dubbed the 'bush brotherhood'. These sentiments tend to echo the thoughts of writers like Adam Lindsay Gordon and Henry Lawson about the equalising effects of life in tough Australian environments. As discussed in chapter 3, the notion of 'mateship born of struggle' became prevalent in the heady days of the 1890s when nationalism and national identity were high on the agenda as the country headed towards federation. Of course, the bush brotherhood was a very masculine notion and Dunphy and his colleagues eventually came under fire – especially from the redoubtable Sydney feminist and lawyer Marie Byles – for forming what appeared to be a secret men's society that excluded women. By 1927 the pressure on the MTC to open the door to women members had becoming quite strong and the matter was formally discussed in a club meeting. Not wanting to open the door too widely, Rigby suggested the idea of probationary members who might eventually become full members. But a decision was taken instead to sponsor the formation of a separate, more inclusive club, the Sydney Bushwalkers.

THE SYDNEY BUSHWALKERS AND MARIE BYLES

Marie Byles joined the Sydney Bushwalkers in 1929. Immediately she convinced her new colleagues to take up a new conservation campaign – for the establishment of a Bouddi National Park to the north of Broken Bay in New South Wales. This was an area of special significance to Byles. When she was a child her parents had owned a weekend cabin at Palm Beach and she later recalled how she often sat looking across the bay to the wooded headlands and hills and secluded beaches to the north. In 1922 she decided to visit the area and took along a group of university friends. The beauty of the area exceeded her expectations and, on this trip, she decided to push for the area to become protected reserve. Already, Sydney people were building holiday homes on the Central Coast and Byles feared that Sydney would spread in that direction. When she became prominent as a woman solicitor, she used that profile to get articles published in Sydney news-papers about the proposal. But she needed help. She got the bushwalkers to write separate letters to the Lands Department asking them to act on the idea. Byles got the Federation of Bushwalking Clubs, formed in 1932, to write to the department and it responded by saying it would send a District Surveyor to the area to investigate. The surveyor became a strong advocate of the idea and, on his recommendation, an area of 650 acres was immedi-ately set aside. In 1936 other large areas were added to the park.

Byles was undoubtedly the best-known woman bushwalker in the early years but she was not alone. Others took the opportunity to join the Sydney Bushwalkers or other new clubs. One of the most adventurous was Dot English (later Butler) who eventually established her own public profile as the 'Barefoot Bushwalker'. She told the authors that she enjoyed bushwalking so much in the early days that she could hardly wait for the weekends to come.[8] As soon as she finished work in the city on a Friday she would head for Central Station, put her skirt and shoes in a locker, put on a pair of shorts and meet her colleagues for another weekend adventure. When we met her in 1996 she was still living her adventures in her mind. Only two years earlier, at the age of 84, she had made the news again by climbing the Three Sisters and abseiling back down. Her stories were punc-tuated by strong belly laughs.

The barefoot Butler sometimes teamed up with another prominent adventurer to tackle some difficult climbs. The Katoomba doctor Eric Dark (husband of the writer Eleanor Dark) was known in the Blue Mountains for being outspoken on social issues but, according to Butler, locals also joked that he was president of the 'Katoomba Suicide Society' because he liked to do dangerous things. In 1936 Butler and Dark became the first to climb Crater Bluff in the Warrumbungle Mountains in north-west New South Wales and Butler told the authors that she always enjoyed climbing with Eric 'because he was such a gentleman. He always climbed in knickerbockers, long sleeves and a tie. And there I was in my shorts and bare feet.' An odd couple indeed, especially at a time when women were generally thought to be too delicate for anything more than horseriding . . . side-saddle.

Soon after she joined the Sydney Bushwalkers, Butler was drawn into conservation work. She remembered collecting signatures on the

Garawarra petition on a number of occasions while travelling by train to the Royal National Park. She also got actively involved with Marie Byles in the Bouddi campaign. Butler also had an opportunity to do conservation work in her job as assistant to the Sydney branch manager of the Melbourne *Argus* newspaper, Walter Trinick. 'He was a conservationist long before the term ever became fashionable', Butler said. Trinick discovered that every public service employee could become an Honorary Ranger under the Wild Flowers and Native Plants Protection Act, so he formed the Rangers League to encourage people to do just that. He got Butler to write to all the editors of house journals in the public service departments to promote the new organisation and before long they had 700 members. Little was expected of League members because the idea was largely to foster their own interest in nature, but they did have occasional meetings and public activities. One activity that really took off was an annual exhibition held when Sydney's wildflowers were at their peak. According to Butler, the distinctive cabbage tree palms of the Illawarra district in New South Wales would be a rare sight today if the Rangers League had not campaigned to have them declared a protected native species. She also suggested that an important success of the League was to have the term 'wildflowers' replaced in official language by the term 'native flowers'.

Tragically, Butler paid a high price for her adventurous life. Three of her four children died in separate accidents in the bush. As a teenager her daughter Wendy caught her foot between rocks in the Kowmung River and drowned. Her son Norman died from snakebite in a treehouse he had built at a rural commune and his twin brother Wade died many years later, disappearing in a remote part of Tasmania. However, Butler was very philosophical about these tragedies and did not regret having encouraged her children to be adventurous. She certainly had no regrets about her own life. Asked to nominate highlights she said that a trip to the Andes in 1969 had been very special, but if she had to nominate a favourite place she thought it might be the Kowmung River area where many of her early adventures took place.

When we interviewed Butler a close friend and fellow bushwalking pioneer, Alex Colley, was also present. In the 1930s Colley and Butler (then English) had been members of a sub-group within the Sydney Bushwalkers, 'The Tigers', who enjoyed going on trips together. At 87, Colley looked even fitter than Butler. He was still playing tennis once a week and going two days a week to continue his voluntary work for the Colong Foundation for Wilderness in the city. When we were discussing the early days, Colley produced some old photos, one of them showing him flanked by Butler and wife-to-be Norma ('monopolising the girls', Butler joked). 'We were all pretty adventurous', Colley said, 'but Dot was something special. She had no fear at all.'

Colley got very involved in the campaign to stop the mining at Colong Caves when that issue broke in 1968. Echoing Butler's comments about the affection she felt for the Kowmung River area, he said that most bushwalkers felt personally betrayed by the decision to mine at Colong, which is in the Kowmung Valley. 'If you had a big, roaring limestone quarry there and a pipeline and a road, all the way into Camden – well, it was going to wreck all that beautiful country. So we started to fight that.' As mentioned earlier, the formidable bushwalkers won the day on Colong and, Colley recalled, they then had to decide whether or not they should wind up the committee they had formed. 'Milo Dunphy [son of Myles] had already got us started on another issue, which was to stop a pine plantation on the Boyd Plateau and we won that one as well. And then Milo argued that we shouldn't stop there because there were many other campaigns we could take on.'

According to Colley, Milo Dunphy became the real inspiration for the Colong Committee (later called Colong Foundation). Looking beyond the Blue Mountains he got the organisation started on a campaign to protect rainforest in the Border Ranges (on the border of New South Wales and Queensland) and to become one of the first organisations to call for the establishment of the Kakadu National Park in the Northern Territory. In 1972 he also became founding director of the newly established Total Environment Centre in Sydney, where he continued to work until shortly before his death (from cancer) in 1996. Of course, Milo was virtually born a conservationist. His first journey, being pushed in a pram by his parents

'The Tigers', an adventurous group of the Sydney Bushwalkers in the 1930s: Dot Butler [then English is third from the right] and Alex Colley is fifth from the right (Courtesy Alex Colley)

from Oberon to Kanangra Walls when he was just 20 months old, has achieved legendary status.

MILO DUNPHY

There were some remarkable similarities in the careers of Milo and Myles Dunphy. Milo followed in his father's footsteps as both conservationist and architect (although, unlike his father, Milo eventually abandoned his career as an architect to become a full-time conservationist). However, there was one important difference. As Meredith put it:

> If there was a difference between father and son, it was one of style. Milo rejected his father's gentlemanly approach, the polite letters, the friendly contacts with bureaucrats. Milo had the fiery zeal of a radical; he made a lot of noise, ruffled the bureaucratic feathers, rattled politicians. In this he was as much a man of his time as his father had been of his.[9]

Colley urged us not to neglect the many unsung heroes of the conservation movement in telling this story. Early bushwalkers like Alan Rigby and Jim Somerville were tireless in their voluntary work for conservation organisations and, of course, we would add the name of Colley himself, who had contributed decades of voluntary work to the Colong Foundation. But Colley agreed that Myles Dunphy deserves all the credit he has received.

> He was actually, in my opinion, a genius, if you consider his infinite capacity for taking pains. He did a terrific amount of work. He had an encyclopaedic memory and, due to the meticulous work he put into all his proposals, they were always listened to and often carried out.

According to Colley, Dunphy was heavily influenced by the ideas of John Muir and the organisation he formed, the Sierra Club. Although he never visited the US, he did keep up an active correspondence with members of the Sierra Club. There are, of course, some similarities in the colonial experience of Australia and the US. In both cases, there were still some relatively undisturbed natural areas left when people like John Muir and Myles Dunphy began campaigning for wilderness protection. In both the US and Australia, the conservation pioneers emphasised the 'preservation' of 'pristine' areas that had not yet been degraded by human activity. Some prominent Aboriginal Australians have argued that the very term 'wilderness' represents a continuation of the concept of *terra nullius* – the idea that Australia was essentially unoccupied by people when the English colonisers arrived.[10] They point out that over the vast expanse of time in which Aboriginal people had lived in this land, they had formed relationships with every part of it. They 'used' the land less intensively, yet more extensively, than the colonisers. Non-indigenous Australians, therefore, display cultural insensitivity when they suggest that 'wild places' have remained people-free and 'pristine'.

Although there is debate about the extent to which the indigenous people altered the Australian ecosystems – e.g. by use of 'firestick farming' or by hunting pressure on the megafauna[11] – there is little doubt that there were significant human impacts and when people are now excluded from areas their absence can cause a chain reaction.[12] So, the notion of pristine wilderness pioneered by people like Dunphy was rather simplistic. It can even be argued that a conceptual separation between people and wilderness represents a continuation of a 'frontier mentality' in both Australia and North America; in that wilderness came to be associated with areas that were too rugged or too remote to be 'settled'. At the edges of settlement, wary pioneers were only too aware that the wilderness also had the capacity to reclaim landscapes that were left 'unguarded'. However, it is also fair to say that Dunphy (in Australia) and Muir (in the US) were reacting in their own ways against a frontier mentality that regarded wilderness as enemy. In collecting local stories for his work in proposing new names for places, Dunphy became interested in the silences created by the elimination of the Aboriginal inhabitants, and he knew that many wild places contained sad memories.[13] Certainly people like Muir and Dunphy created a new mythology of wilderness by writing passionately about its capacity to regenerate the human 'soul'. Yet, this may have been a necessary antidote to earlier negative conceptions and the beginning of a more sophisticated discourse about what wilderness is and what it might mean to people.

Without abandoning the goal of maintaining and expanding the system of national parks, conservationists in Australia are becoming more interested in other conservation strategies that have been tried in other countries, and they are looking beyond the preservation of wilderness and 'mature' ecosystems (Prineas, 1998). At least some are taking up the challenge to learn more about the ecological philosophies and land management practices of the indigenous Australians (Figgis, 2000). This might suggest that the legacy of pioneers like Myles Dunphy is becoming less significant. However, it is important to remember that such pioneers began their work while it was still possible to preserve significant tracts of wild country. In establishing the idea of an extensive system of national parks, they initiated a successful strategy that has created a base from which new strategies might be launched.

Conservationists outside New South Wales often quibble about the attention paid to Dunphy and the Sydney bushwalkers. Queensland conservationists, for example, say that Romeo Lahey was ahead of Dunphy in almost everything the two men did.[14] Victorians might point to Donald

Macdonald or Alex Chisholm (see chapter 5). In his autobiography, Chisholm has said that he was strongly influenced as a 15-year-old by 'The Woodlanders' when they publicly promoted the ideas of Henry David Thoreau (see chapter 5). The story that emerges is that the conservation movement started in several places around the same time because a growing number of nature lovers decided it was time to challenge the national 'habit' of clearing the land. However, it took strong, highly committed individuals such as Myles Dunphy and Romeo Lahey to tackle a challenge as big as that and make progress against the prevailing public sentiments.

THE QUEENSLAND CAMPAIGNERS: JUDITH WRIGHT AND
JOHN SINCLAIR

Myles Dunphy's legacy has been well guarded by the bushwalking fraternity to which he belonged. Romeo Lahey's direct influence was primarily channelled through the organisation that he formed – the National Parks Association. However, he also influenced, more indirectly, a relative (by marriage) from the New England district, who was inspired to play the leading role in the establishment of the New England National Park in 1931. That relative was Philip Wright, father of the poet and conservationist Judith Wright who has said that she, in turn, was inspired by what she heard from her father about the work of 'Uncle' Romeo. This was one incentive for her to become a conservationist, but her early poetry and other writings show that the greater inspiration was the New England countryside in which she grew up (see chapter 4). From the 1940s onwards her poetry is replete with nature themes and imagery. But it was not until the late 1950s, when the burden of caring for daughter Meredith began to wane, that she was able to become more actively involved in conservation work. In 1958 she teamed up with close friend Kathleen McArthur to form a Queensland branch of the Wildlife Preservation Society. Wright was the organisation's founding president and she remained in that role until she moved to Canberra in 1973. What initially motivated her to become a more active conservationist was her concern for rainforests in the area surrounding her home at Mt Tamborine, where new housing developments were opening up. Mt Tamborine was the site of Queensland's first national park but the area's proximity to Brisbane made it vulnerable. Wright enjoyed her best years with partner/husband Jack McKinney at Mt Tamborine and it was where Meredith was born, so she felt strongly about its future.

Wright had also been introduced by McArthur, a wildflower enthusiast, to the delights of southern Queensland coastal flora on the sand dunes behind Cooloola Beach. She was horrified to hear from McArthur that those dunes were threatened by sand-mining. The sand-mining threat also extended to Fraser Island, which had intrigued Wright ever since the time she read in her grandfather's journals about an enjoyable encounter he had with Aborigines while sailing on a boat just off the island's coast. Wright had visited the island herself during a long journey through Queensland with McKinney in 1948 and, although the Aborigines had by that time been long removed to the mainland, she felt the echo of their presence and fell in love with the extraordinary beauty of the world's largest sand island, with its superb lakes and thick forests. After her first visit to Cooloola, Wright took McKinney to the area and they came across a small cottage for sale at Boreen Point, which they could just afford to pay off as a coastal retreat. So Wright had a personal stake in two places that were being threatened by 'development'.

Looking back on the early days of the WPSQ, Wright has written:

> The very words 'conservation', 'ecology' and 'pollution' were unfamiliar. The problem of soil erosion – largely because of the publicity given to it through the dust-bowl disasters of the United Sates – was known and discussed. But few people had read or heard of famous conservationists like Aldo Leopold, who in 1933 wrote: 'Civilisation is not . . . the enslavement of a stable and constant earth. It is a state of mutual and interdependent cooperation between human animals, other animals, plants and soils, which may be destroyed at any moment by the failure of any of them'. (in Brady, 1998)

Wright and her small committee had taken on a huge challenge in a state that was buzzing with excitement about its 'development opportunities'. During the 1960s they would be pitched into some fierce battles over sand-mining and various threats to the Great Barrier Reef. But first they had to learn more about the movement they had become part of. As president, Wright started to receive correspondence from conservationists and organisations in Australia and other parts of the world. She started to feel less isolated in her fears and hopes. The new organisation set out to build branches across the state.

In 1963, Brian Clouston, the proprietor of the Jacaranda Press (where Wright was engaged as a poetry editor), suggested to her that there was a need for a decent nature magazine in Australia. He was sure that it would be used in schools (distributed through state Education Departments) and would also find a public market. He even suggested the title *Wildlife*. Wright thought it a good idea and felt it would give the WPSQ a much higher public profile. So, with her trusted ally McArthur, she began work on the first edition, which appeared in June 1963. With Clouston's support they took the high-risk strategy of producing an expensive, glossy magazine with striking colour photographs. The first edition included a broad range of articles with contributions by some acknowledged experts in their fields. Wright contributed four articles, including one on the life and work of her 'uncle' Romeo Lahey.

The magazine hit the spot in terms of public response, but the Queensland Education Department refused to come to the party with guaranteed distribution. Wright was supported by a team of capable and dedicated volunteers and sales of the magazine steadily increased. But without institutional support, such a high-cost publication was always going to struggle. Jacaranda Press was too small a business to continue subsidising it and so Wright was forced to spend much of her time looking for other sponsors. Eventually she struck gold when a submission to the Myer Foundation in Melbourne resulted in a return phone call from Baillieu Myer, an admirer of Wright's poetry, who said he would be pleased to meet her on a forthcoming visit to Brisbane to discuss ways of making the magazine viable. The result of their meeting in July 1965 was an agreement by the Myer Foundation to pay a regular subsidy for the magazine 'every six months for two years or as long as the WPS would continue to publish it' (Brady, 1998, p. 233). With that achieved, Wright asked to be relieved as editor. By this time her husband's health was very poor (he died in 1966) and her own health was beginning to suffer. She was badly inconvenienced by growing deafness. At first she was replaced by one of the volunteers, Stan Breaden, but when the task proved to be beyond him, Wright wrote to the well-known naturalist Vincent Serventy who had earlier written to her to express his enthusiasm for the magazine. As mentioned in chapter 5, Serventy took on the task for the next 20 years.

In the second half of the 1960s, Wright was on the horns of a dilemma. For personal reasons, and because she wanted more time for her own writing, she wanted to reduce the demands on her time. But the conservation battles were hotting up in Queensland and her reputation as a conservationist was spreading nationally. In 1966, she was approached by noted retired CSIRO scientist Francis Ratcliffe to serve as a councillor on the newly established Australian Conservation Foundation, established in Canberra in 1964 with a grant of £1,000 from Prime Minister Robert Menzies. A little reluctantly, she accepted the invitation because she knew it would give a stronger base from which to pursue her interests in Queensland.

As a conservationist, Judith Wright is probably best remembered for her leading role in the fight to protect endangered sections of the Great Barrier Reef.[15] It was a struggle that continued until the whole reef was incorporated into a protected marine park in 1974. For Wright, the battle began in mid-1967 when the WPSQ received a report from its Innisfail branch saying that water pollution from sewerage discharge was starting to pose a significant threat to areas of the reef. Soon afterwards an even more urgent distress call was issued by Melbourne-born artist John Busst, who had taken up residence on Bedarra Island near Innisfail, after he noticed a limestone mining application for an area of the reef advertised in the local newspaper. Wright made contact with Busst and with members of the Queensland Littoral Society (which included some authoritative marine scientists) and launched into an emergency campaign. Busst had been at school in Melbourne with Prime Minister Harold Holt, a keen diver. This gave him some access to the government in Canberra and, although Holt disappeared at sea near Melbourne late in 1967, his successor John Gorton also took a personal interest in the fate of the reef. Wright was able to convince the ACF to make protection of the reef a national issue (despite some wavering by those who thought that some mining on the reef might be acceptable) and this helped get national media attention. Wright's profile as a writer made her an obvious focus for the media and she was interviewed for newspaper articles in various states and for ABC television's national current affairs program *This Day Tonight*.[16]

The campaign to save the Great Barrier Reef had early success when the proposal to mine Ellison Reef (the proposal that Busst had noticed in the Innisfail paper) was withdrawn. But thereafter it became a long battle against a very hostile Queensland government. At one point it emerged that Premier Joh Bjelke-Petersen had shares in a company that was looking for oil under the reef, but he dismissed public protests about this by saying that all 'sensible' Queenslanders admired a premier who was committed to economic development projects. During a 1969 debate on another conservation issue, the minister for mines had said that he 'would rather see an attractive orchard laid out by man than scrub left standing' (Brady, 1998, p. 360). However, in February 1970 the fate of the reef became a pivotal issue in a by-election for the Gold Coast seat of Albert, normally a safe seat for Bjelke-Petersen's National Party. When the National Party did badly the government responded by setting up a Royal Commission into all the issues related to the reef. National sentiment had clearly swung against any developments that might damage the reef, and Prime Minister Gorton offered financial assistance to the environmentalists to help them put their case to the Queensland Royal Commission. The findings of the Royal Commission did not appear until 1974 (and even then they were ambiguous). But it mattered little because the federal government led by Gough Whitlam had intervened in the meantime and declared the whole reef a marine national park.

The WPSQ was also involved in a successful campaign against sand-mining at Cooloola, which also began in 1967.[17] Kathleen McArthur played the leading role in this campaign and it focused on a grassroots campaign of lobbying local members of parliament. The strategy paid off when, to the surprise and consternation of Premier Bjelke-Petersen and his minister for mining Ron Camm, a joint meeting of National Party and Liberal Party parliamentarians voted against the mining proposal because backbenchers had become nervous about the likely electoral impact of the issue.

The real strength of the WPSQ under the leadership of Wright and McArthur was that it built a strong network of local branches. When Wright was interviewed in 1988 for biographical records to be kept by the National Library in Canberra she was rather pessimistic about public attitudes towards conservation in general.[18] But she was very proud to say that in recent local government elections in Queensland, branches of the WPSQ had successfully campaigned against a range of mayors and councillors who had a poor record on conservation. She saw the election of many pro-conservation candidates across the state as a turning point in Queensland politics. Wright was modest about her own personal contribution to the sea change in political culture in Queensland. She was a prominent female 'greenie' at a time when macho politicians were worshipping at the altar of 'development'. Her courage won her many friends, as well as enemies, and she showed that a love of nature can be self-sustaining. She combined the sensitivity of a poet with the fierce determination of a committed political activist.

During the 1980s, Wright became increasingly concerned about the tendency of nature preservationists to counterpose the interests of 'pristine nature' and the interests of Aboriginal people. In 1990 she publicly condemned the policy of The Wilderness Society on Aboriginal land rights, describing their position as 'a confirmation and endorsement of the *terra nullius* judgement' (Strauss, 1995, p. 23). In saying this she was precisely anticipating the critique of wilderness preservation made by Aboriginal academic Marcia Langton,[19] which provoked considerable discussion in conservation circles in the 1990s. In 1991, Wright even resigned from the organisation she had helped to create – WPSQ – because it had adopted a 'weak' position on Aboriginal land rights.

At the height of the Great Barrier Reef campaign in 1968, Wright wrote a powerful essay titled *Conservation as a Concept* (in *Quadrant*, 12.1, pp. 29–33). In calling for a new sense of responsibility for 'the maintenance of this planet and its elemental and biotic systems' she provided a precise and powerful summary of the most disturbing manifestations of environmental degradation. However, the purpose of the essay was not to plunge her readers into despair, but rather to encourage them to explore 'a new kind of creative relationship' with nature. This would require:

> . . . a renewed humility and a revival of imaginative participation in a life-process which includes us, and to which we contribute our own conscious knowledge of it, as part of it, not as separate from it. (p. 33)

This perspective suggests that environmental activism can be an exercise in self-renewal as much as a pitched battle against 'the enemy'. It probably shows how Wright managed to keep her energy up for the fight for the GBR. No doubt it helped her face the prospect of another big battle that loomed at the same time – the fight against sand-mining on Fraser Island. But on this issue Wright was able to take a back seat because the struggle threw up another highly effective conservation advocate in John Sinclair.

Sinclair was born in Maryborough, a coastal town opposite Fraser Island, in 1939 (Sinclair and Corris, 1994). He left school at 15 to become a 'grease monkey' in his father's garage, but after a year of that he convinced his parents to support him through a course at the Queensland Agricultural College in Gatton. Coming out of college he thought he would surely get a chance to start out as a farmer on a land grant somewhere, but when that did not eventuate he returned to Maryborough to take up a job with the Department of Education as an adult educator in 1962. He now had the responsibility of supporting his new wife, Helen, and one of the first things he did on his return to the town was to take Helen to visit the island that looms large in the consciousness of Maryborough residents. From the mouth of the Mary River, which runs through Maryborough, it is not far by boat across to Fraser Island, which stretches out along the coast like a detached section of the mainland. When Sinclair was young his parents often spoke fondly of their honeymoon on the island in 1935, but it was not until 1955 that he got the chance to visit the place for himself. This opportunity came when a friend of his father needed help to build a house on a block of land he had secured on the island and young Sinclair offered his services for a week. Many years later, he was still able to recall that first trip to the island in precise detail, especially the drive across the island to the 'back beach' (Sinclair and Corris, 1994). The vehicle was driven by a schoolmate, Andrew Postan, son of a wealthy Fraser Island logging contractor, who pointed out the rainforest trees with a 'proprietal air'.

Having settled back in Maryborough, Sinclair followed a family tradition by joining the conservative Country Party (later renamed the National Party) and he rose in its ranks to become president of the Maryborough branch and member of the party's state management committee. At the same time his job as an adult educator put him in touch with a different side of politics when he organised visits to the town by Judith Wright and another poet and conservationist Nancy Cato. In 1967 he joined the Maryborough Field Naturalists' Club, which turned itself into a branch of the WPSQ the following year. Sinclair quit the Country Party in 1970 in protest over the government's support for sand-mining at Cooloola. He was even more enraged that his treasured Fraser Island was facing a similar fate and so he joined with conservationists from Hervey Bay and Bundaberg to form the Fraser Island Defence Committee, soon changed to Fraser Island Defence Organisation (FIDO) – the 'watchdog of Fraser Island'.

Sinclair's experience in state politics proved useful to the campaign because he knew how to make it an issue in the 1971 Merthyr by-election, directly challenging Bjelke-Petersen during one campaign appearance. He also took the fight to Canberra by going there to lobby federal politicians (from all parties), and in 1974 he managed to convince the *National Times* to publish a feature article in its colour supplement under the title 'Paradise in Peril'. He also challenged the validity of leases granted to the companies Murphyores, Dillingham, and Queensland Titanium in the local mining warden's court – a tactic that bought some time in the processing of their applications and which attracted some media attention.

As mentioned earlier, Premier Bjelke-Petersen and Minister for Mines Ron Camm were taken aback when the government's support for sand-mining at Cooloola Beach was overturned in a joint meeting of the coalition parties in 1970. However, this only made them more determined to ensure that the rich deposits of Fraser Island would be made available to the miners. They announced that the mining would definitely proceed even before the findings of the mining warden's court had been released. Sinclair and his colleagues were encouraged by a change of government in Canberra at the end of 1972 because the new minister for the environment, Moss Cass, became a firm opponent of the mining. He submitted a motion of opposition to the development to a caucus meeting of Labor parliamentarians and it was accepted with little opposition. However, the pro-development Minister for Minerals and Energy Rex Connor had missed the meeting because of illness and when he heard the result he rang Prime Minister Whitlam and threatened to resign. Whitlam and some of his senior colleagues decided the issue did not warrant losing Connor from cabinet and so when the issue was put to caucus for a second time the vote went 42–41 in favour of mining.

By 1975 mining had begun on the island and FIDO was fast running out of ways to stop it. A challenge in the Queensland Supreme Court against the decision of the mining warden to approve the mining applications failed, so the conservationists appealed to the High Court in Canberra. There, Justice Ninian Stephen found fault in the decision of the mining warden in regard to the leases granted to Queensland Titanium and that company had to reapply for some of its leases. But this was only a partial delay and not a real victory for the conservationists. Fortunately, Moss Cass had not given up the fight when he was outmanoeuvred by Connor in caucus. Instead he used his own powers as a minister to set up a Commission of Inquiry into the development and he gave FIDO a grant to employ a biologist to develop an environmental plan of management for the island. The Queensland Department of Education refused to grant Sinclair leave to appear at the inquiry but this was overcome by the issuing of a subpoena.

When the Commission finally released its report in 1976, the Whitlam government had been replaced by a Liberal–Country Party coalition government headed by Malcolm Fraser. Although the Commission's report argued that all mining should be stopped because the island's natural heritage was of national and international significance, few believed that the new government would accept this recommendation. It was seen as being even more pro-development than its predecessor. But the government stunned the nation when it announced that it had accepted the Commission's findings and would give the mining companies just eight days to wind up their operations. Most of the island would be declared a national park. Maybe the prime minister had a soft spot for an island that bore the same name. More likely, he and his colleagues could see that national opinion had swung firmly in favour of the preservation of the island's unique heritage. It was a huge victory for the conservation movement.[20]

Sinclair's work for Fraser Island came at considerable personal cost. His marriage broke down and he finally succumbed to the harassment of the Department of Education by resigning his job. The Sinclairs could trace their ancestry back to pioneer settlers in the Maryborough district, but now members of his family were copping abuse from former friends and colleagues in the town. Sinclair moved to Sydney to continue working for the island without the harassment of a largely hostile town. Unexpectedly, at the end of 1976, the Nobel-Prize winning novelist Patrick White, who set two of his novels (*The Eye of the Storm* and *A Fringe of Leaves*) on Fraser Island, nominated Sinclair for the title of 'Australian of the Year' in a competition hosted by *The Australian* newspaper. On New Year's Day 1977, *The Australian* announced that Sinclair had been awarded the honour – just one year after it had gone to his Queensland nemesis Joh Bjelke-Petersen.

The campaign against sand-mining on Fraser Island is sometimes seen as being a watershed in the birth of the 'modern' conservation movement. Others say that the campaign to stop the mining of the Colong Caves pioneered most of the tactics used in subsequent campaigns. Still others say that the failure to stop the flooding of Tasmania's Lake Pedder fired the subsequent, successful, campaign against the flooding of the Franklin River and that this announced the arrival of conservation as a major political force (see chapter 10). The authors would argue that the campaign to stop degradation of the Great Barrier Reef rivals all of the above. Drew Hutton and Libby Connors (1999) have argued that major public campaigns for the conservation of specific areas in Australia, from the Great Barrier Reef to the Franklin River, marked a particular phase in the evolution of the Australian conservation movement. In their account, this phase was followed by another in which a 'more professional' approach to lobbying the major political parties in Canberra held sway.

If we accept that the first wave of conservationists, operating in the last decades of the nineteenth century were essentially engaged in preserving 'scenic' places for hunting and collecting activities that had little to do with the preservation of functioning ecosystems, at what point can we say that an ecologically minded conservation movement first emerged? And what is the relationship between the conservation movement campaigning for national parks in the first half of the twentieth century and the environmental movement that emerged in the late 1960s and early 1970s. Some writers[21] have said that the movement that emerged in the late 1960s was such a radical departure from anything that had gone before that it is not useful to say that there was an earlier environmental movement. However, if we take the transition from Myles to Milo Dunphy as a case study there can be little doubt that the earlier generation did create a foundation on which the 'modern' environmental movement was built. Milo was driven by essentially the same vision that his father wrote about, often very evocatively, in his detailed journals and, as we have seen, Myles Dunphy's ideas were often very radical for his time. We have argued here that people like Myles Dunphy and Romeo Lahey were pioneers in seeing the need to preserve whole ecosystems because they had an empathy for what lay 'beyond the frontier' – even if this also exacerbated the problem of separating people from nature. In this sense, the preservationist movement, which paralleled a similar movement in the US, was itself a departure from the earlier conservation movement in Australia.

However, it is no more useful to try to pinpoint the origins of a movement than to identify the 'father' of a movement. It is probably more useful to say that a mood shift regarding attitudes towards nature started in Australia some time around the 1860s and then deepened, or even changed direction, after the turn of the twentieth century. When such mood shifts occur there is an opportunity for visionaries and pioneer thinkers to articulate the change and initiate new courses of action. The conservation/environmental movement in Australia was well served by strong and visionary pioneers like Romeo Lahey, Myles Dunphy, and Marie Byles. The second generation leaders of the movement they created – people like Milo Dunphy, Judith Wright and John Sinclair – went on to play a significant part in the emergence of what we are calling 'second wave environmentalism'. As it emerged, the nature conservation movement in Australia carried legacies of human-centredness and a frontier mentality. It fostered some traditions and perceptions that were subsequently challenged by 'modern' environmentalists and indigenous leaders. But it also created the discourse about conservation that enabled such debates to occur. The conservation pioneers did their best to nurture a conservation ethic in Australian society.

CHAPTER 7

WORKING IN THE BORDERLANDS:
Australian Innovations in Ecological Science

INTRODUCTION

Ecology as science arrived in Australia from Europe, initially as the study of plant communities, in the early years of the twentieth century. Some pioneering work in plant ecology was done in South Australia. However, agricultural industries became the driving force for a more serious commitment to ecological research from the 1920s onwards as efforts were made to understand the population dynamics of various 'pests' that were interfering with crops and orchards. In the same year, 1927, the Waite Institute for Agricultural Research in Adelaide and the Council for Scientific and Industrial Research in Canberra, were formed and both organisations fostered some internationally significant research on the population dynamics of insect communities, in particular. By the 1950s, Australia had produced three world-renowned insect ecologists in Alexander Nicholson, Herbert (Bert) Andrewartha and Charles Birch. In 1929 the augustly named Empire Marketing Board also brought the English animal ecologist, Francis Ratcliffe, to Australia and he went on to play a role in public education as well as in ecological research.

Although there were clear economic/resource management incentives for taking an ecological approach to the study of Australian plant and animal communities, the academic study of ecology continued to be marginalised by a more traditional, discipline-based, approach to the study of the 'natural sciences'. Even as ecology was gathering strength in applied scientific research, in the academy it remained as an orphan child, rather unloved by the family of 'pure' scientific disciplines. If it was difficult to win acceptance for ecological studies of plant and animal communities, it was even more difficult to win academic acceptance for the application of ecological ideas to the study of human communities, as the innovative academic Stephen Boyden discovered during his career at the Australian National University.

Those who have campaigned for ecology to be accepted as a 'legitimate' scientific discourse have sometimes felt a need to concede ground to their more conservative academic colleagues. When the pioneer ecologist Charles Birch, for example, retired from his position at the University of Sydney to indulge in a long-standing passion for organic philosophy, he felt he had largely renounced the right to call himself an ecologist because he would not be engaged in consistent scientific 'practice', even though he felt his philosophical work was broadly ecological in its approach. While there is a dangerous tendency for people to use the word 'ecology' in loose and rather meaningless ways, the inability of 'tidy' minds to put it in a single 'box' is what has given it both power and a healthy mystique. In academic life, those who want to be bold and innovative thinkers are often condemned to a life in the margins.

However, the study of natural 'systems' indicates that edges and transition zones are often places where new possibilities emerge, as in the hybridisation of species from intersecting communities. Difficult circumstances can sometimes lead to individual and collective innovation, as we can see in the ways that Australian plant and animal communities have adapted to life in some tough environments. Life in 'the margins' may often be difficult and unpredictable, but it can also be a source of creativity, capable of regenerating systems in decline. Unpredictability – the antithesis of tidiness – is being embraced by scientists studying the interplay between chaos and order. In science, as in other fields, those who can adapt to life in the margins may be the greatest innovators.

INTERNATIONAL RECOGNITION

In 1988, the Ecological Society of America took the unprecedented step of awarding its prestigious Eminent Ecologist Award to two men – Herbert G. Andrewartha and L. Charles Birch. Not only were Andrewartha and Birch the first joint recipients of the award, they were the first Australians to receive such international recognition. In a tribute written especially for the award, well-known North American ecologist Daniel Simberloff wrote:

> Both men have ... informed our field to the extent that the 'Andrewartha–Birch school' connotes a widely recognised viewpoint and suggests a distinctive research protocol. *The Distribution and Abundance of Animals* was the landmark synthesis of field population ecology that inspired a generation widely credited with constructing modern ecology ... [They] have been consistent sceptics, continually confronting fashionable models with hard-won field data ... Largely because of their books, Australian systems and Australian ecological research are part of the common vocabulary of ecologists throughout the world ... their names and reputations are as inextricably intertwined in our discipline as are those of

Gilbert and Sullivan or Lee and Yang in other fields. Though either man's independent career is worthy of honour, their interaction has been a highlight of ecology.

The clear implication of this is that Australian scientists have made a significant contribution to the science of ecology internationally, with Andrewartha and Birch leading the way.[1] Not surprisingly, ecological approaches did not penetrate the biological sciences in Australia until well after they had made an impact in Europe. As in both Europe and North America, plant ecology emerged well in advance of animal ecology (McIntosh, 1985). According to the pioneering animal ecologist Charles Elton, this was because plants do not move away when humans move into their habitats. Whatever the reason, the study of plant communities and their interactions emerged strongly in Europe, especially as a result of the work of German botanists, in the 1890s and spread soon afterwards to North America. Raymond Specht has noted (in Kormondy and McCormick (eds.), 1981) that visiting European botanists started using ecological concepts in the construction of vegetation maps in Western Australia, Queensland and Tasmania in the period 1906–1914. An Australian *Journal of Ecology* was initiated in 1913. But, according to Specht, the first Australian botanist to use the word ecology in relation to his own research was A. A. Hamilton, from the Sydney Botanic Gardens, who completed a study of salt marsh vegetation around Port Jackson in 1917. In 1972, R. S. Adamson from the Royal Society of London spent six months working with T. G. B. Osborn at the University of Adelaide and they completed vegetation maps of an arid region near Ooldea and in the Mount Lofty Ranges that year. In the following years, Osborn continued this work and made Adelaide the centre of expertise in plant ecology in Australia.

In 1927, the Waite Agricultural Research Institute was established in Adelaide to conduct research on the problem of insect 'pests'. This coincided with the release in England of the first important book on animal ecology, Charles Elton's *Animal Ecology*, and, under the leadership of James Davidson, researchers at the Waite Institute began using an ecological approach in the study of insect communities. The Council for Scientific and Industrial Research (forerunner of the CSIRO) was also established in 1927 and much of its early work was also focused on the problem of insect pests in the agricultural industries. In 1933, the CSIR entomologist Alexander Nicholson published his first paper on interactions between insect species and, in 1935, he teamed up with physicist V. A. Bailey to present a model for quantifying the fluctuations in insect populations, which attracted a lot of interest internationally.

As mentioned in chapter 5, quasi-scientific natural history clubs and associations had started forming in Australia as early as 1880. These provided a forum for professional and amateur biologists to pursue their interests in the country's unique wildlife. By regularly going into the field, the members of such organisations began to collect valuable information about the habitats and distribution of particular species of fauna and flora. Charles Elton has said that natural historians were the effective pioneers of ecology because they collected information that helped ecologists develop their conceptual models (Worster, 1994). Elton has called ecology 'scientific natural history'. With clearer questions in mind, the plant and animal ecologists who followed the natural historians were able to make more sense of the data they collected. The Australian ecologists who made their mark internationally were working, initially, with insect populations because the control of pests became a major research priority in a country heavily dependent on wealth generated by agriculture. The first was Nicholson working at CSIR in Canberra and he was followed by Andrewartha and Birch, who came together at the Waite Institute in 1939 to begin a collaboration that would last until Andrewartha's death in 1992.

AUSTRALIAN PIONEERS: HERBERT ANDREWARTHA AND CHARLES BIRCH

Bert Andrewartha was already at the Waite Institute when the young Charles Birch arrived there in 1939, fresh out of postgraduate studies in agriculture at Melbourne University. Although Birch left Waite after six years to travel overseas and eventually settle in Sydney, their collaboration resulted in the publication of two books, exactly 30 years apart. *The Distribution and Abundance of Animals* was published in 1954 and *The Ecological Web* in 1984. Both men were initially trained in agriculture and most of their contributions to ecological theory were by-products of their research relating to the control of insect pests in agriculture. Ironically, the motivation for their first book was to challenge the theories that enabled Nicholson to build his international reputation in the 1930s. By the 1950s, these three Australians, Nicholson, Andrewartha and Birch, had provided much of the theoretical base for the cutting-edge research on biological control being carried out at the Riverside campus of the University of California. Yet, as we shall see, they had reached quite different conclusions concerning the factors responsible for the distribution and abundance of insect pests and other species.

In some ways Andrewartha and Birch were unlikely collaborators; Andrewartha being the energetic opinionated field naturalist and Birch the more reserved thinker and experimentalist. Yet, as already indicated, this 'odd couple' would eventually be feted for the strength of their collaboration. By 1945 the suave and urbane Birch was finding the atmosphere of the Waite Institute far too stuffy and isolating and he left for Chicago and Oxford in order to develop his skills as a teacher. When he returned to Australia in 1947 he settled in the cosmopolitan suburbs of Sydney and continued to make many overseas trips. He had been brought up as a Christian and retained a strong interest in the philosophy of life. Eventually he developed a second career as a philosophical writer and won the Templeton Prize (an international award for religious writing) in 1990. Andrewartha stayed in Adelaide and developed his career through research conducted at both the Waite Institute and the University of Adelaide. He was intolerant, according to Birch, of any views he considered to be 'anti-science' and was more at home in the field than the city. Had they not been thrown together early in their careers to study grasshopper plagues in the dry wheat-growing zone of South Australia they might never have found a reason to work together. According to Birch,[2] their theories evolved over numerous evenings spent sharing their ideas while they were out in the field studying grasshoppers.

Not only were Andrewartha and Birch different in temperament and interests, they were rarely in the same location for any length of time. Andrewartha was born in Perth in 1907 and gained his Bachelor's degree in Agriculture from the University of Western Australia before moving to Melbourne University in 1933. After completing his Master's thesis on the biology of apple thrips (a common insect pest) he joined Professor James Davidson at the Waite Institute in Adelaide, to work on the plague grasshopper. Birch was born in Melbourne in 1918 and also studied agriculture at Melbourne University before going to the Waite Institute. When they worked on their two books, Birch was in Sydney and Andrewartha was in Adelaide.

The key ideas in their first book, *The Distribution and Abundance of Animals*, grew out of their joint research on the plague grasshopper. It had been noticed that this species of grasshopper produced very heat-resistant eggs and Andrewartha and Birch were trying to find out if this was the key to the occurrence of plagues. Their first joint publication appeared in 1941 and was entitled 'The influence of weather on grasshopper plagues in South Australia'. However, they came to understand that the abundance of the grasshoppers was influenced by a range of factors, not the least of which was the generous provision by humans of a steady and reliable source of food, wheat. While studying this species of grasshopper, they developed an interest in exploring the way intrinsic (physiological) and extrinsic (environmental) factors interact to determine the distribution and abundance of all species of animals.

The central proposition of the Andrewartha–Birch population theory is that the distribution and abundance of a population depends on the chances individuals within that population have to survive and reproduce, which, in turn, depends on their environments. The environment is regarded as being made up of four directly acting and interacting groups of

Herbert Andrewartha
(Historical Records of the Australian Academy of Science)

factors: resources, mates, malignities and predators. These, in turn, are influenced by more removed factors, and a map of these interrelationships they called an 'envirogram'. They also recognised that local populations may differ widely in both their genetic make-up and their specific environmental conditions, and that this variability helps to spread the risk and increase the stability of populations.

These ideas challenged the views of Nicholson, and others, that populations are regulated only by density-dependent factors such as competition and predation. In contrast, Andrewartha and Birch showed that any component of the environment (density-dependent or density-independent) can play a part in determining population distribution and abundance, and that through careful field studies each of these can be assigned a probability of influence, which will change with both space and time.

The publication of *The Ecological Web* in 1984 marked the end of a celebrated (if rather spasmodic) partnership. Andrewartha was still recovering from a serious stroke that he suffered in 1975 when they wrote this, and he died at age 84 in 1992 (four years after receiving the Eminent Ecologist Award). Birch retired as Challis Professor of Zoology at the University of Sydney the year before the *The Ecological Web* appeared. This gave him the time he needed to further develop his 'second career' in philosophical writing. As early as 1965 Birch had published a reflective book entitled *Nature and God,* and in the late 1970s he began a collaboration with the US philosopher John Cobb Jr (a former student of Birch's long-time mentor Charles Hartshorne – see below). They co-authored the book *The Liberation of Life: from cell to community* in 1981. After his retirement from Sydney University, Birch published two related books, *On Purpose* (1990) and *Feelings* (1995) and in between (1993) came *Regaining Compassion For Humanity and Nature,* which was written for a broader audience. He was awarded the international Templeton Prize after the appearance of *On Purpose.*

Birch's philosophical work places him firmly in the organicist school pioneered by the English philosopher Alfred North Whitehead, who trained at Cambridge but spent the latter part of his career at Harvard University. Birch has explained his debt to Whitehead and his followers in his 1999 book *Biology and the Riddle of Life,* perhaps his most comprehensive contribution to philosophy.

As a young scientist working at the Waite Institute, Birch found the mechanistic world-views of most of his colleagues deeply disturbing. This was partly because he still considered himself a Christian, but also because their view of life failed to inspire him emotionally. 'It had nothing to say about my feelings, which were the most important part of my life,' he wrote (1999, p. 138). Birch wanted to read work by biologists that took a broader view of life and he remembered a lecture he had attended while studying zoology at Melbourne University, given by Professor W. E. Agar, in which Alfred North Whitehead's ideas had been mentioned. 'I didn't understand a word of the lecture', Birch told the authors, 'but it encouraged me to write to Agar to ask his advice about how to learn about Whitehead.' Agar wrote back recommending a book called *The Philosophy and Psychology of Sensation* written by Whitehead's most distinguished student, Charles Hartshorne. To Birch's surprise, Agar also mentioned a book he had written himself called *A Contribution to the Theory of Living Organisms* and it was the opening line of this book that convinced Birch he was now on the right track. It read: 'The main thesis of this book is that all living organisms are subjects' (Birch, 1999, p. 139). Birch was attracted to Whitehead's ideas because they were grounded in scientific thinking but were addressing 'interesting' questions that most scientists preferred to ignore. As Birch (1999, p. 8) explained, Whitehead first rose to public prominence after working for ten years with that other 'great polymath of the century', Bertrand Russell, on the classic book *Principia Mathematica*. However, at the end of this project the two authors parted company. When Russell said, famously, that 'either life is matter-like or matter is life-like', he chose the former whereas Whitehead chose the latter, to become an organicist. Russell set out to demonstrate that the complexities of life could be reduced to some simple understandings while Whitehead wanted to dwell in complexity and mystery. Whitehead once introduced Russell at a lecture at Harvard by saying: 'Bertie says that I am muddle-headed. But I think he is simple-minded' (Birch, 1999).

In 1929, Whitehead irritated traditional scientists by suggesting that 'In the real world it is more important that a proposition be interesting than that it be true.' By this he meant that the pursuit of interesting ideas might lead to a degree of understanding of the complexities of life while a simple pursuit of the 'truth' would lead to a narrowing of the focus. Whitehead himself pursued the interesting notion that matter is life-like by suggesting that all entities have experience from the past as well as an orientation to the future and this gives them a degree of self-determination, or subjectivity. In 1938 he wrote 'Life is the enjoyment of emotion derived from the past and aimed at the future.'

Birch began *Biology and the Riddle of Life* by saying that whereas books on biology once routinely began with a definition of life, one searches in vain for any attempt at this in 'modern texts'. However, for Birch the question 'what is life?' is the most interesting one in biology. For this reason he believes that work being done on questions about consciousness and the way that non-human organisms experience 'feelings' is the most important, yet it rates poorly with funding bodies and in the media alongside 'mechanical' research in areas like genetic engineering. For Birch,

Whitehead not only returned the focus to the broad and interesting questions, but he also launched a counterattack against those who predicted that biology would one day be reduced to a study of the physics and chemistry of living organisms. In 1925 he wrote that 'Science is taking on a new aspect which is neither physical nor purely biological. It is becoming the study of organisms. Biology is the study of larger organisms, whereas physics is the study of smaller organisms.' Birch said that he wholeheartedly agreed with a comment attributed to world-renowned physiologist Sir John Eccles when he reportedly said 'I'll be reduced to physics and chemistry only when I'm dead' (1999, p. 4).

It is interesting to note that Whitehead himself found much inspiration in the poetry of the great romantic William Wordsworth. Good poetry, of course, has the capacity to engage its readers emotionally as much as intellectually. Whitehead approved Wordsworth's celebration of the

Charles Birch, circa 2000

mystery of nature and concurred with his famous line 'We murder to dissect'. Birch points out that similar insights can be found in the work of other great writers like Shakespeare. But he goes on to argue strongly against some of the more simplistic, romantic views of nature that emphasise notions like wholeness and harmony. 'Life is a struggle against enormous odds', he has written. 'Any ecologist who has studied plants and animals in nature knows this' (1999, p. 92). Many of the simplistic, romantic notions of nature are based on the concept of a 'balance of nature' and Birch hastens to point out that the books he wrote with Andrewartha were an attempt to provide alternative models to the notion of balance. He also points to an article he wrote with Paul Ehrlich in 1967 that criticised the concept of balance in regard to population dynamics. For Birch 'the ecological crisis should prompt efforts to deepen our scientific understanding of ecology rather than a retreat into more simplistic notions and slogans'.

As a result of work being done at the Center for Process Studies at Claremont in California, Whitehead's organicist philosophy has also become known as 'process philosophy'. Whereas many biologists have suggested that consciousness must have arisen at a particular, late, stage of natural evolution, the process philosophers argue that mind and matter have co-evolved in the form of organisms. In two of his key philosophical books, *On Purpose* and *Feelings*, Birch has argued that evolution has been driven by an increasing tendency towards greater subjectivity and self-determination and that life is distinguished by the emergence of feelings in sentient beings. If we acknowledge that all living things have purpose and that many of them can experience feelings then we ought to have much more compassion for them as being subjects like ourselves. Birch argues that we have become separated from nature either by seeing ourselves as superior beings (a product of western 'enlightenment') or by focusing on our own struggle for survival (especially in poor countries). Only by regaining compassion for non-human life can we reawaken our emotional lives and renew a sense of belonging to something much greater than ourselves.[3] For Birch, this is where an understanding capable of uniting scientific and religious insights can emerge.

The Australian ecophilospher Val Plumwood (1993) finds difficulty in the way that process philosophy continues to place humans at the apex of the evolutionary tree. She argues that it remains anthropocentric in that it 'conceives the world of nature as similar to but of lesser degree than the human mind, rather than simply different'. Birch would counter this by arguing that many contemporary ecophilosophers have lost sight of Whitehead's contribution because they have turned their backs on science. Perhaps both are right. What occurs to the authors is that Australia has been able to produce both a distinguished Whiteheadian philosopher (in Birch) and people like Val Plumwood who have played leading roles in the forging of more recent ecophilosophies (see chapter 11).

When the authors met Charles Birch in 1997 he was, at the age of 79, bright-eyed and energetic. He greeted us at the door of his unit overlooking Sydney Harbour with a quizzical half-smile and a warm welcome. Even on a late autumn day the sun was streaming in through expansive windows and sparkling on the water below. An occasional gust of wind set off the sound of metal fittings clanging against the masts of luxury yachts moored in an adjacent bay. In the distance, in silence, ferries and other working boats made their way across the harbour. A writing desk held pride of place in the lounge-room and crowded bookcases lined the walls. The interviewee worked very hard to answer our questions fully and accurately, only occasionally allowing his famous smile to flitter across his face. His concentration on the task was broken only by his offer of a cup of tea, served in fine china cups.

'I'm really not sure where you would start with the history of ecology in Australia', he said as he sought clarification of our project:

> There was a Joe Wood in Adelaide who was a sort of ecologist because he made the first vegetation maps of South Australia which we later used. But I don't know who you would call the first ecologist. I know the Waite Institute played a role because it focused on practical work related to agriculture. I think that a background in agriculture [which Birch and Andrewartha shared] was the best introduction to ecology in those days because you got into soils, botany, climatology, statistics and all that as well as the animals. The people who were teaching zoology at that time were very traditional. There had been a plant ecologist, Professor Ashby, at Sydney University before I got there [in 1948], but going there was like going to a foreign land after being at Waite and then at Oxford with [Charles] Elton.

How did Charles Birch get interested in biology?

> Oh, that's very simple really. When I was at Scotch College in Melbourne I had a biology teacher who was probably the best in the state and she got me interested in the question: 'What is life?' She got me interested in living things and I became a beetle collector. Then my mother noticed this interest and she started to feed me with books. One book that really inspired me was Haldane's *Possible Worlds.* After I read that I said to myself: 'Oh, I want to be like Haldane.' I met Haldane in 1947 and after that I had correspondence with him while he was in India. He was quite nutty you know. When I first met him he introduced me to Helen Spurway, who was a member of his staff, and he introduced her as his divorced wife. I thought that was a bit funny. I can still see him now; all dressed up in his Indian clothes. He wrote to me once and told me I should go to India and become a Hindu. But I'd given up wanting to become like Haldane by then.

On his own admission, Birch applied for a scholarship to study and work at Waite primarily to avoid going to war. At the time it was a case of sign up for the army or do work of national importance, and the work at Waite on finding out what was happening to the piles of wheat that were awaiting transportation overseas was considered to be important enough to avoid conscription. Because of the war, there was a shortage of ships to transport the wheat and billions of tons were being stored. Since there was too much for the silos, stockpiles were made on the ground and then sheds built over them. But the wheat was going soft inside and insects were chewing the husks, turning it into flour. Birch was given the task of finding out more about the grasshoppers that were causing much of the damage. This was his entry into the ecology of insects.

> Waite was a funny place to start my career. It was set up when Mr Waite, a wealthy pastoralist, donated his property and James Prescott and James Davidson set it up along the lines of a similar institute, the Rothamsted Experimental Station in England. Prescott had been at Rothamsted and he was one of those stern Englishmen who like to look down their noses at you. When I first met him he said to me: 'You're only the second research student we have had and the first one was a failure.' Prescott would hate it if you said hello to him twice in the one day and so people rarely spoke to one another. You would walk down this really long corridor and pass people without even saying hello.

Birch had no choice about working with Andrewartha because he was assigned to be the supervisor for Birch's Masters thesis and they were put to work on joint projects. If they had not been thrust together in this way they may not have become natural allies. But Birch came to have great respect for the older man's ability to generate ideas and to express his ideas simply. 'Andrewartha was a very practical man', he told us. 'He was suspicious of me at first because he thought that some of my ideas were anti-science, but we got over that.'

In being put together with Andrewartha, Birch began research and a collaboration that would strongly influence his career. But he also reached the conclusion that he did not want to be a full-time researcher. By the time he had been at Waite for about six years, he 'started to realise that about half the people there were really shunning society because they had great difficulty with human relationships'. However, James Davidson was not one of these and he actively encouraged Birch to pursue outside interests. One of those interests was to find out more about philosophies that underpin the study of biology and it was this interest that led him to write to Professor Agar in Melbourne about Whitehead. 'I may have been the only person to read Agar's book on this subject', Birch joked, 'but it confirmed my interest.'

After he took the decision to leave the Waite Institute in 1946, Birch headed for Chicago University because some excellent zoologists were working there and he now knew that he wanted to become a teacher rather than purely a researcher. However, the other attraction of Chicago was that Charles Hartshorne was there (in the philosophy department) and Birch made a point of going to meet him. To Birch's delight, Hartshorne showed surprising interest in a visitor from Australia because, as a keen bird-watcher, he loved to visit the land that is host to such an extraordinary diversity of birds. The two men established a rapport that would last five decades, with Hartshorne celebrating his 100th birthday in 1997.

The year that Birch spent at Chicago University in 1946 was pivotal in his life. In the field of zoology he was exposed to the ideas of Professor Sewall Wright, a leading exponent of the neo-Darwinian theory of evolution, who also happened to share some of Hartshorne's ideas. Then, as well as Hartshorne in philosophy, there was a cluster of Whiteheadian theologians in the school of divinity. During a conversation with Hartshorne, Birch asked him if he could recommend anyone else who might help him develop his own ideas and Harthsorne recommended 'my most brilliant student', John Cobb, who had begun his long and distinguished association with the Center for Process Studies at Claremont, California. Birch made contact with Cobb and began an exchange of ideas that finally resulted in the publication of the joint book *The Liberation of Life: from cell to community* in 1981.

At the end of his year in Chicago, Birch got the opportunity to go to Oxford to study at the Bureau of Animal Population set up by Charles Elton, the 'father' of animal ecology.

> Oxford was deadly dull after Chicago. The English were so strange. They hardly ever talked to each other. There I was, dying to meet all the famous names in ecology. I wanted to go to Newcastle to meet George Varley. Alastair Hardy [a famous marine biologist] arrived at Oxford while I was there. But, you know, when the summer came around I said to Elton that I would like to travel around and meet famous scientists to see what they were doing he said to me: 'Oh, what would you want to do that for? The only thing that is happening is here.' Even Elton didn't encourage me to take a close interest in his work. By then he had been working in Wytham Woods for years so I said that I would like to come with him one day and observe the way he worked. But he said he would prefer that I didn't come because 'I don't know what I am doing'. Elton was a terrible lecturer. But he was very, very bright. He could see through the mess of other people's ideas. I still think his book *Animal Ecology* [which was published in 1927] is one of the best things ever written on ecology.

Birch was more inspired by his time at Chicago because the leading scientists were much more interested in sharing their ideas. A group of them,

Allee, Emmerson, Park, Park and Schmidt (Birch rattled off the names as if they were those of the partners in a law firm), would have breakfast together every Sunday and because of their initials became known as 'the Great Apes' (actually APPES). They wrote a book called *Principles of Animal Ecology* that Birch described as 'probably the most boring book on ecology ever written', but he found their discussions and seminars fascinating.

Between 1946 and 1960, Birch had four stints in the US and another at the University of Sao Paolo in Brazil. He began his career at the University of Sydney in 1948, becoming a professor in 1958. He was first appointed to the position in Sydney by Professor William Dakin, who had been inspired by Charles Elton's 1927 book about animal ecology to give lectures on the subject at that university in the 1930s.

Dakin was a great pioneer in his own right. His long-time research assistant Isabel Bennett reminded the authors[4] that he had arrived in Australia after a 'brilliant career' in Europe to become the first professor of biology at the University of Western Australia in 1913. He not only pioneered the teaching of ecology at the University of Sydney in the 1930s but also conducted the first serious study of Australian plankton at the same time and successfully lobbied the CSIR to establish a Fisheries research section (established in 1937).[5] In 1937 Dakin began a very popular weekly broadcast on ABC radio called 'Science in the News' and successfully campaigned to have biology included in the curriculum of the New South Wales High School Leaving Certificate. He is probably best remembered for an extensive study of life in the intertidal zones of Australia's temperate seashores, first written up as a CSIRO research paper in 1948 under the title 'The Ecology of the Intertidal Zone', but subsequently used for a textbook that went through 11 editions from 1948 to 1997.

Dakin died immediately after retiring from the University of Sydney in 1950, with the survey of the coastal regions in temperate Australia only partly completed and it was left to Isabel Bennett to complete this Herculean task. In finishing the professor's unfinished work, Bennett was able to establish herself as the foremost expert on the ecology of Australian seashores. The text on Australian seashores was so well regarded that in 1960 the Lansdowne Press commissioned her to write a similar text on the corals of the Great Barrier Reef where she was able to witness, first-hand, some of the human-induced degradation of parts of the reef. When her book on the corals was finally released in 1971, it was launched in Brisbane by Judith Wright. What is impressive about Bennett's career is that she had no scientific training at all when she first went to work for Professor Dakin. When interviewed, the 93-year-old Bennett told the authors that she first met the professor on a cruise ship to Norfolk Island in 1932. Having been born in Queensland, she moved to Sydney to take up a job in the tax office but was out of work when her mother urged her sister and her to go on the inaugural Norfolk Island cruise. Bennett found herself in the cabin adjacent to that occupied by Dakin and his wife and when the professor learnt that she was out of work he promised to create a position for her. That was the beginning of a partnership that continued up to and beyond Dakin's death. Bennett urged the authors to give him his due as a pioneer thinker.

Interestingly, Birch told the authors (in May 1997) that when he arrived at the University of Sydney in 1948, Dakin had not encouraged his interest in ecology, telling him instead to concentrate on teaching comparative physiology. Bennett suggested that this might have been because Dakin had appointed him specifically for that purpose, yet Birch was able to get around the problem by introducing ecological concepts into his teaching of comparative physiology. Bennett, who was working for Dakin when Birch arrived, said that the new man's interest in ecology was well known to his new colleagues, who jokingly gave him the nickname the 'flourbag ecologist' as a result of the work he had done with Andrewartha at the Waite Institute.

Birch told the authors that he may not have kept in touch with his old partner Andrewartha after leaving Adelaide if not for a third person who was about to leave Sydney University to go to the Waite Institute when Birch arrived at that university. This was Tom Browning who worked with Andrewartha at Waite and who kept Birch informed about their work. 'It was really Tom's idea that Andrewartha and I should put our ideas into a book in order to counter the influence of Nicholson's ideas', Birch said. The publication of the book led to an invitation for Andrewartha and Birch to visit the Riverside Campus of the University of California, where Nicholson had been much revered. Given that the younger Australians were promoting an alternative to his theories on distribution and abundance of species, Nicholson saw their arrival at Riverside as a threat and he set out to protect his reputation. According to Birch, this is what lay behind a fierce argument between Nicholson and Andrewartha at a 1957 symposium organised by the geneticist Theodosius Dobzhansky.

Birch was much less successful in finding colleagues at Sydney University to work with. He told the authors that throughout his long career at the university (from 1948 to 1983) he felt rather marginalised by more senior people who often made jokes about his work. When asked to nominate the people at the university who had most influenced him, Birch quickly replied 'the students!'.

> Some time in the early 1970s some students at the Students Representative Council decided to organise a series of lunchtime lectures on environmental issues. They wanted to have one session on population so they invited Bob May[6] and myself to be the speakers. Bob May was a theoretical physicist who had come to seminars we had organised on population biology. He

thought the mathematics of the population biologists was pretty archaic and he thought he could do something about it. So he developed a strong interest in the field. Bob was a good lecturer and our session went very well. We got a tremendous response with about 300 students filling the Wallis Theatre. I would have to say that was a highlight of my time at Sydney University and I would have to thank the students for it.

Birch also recalled one particular student who had an impact on him. David Mowbray was a postgraduate student in zoology who was also one of a set of triplets who gained media coverage when they refused to register for military conscription at the time of Australia's involvement in the Vietnam war. Birch recalled a time when he had been walking across the campus and noticed a crowd of students gathered around a car. When he got closer he realised it was a protest and saw that David was one of the students lying on the ground blocking the vehicle's progress. Knowing someone so involved in the protest made him pay more attention to what was happening.

David Mowbray talked often to Birch about the Vietnam war and he came to the realisation that he could use his position to speak out against Australia's involvement. And as he started to take a public stand on the war he found that he was drawn into discussions about other social issues of the day. He began to attend the Wayside Chapel in Kings Cross, which was famous for its open door policy and lively, contemporary discussions. He was invited by the Rev. Ted Noffs to run a weekly discussion group, which he did for many years, and there was 'tremendous interest'. 'All kinds of people would walk in and yet we would always get into lively and interesting discussions.' Birch also started to pay more attention to the public discussion of environmental issues. He credits this largely to the influence of Paul Ehrlich, the US entomologist and ecologist who has led the crusade for population control.

Ehrlich was greatly influenced by the first book by Andrewartha and Birch, *The Distribution and Abundance of Species,* so he planned a trip to meet the authors. The meetings went well and from the late 1960s onwards Ehrlich became a regular visitor to Australia, where he established a strong working relationship with Birch. After one such visit, Ehrlich and his ecologist wife, Anne, decided to visit India before returning to the US and it was on that visit that he decided to write the book that became the best-selling *The Population Bomb,* which the Ehrlichs used to launch a movement for population control called Zero Population Growth (ZPG). 'I get on well with Ehrlich', Birch commented, 'and since our first meeting we have met at least once a year. I would say it was Ehrlich who convinced me that scientists should be more actively involved in public debates.' In 1999, Birch invited Paul Ehrlich to deliver the Templeton Lecture at Sydney University, as part of an ongoing series funded by a trust Birch set up after winning the Templeton Prize in 1990.

Birch's long collaboration with Paul Ehrlich was one of a number he developed with biologists outside Australia. While his collaboration with Andrewartha was central to his career, he did not find it easy to build collaborations with other Australian scientists. He had more luck outside the country. For example, while he was at Columbia University in New York in 1953, he began working with the famous geneticist Theodosius Dobzhansky in a collaboration that was mutually beneficial. According to Birch, Dobzhansky is one of four founding fathers of the neo-Darwinian theory of evolution, along with J. B. S. Haldane, Ronald Fisher and Sewall Wright (1999, p. 140), so he was very pleased to have the chance to work directly with him.

> I started working with Dobzhansky because I didn't know much about genetics and he didn't know much about ecology. I met him at Columbia and went with him to Brazil where he was doing some work on a species of *Drosophila* [fruitfly]. He told me he thought it was time he knew something about the 'ecology of this bug', and I thought it was important to learn about genetics because ecologists were being accused of focusing too much on environmental factors.

According to Birch, Dobzhansky enjoyed his introduction to ecology so much that he decided to organise the 1957 Cold Spring Harbour Symposium to bring interested ecologists, geneticists and other biologists together. It was at this symposium, however, that Andrewartha and Nicholson had their fierce argument. According to Birch:

> Andrewartha would get terribly hot under the collar if he thought someone was saying something terribly illogical and so he got stuck into what Nicholson was saying. Nicholson responded to us by saying that 'the trouble with you two is that you are studying very atypical populations on the edge of deserts, which is entirely different from where most animals are'. But we were saying that there was no typical environment because you need to consider a wide range of factors in every case.

Although the argument between the Australians spoilt the atmosphere of the symposium it did not dim Dobzhansky's enthusiasm for ecology.

While at Columbia University, Birch also met Dobzhansky's 'most brilliant student', Richard Lewontin, who would later become the co-author (with Richard Levins) of the influential book *The Dialectical Biologist*. Both garrulous, but one a radical Christian (Birch) and the other a Marxist (Lewontin), the pair started conversations that would continue for decades. Birch recalled that he was invited to be on a panel examining Lewontin on his PhD work and he and the other panel members finished with the impression that they had been interviewed by the candidate. Later Lewontin came for a sabbatical in Birch's department at Sydney University and in 1997 he was invited by Birch to give the Templeton Lecture.

> The trouble with most of the biologists I know is that they are fundamentally mechanists. I mean, I know Ehrlich very well and I respect his work, but he is a complete mechanist. He's like Andrewartha in thinking that philosophy never got us anywhere. And I can't talk about philosophy with Lewontin either. I think he is probably the brightest biologist on the face of the earth. But when I said to him once that I thought it was important to ponder the question: 'what's the point of my life?' He told me that if he ever felt that way he would go to bed and pull down all the blinds until the feeling had passed.

One of the few biologists who shared Birch's enthusiasm for Whitehead was Conrad Waddington from the University of Edinburgh. Birch discovered this fact when he was with him in Rome one day during a break in the meeting of the International Union of Biological Sciences that Waddington was chairing. Waddington asked Birch to go into a bookshop with him because he wanted to buy a book on pruning olive trees. Birch was surprised to find that as soon as he had found an appropriate manual he headed straight for the philosophy section and picked up a book on Whitehead. 'So I asked him if he was interested in Whitehead and he told me he had read every one of his books. He said that it was reading Whitehead when he was an undergraduate at Cambridge that got him interested in biology. He's the only major biologist I know who was led into the field by philosophy.'

When the authors suggested to Birch that his own career seemed somewhat schizophrenic in that he had kept his own scientific and philosophical writings separate, he argued that it was very important that ecology be kept separate and given a solid basis in science.

> I think the word 'ecology' is used much too loosely now. You've got eco this and eco that, and I really don't know what it means, do you? I think ecology needs to have scientific credibility if it is to mean anything.

> But in working in science you end up working with scientists who don't often have an interest in philosophy. Now I don't think I can keep up with developments in population biology, so I have turned my attention to philosophy. But I think my philosophical models are ecological because they stress relationships. There's probably a closer link between Waddington's developmental biology and philosophy than there is between animal ecology and philosophy because animal ecology has been studied in a strictly mechanical fashion.

As indicated, Charles Birch was not afraid to take a stand on matters of public interest and yet he had virtually nothing to do with the nature conservation movement. This seems like a missed opportunity on the part of the conservationists, especially in view of the high-profile role played by another ecologist working in Australia, Francis Ratcliffe, in establishing the Australian Conservation Foundation in 1965.

EARLY EDUCATOR: FRANCIS RATCLIFFE

Francis Ratcliffe was born in British India in 1904 and studied alongside Charles Elton under Julian Huxley in England. He was recruited to work in Australia in 1929 by the Empire Marketing Board as part of its effort to end the damage being inflicted on orchards along the eastern seaboard by the fruit-eating 'flying fox'.[7] It is hard to know what his employers expected of him as an ecologist but it was probably not what they got because Ratcliffe urged fruit growers to look for ways of coexisting with their 'enemy'. By making a study of flying fox needs and habits it should be possible to do two things at the same time: make the fruit-growing areas less attractive to the flying foxes while ensuring that they have alternative habitats away from the fruit-growing areas. Part of the problem, Ratcliffe reported, was that their natural habitats were being destroyed so they were seeking out new opportunities, which humans were conveniently providing. A better ecological understanding of the flying foxes was the alternative to an ongoing war of attrition.

During his two-year study of the flying fox, Ratcliffe apparently fell in love with Australia because, after returning to Britain for a four-year stint as a lecturer at Aberdeen University, he jumped at the opportunity to return to Australia to work with the CSIRO on problems of soil erosion in the outback. This was the first time that this problem had been examined from an ecological perspective and Ratcliffe documented his experiences in the popular book *Flying Fox and Drifting Sands* published in 1938. Ratcliffe's book played a role in bringing into public focus the need for resource managers to take into account ecological limits and to be proactive in conservation measures, especially in fragile areas. His ecological understanding was also evident in his recognition of the need to treat the environment as a system with a consequent need to employ integrated, interdisciplinary approaches in solving problems. Ratcliffe's efforts were influential in the establishment and design of the state soil conservation authorities, the first of which was set up in New South Wales in 1938.

Ratcliffe's leadership qualities, and ability to apply ecological understandings to a wide range of issues, enabled him to raise the profile of ecology within the CSIRO. He became its first Officer-in-Charge of Wildlife Surveys and championed the need for Australia to develop wildlife conservation policies. At CSIRO, Ratcliffe had a significant influence on rabbit control programs introduced in the 1940s. He pointed out that agricultural areas fall into three distinct zones – relatively wet coastal areas, arid pastures bordering the interior, and intermediate areas that comprise the 'wheat belt'. The rabbit problem was most severe in this latter zone and this recognition led to a concentration of effort in the most critical area – a saving of resources and a maximum return on investments. After his retirement from the CSIRO he was invited by Prime Minister Robert Menzies to play the leading role in establishing the Australian Conservation Foundation in 1965 and he worked for the new organisation in a voluntary capacity until just before he died in 1970.

From Francis Ratcliffe through to Andrewartha and Birch scientific research in ecology in Australia was largely geared to the commercial needs of rural industries. The funding for research was often directed through the CSIRO and its charter was to serve industry. CSIRO was a principal employer for ecology graduates interested in research.

STEPHEN BOYDEN: PIONEER IN HUMAN ECOLOGY

One scientist who started his ecological career in Australia by asking non-commercial questions, however, was Stephen Boyden at the Australian National University (ANU). Born in England and initially trained as a veterinarian and then as an immunologist, Boyden migrated to Australia in 1959 and, within a few years, discovered a passion for exploring the link between biological systems and human culture. In 1965 he put his established career in immunology in the John Curtin School of Medical Research at risk by going to the Vice-Chancellor of the university, Sir Leonard Huxley, seeking permission to establish a new unit in which he could pursue his new interests. To his surprise and delight, Huxley agreed to give the idea a trial period of 18 months and a new unit called 'biology and human affairs'[8] was established, initially within the Research School of Social Sciences but subsequently as an independent unit operating out of a small university-owned house.

The unit survived the trial, changed its name to 'urban biology' and later to the Human Ecology Unit. However, as a unit situated in the university's Institute of Advanced Studies it could attract PhD students but not teach undergraduates. In 1970 Boyden began to develop his ideas for a 'human sciences program' that would take undergraduate students from a grounding in either Human Biology or Sociology through two sequential subjects related to biology and culture. This cluster of subjects would then be available to students doing a wide range of degrees. In 1972 he took the proposal to the new Vice-Chancellor, Sir John Crawford, and it was accepted about six months later. The program began with Boyden teaching students in Human Ecology.

Boyden was able to get his opportunity to start a new program crossing boundaries between biology and social science because he had already built a successful career in immunology. He completed a PhD in immunology at Cambridge University with the Australian microbiologist, Professor Ian Beveridge, as his supervisor. Beveridge himself had come to Cambridge with a strong reputation as the scientist who discovered the cause of foot-rot disease in Australian livestock and apart from giving Boyden sound advice about research opportunities he also sparked his interest in Australia. After leaving Cambridge, Boyden landed a job at a prestigious serum research institute in Denmark, where he remained for eight years. When he left that position he had to choose between jobs in Australia and the US. Fortunately for Australia he chose the Australian National University. He was elected as a Fellow of the Academy of Sciences in Canberra in 1965 and used that position to organise a symposium in 1969 on the topic 'The Impact of Civilisation on the Biology of Man', with famous bacteriologist and humanitarian, Rene Dubos, from the Rockefeller Institute for Medical Research, as special guest. Boyden had established a friendship with Dubos many years earlier when he had gone to the Rockefeller Institute as part of his PhD program.

In boldly pursuing his interests rather than a safe career, Boyden was very grateful for the support he got from people like Huxley, Crawford, and the director of the John Curtin School, Frank Fenner. In an interview with the authors[9] he also wanted to acknowledge a wide range of other university colleagues who worked with him in various projects and initiatives. However, when he retired from the university in 1990 he felt somewhat worn down by the effort involved in keeping new ideas alive against the conservative drag of all those who saw no reason to venture beyond the boundaries of established disciplines. After building the reputation of the undergraduate unit, he consciously withdrew from it so that it was not seen as a one-man show and, although the subject offering had survived in a form up until the time of the interview, it had gone through many changes and many homes within the university. A later initiative, the Fundamental Questions Forum, enjoyed an active life for just two years after it was set up in 1988, but it collapsed on Boyden's retirement in 1990. The aim of the initiative had been to draw together a range of experts from the natural and social sciences to discuss pressing contemporary concerns and then to produce some published documents that might contribute to public debate on those issues. Boyden said that many people from ANU and other universities around Australia participated in a lively set of symposiums and small group discussions that he and his colleagues in the program set up. The program resulted in the publication of a range of papers and a book called *Our Biosphere Under Threat: Ecological Realities and Opportunities for Australia* that Boyden wrote with his assistants Stephen Dovers and Megan

Shirlow. However, the project was not around long enough to establish a serious public profile. When we talked to Boyden in 1997, he was keen to contrast that experience with a more recent community-based project he was involved in called the Nature and Society Forum. This project was initiated by a group of people after they attended a talk Boyden gave on World Environment Day in 1991. Its aim was to involve local 'experts' in discussions and projects related to contemporary community concerns. Boyden felt that this project was not fettered by the boundaries of academic disciplines and had a spark of energy missing from some of the projects he started at ANU.

During our 1997 interview, Boyden was friendly and engaged, but also softly spoken and surprisingly self-effacing. By his own assessment, he has been good at creating ideas, but not as adept at communicating them, especially in writing. In his retirement (living on a block of land out of Canberra) he appeared relaxed, but the discussion of his career frequently sent dark clouds scurrying across his sensitive face. In saying that he still wanted to improve his communication skills he seemed to blame himself for many of the difficulties he had in convincing university colleagues of the merit of his ideas. He was clearly not at ease with his own, very considerable, achievements, but he was very keen to tell the story of how his unusual career had unfolded.

> I first got interested in animals when I was a young child living in London because I had an aunt who visited us every week and she always brought me photos of animals that she had cut out of magazines to include in an album she had given me. I can remember that even though I was very young. We moved to the country when I was five and I started to see some of the animals in reality. Then I started looking for unusual animals. I can still remember the excitement I felt when I found a newt under a rock. I got particularly interested in collecting lizards and snakes.

Boyden said that a love of animals encouraged him to become a vet, but, although he enjoyed the work he found that he was becoming more interested in the processes underlying disease than in the treatment of it, so he took up a grant to study human pathology at Cambridge and from there he moved into his PhD project with Professor Ian Beveridge.

> I'd have to rate Beveridge as an influence simply because he was a good supervisor and I got on well with him personally. I also enjoyed working with the director of the serum institute in Copenhagen, Jeppe Orskov, because he shared my love of nature. I got on particularly well with Rene Dubos when I met him at the Rockefeller Institute. He had a house in the hills north of New York and when I visited him there we would go on long walks in the country having a jolly good discussion about all sorts of things.

He was not only a good scientist but a great humanitarian and he had interesting things to say on a wide range of topics. He also had a fascination for mushrooms and I had a similar passion for lizards so when we went on our walks he would tell me all about any mushroom he saw and I introduced him to the world of lizards. Perhaps his influence on me was simply to encourage my enthusiasm. He once told me that the word 'enthusiasm' was one of the most interesting in the English language and he then proceeded to tell me all about its Greek origins.

As mentioned earlier, Boyden was grateful for the support he got from leading figures in the ANU when he took some radical ideas to them. But he particularly appreciated the interest and support that he got from Frank Fenner. Fenner took a personal interest in the Human Ecology initiatives and he was always someone Boyden could turn to. 'Even now, Frank is patron of the Nature and Society Forum', Boyden said, 'so that connection has continued.'

Fenner played a crucial role in the most ambitious project that Boyden ever undertook, which was a broad-scale study of the urban ecology of the city of Hong Kong. This project was born when Boyden addressed a conference on human ecology in Hong Kong in 1972 and made the observation that the city would be an ideal place for a study of the urban ecology of a whole city because it was confined geographically. To his surprise, people from the Geography Department at Hong Kong University said they were prepared to accept the challenge of such an ambitious study provided he would act as a consultant. When he returned to Canberra, Boyden discussed the project with colleagues and was encouraged to participate. However, just as he was preparing to tell the people in Hong Kong what sort of support he could offer them, he received a message saying they had decided against such a big undertaking. At that point Boyden went to talk to Fenner, who was very enthusiastic about the project and keen that it should go ahead even without the people in Hong Kong. As a direct result of Fenner's support, Boyden was able to get funding for the project from the Nuffield Foundation in England and he was joined by two PhD students, Ken Newcombe, a former president of the Australian Union of Students who would tackle the collection of data about factors such as energy and nutrient flow through the city, and Sheila Millar from Britain who took on the task of designing and implementing a biosocial survey of people living in the city.

Boyden also got some unexpected assistance in the project from his teaching colleague Jeremy Evans who bumped into a person from UNESCO at a social event in Canberra and told him all about the project. When Evans told Boyden that UNESCO might be interested in the project he wrote a letter to the organisation to find out if there were any avenues for this. He was surprised to get a letter in response that apologised to him for not having informed him that the project had already been adopted as the first in a new series of studies of human settlements that was part of the broader Man and the Biosphere (MAB) program. Obviously Evans had done a good job of selling the idea at his chance encounter. UNESCO support delivered a small amount of extra funding for the implementation of the study but, more importantly, it guaranteed that the findings would be published and circulated widely. The study was written up in a report that became the first in an MAB series of publications distributed internationally.

According to Boyden, the Hong Kong study became a crucial test of some of his rather abstract ideas and he was pleased with the outcomes. It confirmed the importance of studying the interplay between biological systems and human systems and led to the development of new ideas in urban ecology. One of the key ideas that Boyden developed in this study was that cities have a metabolism that takes two forms:

> First is 'biometabolism', which consists of inputs and outputs of foodstuffs used with the human body, in biological processes. This is the internal metabolism of the human organisms within the system. The other form of metabolism is 'technometabolism' which consists of the inputs and outputs of resources and energy used outside the human body for things like machines, motor cars and so on. In one sense technometabolism is an extension of biometabolism, but the problem with 'modern' industrial cities is that technometabolism has outgrown biometabolism dramatically, making the systems increasingly unsustainable.

Boyden's study of the environmental and human health consequences of rapid urbanisation also prompted him to look at the evolution of cities through the lens of what he has called 'biohistory'.[10] As a result of this he has suggested that human history can be divided into four distinct ecological phases: the hunter-gatherer phase; an 'early farming' phase; an 'early urban' phase; and a modern industrial phase. The jump from each phase to the next has been accompanied by a dramatic increase in per capita consumption of resources and in surges in population numbers. For Boyden, the inescapable conclusion is that such growth surges have not been accompanied by a corresponding awareness of demands being made on the biosphere and if there is to be a fifth ecological phase it must involve a sudden *reduction* in demand for resources. 'I think this is an idea that has caught on quite well judging from the letters I still get from various parts of the world asking me to come and talk about it', Boyden said.

As an international pioneer in the study of human ecology, Boyden has been interested in the way the term has been applied in other countries. He has exchanged correspondence with Torsten Malmberg in Norway and with people in Sweden, Austria and the US. He 'had a bit to do' with Geoffrey Harrison in the Department of Biological Anthropology at Oxford

University. But the person he enjoyed meeting the most was Zena Daysh at the Commonwealth Human Ecology Council in London because 'I always got lots of ideas whenever I visited Zena.'

However, Boyden became increasingly disturbed at the lack of any apparent common ground for people who adopted the term human ecology. 'There is a Society for Human Ecology in the US', he noted, 'but whenever they have a conference the only link seems to be an interest in humans and that is not enough. I dropped the term after a while, except in situations where I could define my own use of it clearly.'

> My own approach was much more deeply rooted in biology than that of the other people I knew about. I have always been particularly fascinated by Darwin's ideas. I keep coming back to read what he wrote because it provides so much to think about. I was also influenced by people like Konrad Lorenz [the animal behaviourist] and I have an interest in socio-biology, although I think they focus too narrowly on particular aspects of human behaviour, like incest and homosexuality. You can't study humans individually because we exist in groups, and the evolution of culture involves so many intangible factors related to prevailing assumptions, values and priorities, as well as individual feelings and emotions. You can only make sense of it if you look at the whole system.

Through his work on the Hong Kong project, Boyden became actively involved in UNESCO's MAB program up until the time of his retirement from ANU in 1990 and he considered this to have been some of his most important practical work. He also thought that the Hong Kong project began to crystallise the ideas that remained his strongest theoretical contributions. In particular, he rated three ideas that he wanted to highlight. The first was the idea of cities having a technometabolism (mentioned above). The second is what he has called the 'evodeviation principle', which says that humans evolved biologically for certain ways of living in certain conditions but now live in radically different conditions. This means that our biology no longer matches the ways we live and the mismatch results in chronic health problems. For example, most people consume much more energy in their food than they expend in physical work and so the surplus energy is stored in the form of fat that might be a reserve source of energy in the 'lean times' that no longer arrive. We become chronically unhealthy because we don't learn what our biological systems were 'designed' to do.

The third idea is that culture affects biology because it can produce evolutionary advantage for groups of people. For example, we can be advantaged by co-operative ways of meeting our biological needs and this led to the creation of towns and cities. But such cultural advances can lose sight of biological imperatives and we develop the illusion that we are masters over nature. Boyden thinks that the difference between biological and cultural evolution has been over-emphasised. Culture does not render biology obsolete.

Like Charles Birch, Boyden has often found that his ideas were better received overseas than in Australia. Of course, he arrived in Australia when few people believed we could be world leaders in anything. He told an interesting story of an experience in the early 1960s when he noticed that plans were being made to build a zoo for Canberra. Seeing little educational value in yet another zoo, he wrote to Prime Minister Menzies suggesting the idea of a Biology Park where whole ecosystems could be modelled. Menzies wrote back asking for more details and so Boyden sent in a detailed submission with personal endorsements from famous biologists including Konrad Lorenz, Nico Tinbergen and Julian Huxley. As a result of the submission he had meetings with government ministers John Gorton and Malcolm Fraser (both future prime ministers) and the only question both men had was 'Is there a precedent for this overseas?' When Boyden replied that it would be a world first they lost interest and his submission disappeared without trace.

Another problem for Boyden was that his greatest strength was also a weakness in the context in which he worked. As a bold and lateral thinker he inevitably crossed the boundaries of academic disciplines and was seen by some as a trespasser and by others as a traitor. Many of his conservative colleagues were probably frightened by his enthusiasm for new ideas and projects and doubted his ability to 'deliver the goods'. Rather than being valued for his obvious strengths he was left with the feeling that he was somehow to blame for not being able to communicate his ideas better. 'I never really convinced my academic colleagues of the merits of my work', he said sadly.

As mentioned, Boyden was very grateful for support given by Frank Fenner over many years. However, one piece of advice that Fenner gave early in their association became a sort of albatross around Boyden's neck. 'Frank always told me that I had to work very hard to earn the respect of my colleagues. I had to show that my ideas were scientifically credible.' In trying to do this he found himself working in a culture that was largely hostile to his ways of working and his longing for credibility undermined his self-confidence. There was an echo of this in the way Birch also discussed his career at Sydney University.

The experiences of Birch and Boyden as scientists at the edge of their disciplines were both negative in terms of lack of funding and recognition and positive in enabling them to think more creatively and holistically about the subjects under investigation. This seems to be a common experience for pioneers in academia who seem to spend most of their careers having to fight to maintain their research and teaching programs until their foresight is recognised (often, sadly, following their retirement or death).

Ironically, as discussed in chapter 5, the amateur movement of field naturalists had earlier felt marginalised by the academic biological sciences (including those who adopted ecological approaches) yet the 'amateurs' were much more effective in public education. For similar reasons, Birch felt that scientists who had an interest in philosophy or spirituality tended to be viewed with great suspicion by a big majority of their colleagues. There are constant processes of marginalisation and 'normalisation' going on in any field, yet the most innovative work is often being done in the margins, because the centre becomes obsessed with order and predictability (seen as necessary ingredients for normalisation). Scientists who have been attracted by developments in chaos and complexity theory[11] like to say that there is a necessary interplay between chaos and order, with innovation occurring at the 'edge of chaos'.

Certainly this is supported by the study of natural 'systems' because the greatest biodiversity and evolution of new forms and functions occurs at the edges, between major biomes, as in the meeting of coastal zones and the vast oceans. Many of us are fascinated by the great diversity of life found in rock pools that are regularly washed by the ocean, yet this diversity emerges from intense struggles for survival in turbulent conditions. Those brave souls in academia who dare to think about the world in new and different ways (with an inherent critique of the status quo) also have to endure turbulence. Often those who have the 'good fortune' to receive early recognition and funding will have their creative impulses tamed and their critical voices silenced. True pioneers may be those who are willing and able to live in the insecurity of the borderlands.

CHAPTER 8	# THINKING LIKE AN ECOSYSTEM: *Australian Innovations in Land and Resource Management*

INTRODUCTION

Scientific ecology gained its impetus in Australia from the desire to control 'pests' that were making life difficult and expensive for farmers (see chapter 7). Some of the ecologists who came to work on the problem of pests suggested that a better understanding of the needs and habits of these 'enemies' of agriculture was needed in order to design forms of peaceful coexistence between humans and the offending insects and other animals. However, no amount of ecological knowledge about the 'enemy' was enough to convince humans that insects and other 'humble' creatures might also become their teachers. In a society already committed to quick-fix, high technology 'solutions' to the problems posed by recalcitrant nature, the idea of learning from nature seemed to be both too primitive and too radical to consider. Consequently, when P. A. Yeomans, a mere farmer, set out in the mid-1940s to demonstrate that careful observation, field-scale experimentation and patient persistence had enabled him to profitably work with, rather than against, natural processes, he was viewed sceptically, if not with outright hostility, by some agricultural 'experts'.

The Keyline approach to farming that Yeomans conceptualised and refined, on the basis of his ground-breaking landscape design experiments on his farm near Sydney, achieved outstanding results in a matter of years. Although many people flocked to his farm when first hearing about his results and he gained some notoriety, the advocates of input-dependent conventional farming methods – the ongoing source of widespread environmental degradation – soon managed to quarantine the industry against Yeomans' influence and his star quickly faded. His legacy has been kept alive, however, primarily by two of his sons and the small, but significant, alternative farming community in Australia. In the late 1970s, Keyline received an unexpected boost, as it was a major source of inspiration for the development of Permaculture, which was pioneered in Tasmania by David

Holmgren and Bill Mollison. This design-intensive and species-rich approach to landscape management has gone on to achieve a much higher public profile, both inside Australia and internationally, than Keyline had been able to achieve on its own. Much credit for this must go to the irrepressible extroverted performances of Bill Mollison, who has managed to ride a wave of public acclamation as the best-known advocate of Permaculture. It was, however, the more low-key Holmgren, who was primarily responsible for the design concepts and who has continued to systematically test his ideas in practice in order to refine the design principles upon which Permaculture is based.

Whereas Keyline and Permaculture concepts of landscape design have focused on redesigning the ways in which humans intervene in natural processes to produce food and other resources, the bush regeneration movement has concentrated on reversing the steady decline of native plant and animal communities that has been an ongoing legacy of colonisation of Australian nature. Sisters Joan and Eileen Bradley, living on the northern shore of Sydney Harbour, stumbled into what became their life's work in the mid-1960s by taking a stand against invasive weeds in an area of remnant bush near their home. Intuitively at first, they gradually developed a radically different approach for changing the odds in favour of the native species. Subsequently, in the process of publishing their results and ongoing observations, they became increasingly conscious of the ecological principles that provided the foundation for what was rapidly becoming known as the Bradley Method of Bush Regeneration. Like Yeomans before them, the Bradleys had relied heavily on careful observation and a willingness to patiently test emergent ideas, even if they contradicted conventional wisdom. More recently still, the Sydney academic and eco-village advocate Ted Trainer has injected a tone of urgency into discussions about local ecological redesign work by pointing out that global 'development' agendas are making this work even more necessary and yet more difficult.

From Yeomans through to Trainer, the advocates of ecological redesign have repeatedly stressed that we must abandon our out-dated, life destroying notions of mastery over nature and learn how to think like ecosystems. Instead of responding to regular environmental crises with reactive, high-technology and disruptive interventions, they advocate forms of interaction that enhance the regenerative capacities of nature's cycles and processes. Together they offer a radical critique of conventional resource management practices, centralised planning and market-based policies and decision making, yet they also hold out the promise of human and environmental renewal, through the promotion of sensitive collaboration in place of blind competition.

The interventionist philosophy that underpins most ecological redesign work stands at odds with the preservationist tradition for nature conservation discussed in chapter 6. Its focus is fundamentally anthropocentric and, because it involves much more blatant manipulation of landscapes than was evident in Aboriginal land management practices (see chapter 9), its margin for error is also much greater. Yet the practice of ecological redesign, especially its emphasis on learning from nature, has much in common with the Aboriginal notion of 'caring for country'. Taken together, we believe that ecological redesign *and* 'caring for country' can offer a refreshing alternative to the still dominant market-based policies that remain uninformed by both ecological wisdom and concern for longer-term and broader outcomes.

THE 'KEYLINE' CONCEPT: P. A. YEOMANS

During a family camping trip in the desert in 1931, three-year-old Neville Yeomans wandered off and could not be found. His parents were, naturally, very concerned; the landscape was unforgiving and a young child would not last long without water. Discussing this incident 66 years later,[1] Neville could remember feeling scared but, even more significantly, he remembered being found by an Aboriginal man who took him straight to a source of water in this dry landscape. Neville's father P. A. Yeomans was obviously relieved that his son was returned to him unharmed, but he was also deeply impressed by the local knowledge of the man who had found him. According to Neville, it was probably this incident that gave his father his enduring interest in the movement of water through Australian landscapes, because he could see that an understanding of this would be a huge advantage for people living in the driest continent on Earth.

To study the movement of water it is best to go outside in the rain. Neville recalled that just as other folk sought shelter when it rained, his father would take him and his brother Allan out to study the pattern of water infiltration and movement on the grassy slopes near their home. What Yeomans senior discovered through such patient observation was that there is a line across the slope of a hillside where the watertable is closest to the surface. The ground along this line looks wettest and is reflective when it rains heavily. This is the line where the slope changes from being convex or steepest above to concave below (the line on a map where the contours go from being close together to further apart). It is the line just below which it makes most sense to locate the highest irrigation dams within the landscape, because this is where the run-off water from above can most effectively be collected and subsequently used at the most appropriate time to irrigate the more gently sloping land below. Yeomans called this line the Keyline. The optimal locations for dams along the Keyline are where it crosses the drainage lines within primary valleys, and he called these the Key Points. Yeomans first outlined his ideas about water movement and how to detect Key Points in a book entitled *The Keyline Plan* in 1954.[2]

Neville noted that his father was obsessed with the idea that no water should be wasted as it moves across the landscape. He joked that he suspected that 'dad probably would have been happy if not a single drop of rain that fell on the land reached the sea'.

The name Percival Alfred Yeomans, or 'P.A.' as he preferred to be known, is certainly not a household name in Australia. Yet when the author, Stuart Hill, visited Lady Eve Balfour (colleague of Sir Albert Howard, the father of organic farming, 1940) on a visit to England from Canada in the mid-1970s, she urged him to contact, or preferably visit, Yeomans in Australia. She claimed that he was making the greatest single contribution to the development of sustainable farming in the world. Unable to make the long journey to meet Yeomans, Hill immediately wrote to him and was delighted to receive, with his reply, a copy of his recently published book, *The City Forest, the Keyline Plan for the Human Environment Revolution*. He also enclosed a copy of a *Farm Manual*, which explained the use of a plough Yeomans had invented, which was then known as the 'Bunyip Slipper Imp with Shakerator'.[3] In 1974 this plough had won the 'Prince Philip Prize for Australian Design', awarded by the Industrial Design Council of Australia. A particularly innovative feature was its vibrator attachment (the 'shakerator'), which not only shattered clumps of compacted soil but, in so doing, reduced the resistance to the passage of the 'slippers' through the ground. This saved farmers fuel costs because they could use lighter weight, lower horsepower tractors for ploughing compacted soil. However, despite this award, Yeomans was clearly far ahead of his time in Australia and, although many visitors came to his farm, his main ideas had relatively little acknowledged impact on the farming practices of the day and subsequently. Instead, the land continued to suffer under the impact of what Yeomans called the ongoing 'bastardisation of agriculture' (1971, p. 88). Whereas Yeomans set out to study complex natural systems in a holistic way in order to find ways of working *with* nature, the dominant farming practice was to try to simplify and control nature by the use of powerful machinery, fossil fuels and synthetic chemicals. We have paid, and continue to pay, a very high price for this in widespread soil erosion and desertification, soil salination, long-term declines in soil fertility and a dependence on regular curative interventions to control the inevitable outbreaks of pests and disease. Such unsustainable farming practices have been partly responsible for the increase in farm bankruptcies, dependence on fickle government subsidies, the demise of the 'family farm' and the decay of many rural communities. Increasingly, we are seeing state governments, under political pressure from farmers arguing over access to water taken out of rivers like the Darling, Murray and Snowy. Widespread application of Yeomans' ideas in the 1950s may have prevented many of these outcomes. Interestingly, although ignored by most agriculturists, some of his ideas were central to the development by David Holmgren and Bill Mollison of Permaculture. Holmgren told the author that he considered that 'Yeomans has made Australia's greatest contribution to sustainable land use, because he introduced the practice of design to what previously had been just husbandry and cultivation'.[4]

Because Australia's future will increasingly be shaped by issues relating to access to water – for drinking, irrigation, industry, and ecosystem maintenance – Yeomans' contributions to the ecology of water management may yet be acknowledged. This is unlikely to happen, however, as long as we continue to look for 'magic bullet' curative solutions to our problems, the most recent of which are being forged in the narrowly conceived biotechnology and genetic engineering programs that are currently attracting generous research and development funding. In contrast, it remains virtually impossible to obtain support to study complex, whole-system design approaches, despite their proven record of success. Although Yeomans had to sell his 'experimental farm' near Richmond, New South Wales, in 1964 to pay death duties incurred when his wife died, it still stands out as a patch of well-watered green in the midst of much browner and less productive landscapes.

P. A. Yeomans was born in Harden, New South Wales, in 1905. Starting his own family in the Depression days of the late 1920s, he struggled to find reliable work and tried his hand at many jobs before discovering an aptitude for being an assayer and valuer of gold and tin mining projects. This was reliable work in the 1930s because, with the Depression still fresh in their minds, there were plenty of hopeful prospectors roaming the country looking for instant wealth. Bogus claims of gold and tin 'discoveries', that would instantly increase the resale value of mining leases, were rife, and Yeomans' job frequently involved distinguishing between genuine and false claims. In this job he developed a sensitive eye for landscapes, learning how to pick country that might yield up valuable minerals by being able to recognise the kinds of plants that are likely to grow in such soils. With his wife Rita and two eldest sons, Neville and Allan, in tow, he was constantly on the move throughout eastern Australia and New Guinea, going wherever his skills were in demand. It was during these travels that he was able to observe a diversity of landscapes and appreciate the importance of access to water, so essential for mining as well as agriculture. He recalled observing that innovative miners (probably Chinese immigrants) had built dams to capture water needed to separate the gold from the rock and to disperse the tailings (Doherty, 1996). Whereas these miners had quickly found ways to gain an abundance of water from their makeshift dams, built largely out of logs, he was surprised

P. A. Yeomans opens a gate to release water into a keyline irrigation channel
while demonstrating his methods to visitors to his farm in the 1950s.
(Ken Yeomans)

to find that the neighbouring farms lacked any dams, and that their
productivity was severely limited by this lack of water. Such experiences
certainly influenced his recognition of the potential of farm dams for irrigation and water management, and consequently for increasing Australia's
agricultural productivity. Later on he established himself as an earth-
moving contractor, and during World War II his company was busy
supplying open cut coal to the Joint Coal Board.

Partly motivated by tax benefits in 1943, Yeomans purchased two
adjoining blocks of poor unproductive land (1,000 acres) near North

Richmond, and his brother-in-law, Jim Barnes, became the farm manager. They called the farm 'Yobarnie', combining parts of their two names, and at first they set about implementing conventional soil conservation practices, imported largely from the US and based on the work of Dr Hugh Hammond Bennett (1939). However, they were unsatisfied with the results and they began to experiment with their own ideas. Towards the end of 1944, Barnes tragically perished in a grass fire that got out of control. This was, of course, a huge personal tragedy for Yeomans, but it also made him more determined to find more effective ways of using water, because now he also wanted to find ways to fireproof his farm. So he stepped up his efforts to implement the Keyline system of landscape management that had grown out of his observations of water movement. Professor J. Macdonald Holmes, Head of Geography at Sydney University, who visited Yeomans regularly and wrote a booklet (1960) about the geographical and hydrological aspects of Keyline, picks up the story here. Following the fire, he recalled that Yeomans had:

> . . . invited the heads of the then recently formed Soil Conservation Department to visit him . . . Their comments and advice about the property were considered by Yeomans to be negative and pessimistic, and so he decided to proceed alone with a general policy of soil conservation which had, for those days, a new twist. Whereas the approach of soil conservation is generally applied to getting rid of water safely, he decided that it would be a better proposition to hold, store and use for irrigation as much of the run-off as possible . . . (1960, p. 35)

> Keyline overall planning had a new psychology. Its approach to land is no longer the idea of conservation. . . . The object of Keyline is development rather than protection . . . [also] Keyline means a new philosophical outlook. Land is to be lived on and enjoyed, not to be lived off and destroyed by wrong practices in an effort to join a precarious livelihood . . . I see Keyline as an abiding, self-renewing, highly productive enterprise, and a very satisfying adventure. (1960, pp. 45–46)

The farm began to thrive and Yeomans was able to buy the adjoining property, which he called 'Nevallan', a combination of the names of the two sons he had at the time (Neville and Allan), who were then in their teens. Ironically, it was Yeomans' third son, Ken, who was not born until 1947, who took most interest in his pioneering farm design work, continuing to promote his father's ideas up to the present.[5] Allan became an engineer and further developed the Yeomans Plough, and Neville went on to become a psychologist, and was one of the first in the world to pioneer 'therapeutic communities' (at 'Fraser House' in Sydney). This work, which was also inspired by his father, he regarded as the development of 'Cultural

Keyline'. He died not long after being interviewed for this book, but his work continues (see http://www.laceweb.org.au/htm).

As mentioned above, Yeomans began his approach to farm design by simply observing the movement of water through the land. From such a humble beginning he probably had little idea where his quest would take him. His passion for water is clear, however, in the following statement.

> I was always interested in water control, and whether experimenting with 'wild flood' or contour furrow irrigation or getting oneself saturated watching run-off in heavy rainfall, the flowing water seemed to hold many of the answers to the questions of land. (1958, p. 262)

He was convinced that Australian landscapes as a whole would benefit if the water that fell as rain could be encouraged to travel the maximum distance before ending up back in the sea:

> The floodwaters from prolonged heavy rains, which now go to sea within a few days, would still be in the soil and in the farm dams months later. Some of the water would remain there for a year or more. During this time the increased soil moisture would be feeding ground water supplies, which flow as springs to feed creeks and rivers. Therefore, river flow would be more constant. Then the continuous but slow seepage from farm dams would be adding to these underground supplies. This would be clean and clear, as well as constant. The present silting up of the rivers would cease and the constant flow of silt-free water would speedily regenerate them. (1958, pp. 9–10)

His ideas were not based solely on observation and conjecture because he was able to use his own farm as an experiment. As a result, all of his ideas were tested for their application, effectiveness and economic viability. So, for example, it was one thing to build dams at the key points along the keylines of natural water flow, but quite another to then encourage moving water to take a longer passage towards the creeks. For this purpose he designed the Yeomans Keyline Plough, which involved redesigning the traditional chisel plough to enable him to create parallel channels about 45–50 cm apart below the soil surface and roughly in parallel with the contour of the Keyline. Because the contours are not parallel as one moves down the slope, the plough channels progressively veered slightly down the slope at a shallow angle, thereby enabling both rain and irrigation water to distribute as widely as possible across the landscape and onto the ridges. By repeating these cultivations once a year (ideally when the weather is warm and rain is likely to follow) over the first three years at increasing depths (e.g., 20 cm, 30 cm and 40 cm or more, depending on the particular conditions and depth of the existing soil), and mowing or heavily grazing the pasture, which results in the death and regrowth of the roots of the pasture plants, Yeomans was able to rapidly transform pale, dead, non-productive soils into dark humus-rich productive soils that were crawling with earthworms where few had been seen before. This also resulted in a rich diversity of other beneficial soil-forming micro-life. The outcomes of these innovations were the ongoing maintenance of the soil, without the need of supplemental chemical fertilisers, no pest outbreaks, healthy productive livestock and no veterinary bills, and hence a profitable farm system. Of equal importance today is the phenomenal capacity of this system of soil management to capture carbon dioxide from the atmosphere and fix it in the soil as organic matter, thereby reducing the threat of global warming. A forthcoming book by P. A. Yeomans' son Allan will document the potential of this system to make the greatest contribution among the alternative strategies for removing from the atmosphere the extra carbon dioxide resulting from the burning of fossil fuels.[6] He claims that a one per cent increase in organic matter content of our agricultural soils would remove all of Australia's elevated carbon dioxide contribution. Like all other carbon fixation strategies, however, it will only buy time, during which we must reduce our burning of fuels and excessive production of carbon dioxide and convert to a solar-based economy.

Through a combination of theory and practice, Yeomans came to arrange the factors that must be considered in farm design into what he called a 'hierarchy of permanence'. Three factors – climate, land shape and water supply – were listed as the 'inseparable trinity of landscape design', while the other main factors to consider – farm roads, trees, farm buildings, subdivision (fences), and soil – were regarded as being more 'negotiable'. Many would be surprised to see soil at the bottom of this list, but this reflected his finding that soil can be made fertile very quickly by his system of cultivation, irrigation and pasture and livestock management. Thus, this does not represent a hierarchy of importance. Indeed, the central focus of his design and management was to improve, in his terms, 'the climate of the soil' for the life within it, as the sustainability of the whole farm, he considered, depends on having a live vibrant soil. What the hierarchy seeks to do is indicate where a designer can be most effective when planning or redesigning a farming system.

In his pragmatic way, Yeomans made extremely significant contributions to soil science. Whereas conventional soil science usually assumes that it has taken 300 to 1,000 years to develop each inch (2.5 cm) of topsoil, Yeomans showed how to work with nature's processes of decomposition to make many inches in as little as three years. He did this by creating a deep profile of optimal conditions (oxygen, moisture and dead organic matter) for the soil's rich decomposer and soil-forming community, by means of his innovative system of tillage, irrigation and pasture management. This provides the decomposers and soil-builders with a regular supply of substrate, the dead roots of the pasture plants that are added every time the 'browse' is removed by grazing or mowing, and by adequately resting the area for recovery in between the periods of grazing. For Yeomans, the evidence of success was plain because earthworms became increasingly abundant over the three years of his cycle, and the depth of transformed dark soil was increased every year. Thus, in addition to being one of the first users of the chisel-type plough in Australia, which he introduced into Australia, Yeomans was also a pioneer in the introduction of André Voisin's system of Rotational or Cell Grazing (Voisin and Lecomte, 1962). Voisin is regarded by many as France's greatest agronomist. The extent to which chisel ploughs, rather than the more destructive mouldboard and disc ploughs, are currently used in Australia is one example of Yeomans' extensive, yet hardly acknowledged, influence on Australian agriculture. However, his whole system redesign approach is yet to be widely adopted.

Yeomans owed his success to being an astute observer, a tireless experimenter, and a critical pragmatist. He was also an avid reader, and had made extensive studies of many other pioneers who had preceded him. He was, for instance, influenced by the writing of Charles Darwin (1882) on the importance of earthworms, and by Edward Faulkner's (1943) treatises on the folly of mouldboard ploughing. He was familiar with Friend Sykes' (1946) ideas on pasture management, Andre Voisin's (Voisin and Lecomte, 1962) system of rotational grazing and understanding of pasture ecology, and Sir Robert McCarrison's (1944) and Sir Albert Howard's (1947) understanding of the relationships between the health of soil, plants, livestock and people. He also travelled overseas (with his son Allan) and learnt from his own observations of Louis Bromfield's (1947) famous farm in Ohio, the long-term plots of Dr William Albrecht (1975) at the University of Missouri and the Rothamsted Experimental Station plots in Harpenden, England.

Yeomans had also opened his farm to visitors, and regularly showed busloads of interested farmers and scientists what could be achieved by his methods. Perhaps the most important of these visitors was Sir C. Stanton Hicks, Professor of Human Physiology and Pharmacology at the University of Adelaide. He had taken the time to study the Keyline system in detail and, based on his extensive comparisons with developments in the rest of the world, declared it 'the most interesting development in world agriculture'. He subsequently became one of the founders of the Keyline (Research) Foundation, which unfortunately was never able to follow through on Yeomans' vision of fully investigating Keyline as 'a complete plan for agricultural land development'. It did, however, produce six issues of its magazine, *Keyline,* and in the first issue Hicks (1955) recorded his version of the development of Keyline. He noted that Yeomans began by using conventional farm planning methods but was very disappointed by the results:

Despite water conservation in dams and on contour ledges, and despite the control of erosion and the better growth of pasture, the latter did not retain its vigour and stock did not put on condition . . . The turning point came in 1950 when his review of the high cost of all that he had done impelled him to make a new decision.

This is the classical story of discovery. The new decision . . . was based on what he had been doing for seven years, but it aimed to simplify the procedure and reduce cost . . . [he searched for] a blueprint, to which a farmer could work on any given property . . . So the Keyline was born. Not difficult contours to be run and ploughed, but one contour only to be the guide-line for the man on the tractor . . . The results . . . were as startling as they were unexpected . . . The moisture, which was absorbed in the loosened earth, therefore tended to migrate outwardly onto the ridges. And growth naturally followed it. . . . The bare ridges became covered with lush pasture. Erosion in the valleys ceased. Earthworms, which had never been seen, appeared in their myriads. Soon bare loose red shale became submerged in rich black soil. . . . In three years he has produced four inches of friable black soil where bare weathered shale or sandstone so recently comprised the barren soil . . .

The Keyline Plan is simple to put into effect. It is inexpensive. Results are rapidly produced and the land values improve quickly . . . It is a complete plan [that] . . . deals with . . . water and air in the soil. It places the dams in the most effective situations and constructs these by the cheapest technique . . . The plan appeals to me as the basis for renaissance in Australian land use. Through its application we may well hope for the much needed extension of rural population in Australia, that will rapidly become profitable even if poor to begin with.

For many, however, Yeomans' ideas seemed complex and, for some, the descriptions of his techniques were too difficult to follow. He was also a man who did not suffer fools gladly and it took exceptional persistence on the part of interested farmers to fully grasp and implement his Keyline system. The author (Stuart Hill) was among those who initially failed to completely appreciate the significance of his ideas from his writings. Subsequently, however, he was instrumental in getting two of Yeomans' books that were out of print, *The Keyline Plan* (1954) and *Water for Every Farm* (1965), republished in North America in the 1970s, thus helping to revive interest in his work. New ideas, radically at odds with prevailing wisdom, often take time to establish their efficacy, even for those who might be most sympathetic to them. However, there is a more significant and pernicious reason for the difficulties Yeomans experienced in trying to change farming practices, and this relates to the fact that his ideas posed a direct threat to the powerful pharmaceutical and chemical companies that were increasingly becoming the major players in world agriculture. Such companies, selling their chemicals to farmers, offered seductive quick-fix 'solutions' to the problems farmers were experiencing in implementing the traditional input-output model of agricultural production. A demonstration of increased production without the need for purchased inputs would certainly have been viewed as bad for agribusiness. As such, it would have been subjected to all the usual 'damage control' strategies by the threatened industries, government departments and 'colonised' academics, i.e., ignore as long as possible, then ridicule it, describe it inaccurately, conduct fake experiments, make false comparisons and then dismiss it.

Yeomans was driven by a force that he probably did not fully understand himself, because once he started down the path of whole-system planning and redesign, informed by a deep understanding of natural ecosystemic functioning, he could not stop. The methods he pioneered required intense knowledge and skill development. They demanded a much greater commitment than the 'magic bullet' approaches offered by the advocates of the prevailing farming methods. Perhaps the costs of non-sustainable methods are more apparent now than they were in the 1950s and, of course, the cost of 'addictive' and non-renewable products can only continue to increase. Daily it becomes more evident that we must replace our dependence on disruptive input-driven production systems with knowledge-based ecologically sustainable systems that foster the development of wisdom and skill. If Yeomans was way ahead of his time, we can only hope that his time will still come.

'PERMACULTURE': DAVID HOLMGREN'S CONTRIBUTION

Despite its marginalisation by conventional agriculturists, Yeomans' approach to ecological design was a main source of inspiration for the development of 'Permaculture'. The birth of this movement dates back to 1972 when Bill Mollison, a psychology lecturer and well-known 'identity' at the University of Tasmania, and David Holmgren, a student in the Environmental Design Course at the College of Advanced Education in Hobart, began an unlikely but highly productive collaboration. The extroverted Mollison (1996) has gone on to establish an international reputation as the 'father' of Permaculture; giving inadequate credit to both Holmgren and Yeomans. Holmgren's story is certainly less well known, but of great importance in tracing the lineage of ideas upon which Permaculture design practices are based.

David Holmgren was born in 1955 and grew up on the Swan River near Fremantle in Western Australia.[7] His parents were widely read political and social radicals who encouraged his creativity and interest in designing and making things. He remembers being taunted for bringing wholemeal sandwiches and dried fruit to school, but he managed to 'survive' by relying on his 'gift-of-the-gab'. He had witnessed the draining of swamps along the Swan River, where he had played as a child, and their transformation into suburbs of Perth. This loss of habitat had instinctively disturbed him. Instead of progressing from school to university, where he felt he could not survive, he embarked on a year of hitchhiking around Australia. In the process he discovered both Tasmania, which he fell in love with, and the innovative Environmental Design Course (at Hobart's College of Advanced Education), in which he enrolled in 1974.

During his first year in the design course he 'dabbled' in the program while sharing a house with other like-minded students. They composted their waste, grew their food organically and baked their own bread. Although Holmgren was developing a strong interest in the relationships between agriculture, landscape architecture and ecology, he found his coursework uninspiring and felt alone with his particular interests. Near the end of his first year at college, he needed a place to stay for a few weeks and heard that Bill Mollison had a spare room in his house. Holmgren, Mollison, and Mollison's second wife Philamena quickly discovered that they had many common interests and they started to engage in spirited, daily discussions. The two men read widely and enjoyed discussing a broad spectrum of ideas; arguing long into the night and experimenting by day in their collaborative project to recover the 'trashed' backyard garden and establish a species-rich cornucopia as their primary source of fruit and vegetables. Mollison was the expansive, divergent thinker and explorer, regularly collecting plants and ideas from near and far. In contrast, Holmgren was the convergent, critical practitioner who did most of the actual work of design and maintenance. Mollison was the mentor with challenging ideas, and Holmgren the questioning, reflective and practical student. It was natural that Holmgren would use this experience as the basis for the thesis required for his college course and, at the same time, to start to draft the outline of their evolving design principles and practices for sustainable land use for a post-industrial era. It was this latter document that was eventually jointly published in 1978 as *Permaculture One*. As Holmgren put it:

> I wrote the manuscript [for *Permaculture One*], which was based partly on our constant discussions and on our practical working together in the garden and on our visits to other sites in Tasmania . . . I used this manuscript as my primary reference for my thesis, which I submitted and was passed in 1976.[8]

For Holmgren, his manuscript represented a very tentative outline of potentially important ideas that needed more testing and refinement, and he was in no rush to publish them. For Mollison, however, it was the answer to the numerous demands he was receiving for a book to back up his growing media commitments, lecture circuit and workshop presentations, at which he was rapidly becoming famous for his outrageous claims about the wonders of Permaculture.

Permaculture, like Keyline, had been built on the outstanding work of earlier pioneers. Holmgren confirmed that Yeomans was a direct influence but there were other influences. These included F. H. King's *Farmers of Forty Centuries* (1911), Lewis Mumford's *Technics and Civilisation* (1934), Russell Smith's *Tree Crops: A Permanent Agriculture* (1977), Howard T. Odum's *Environment Power and Society* (1971), and Sir Albert Howard's *An Agricultural Testament* (1940). Holmgren and Mollison were also influenced by the general literature on organic and Bio-Dynamic farming (Pfeiffer, 1943) and by Geiger's *The Climate Near the Ground* (1965). It was a mixture of this knowledge of relevant literature, direct experiences of Australian environments, their very different but complementary personalities, and a shared passion for ecological and community-based approaches to design that enabled Holmgren and Mollison to develop and articulate the principles and practices that became known as Permaculture. However, it was the receptivity to such ideas by those who had been 'radicalised' by the various social movements of the 1960s, together with Mollison's missionary zeal and thirst for international recognition that enabled their ideas to gain a much higher public profile than those of Yeomans. Not surprisingly, the concepts immediately appealed to the growing 'counter-culture' movement, and particularly to those who were leaving the cities in their search for meaning back on the land. It was also not surprising that they were attacked by many in the farming and agribusiness communities as being impractical and of relevance only for hobby farmers and garden-scale levels of production – of no relevance for Australia's export-based, broad-acre agriculture. As Holmgren had feared, Mollison's rather premature promotion of Permaculture, and his exaggerated claims about what it could achieve, helped to fuel these criticisms. This reaction by the establishment was as disappointing as it was predictable, for, even if the concepts did require more work, it was clear that they were based on some sound ecological wisdom, which conventional agriculture was desperately in need of acknowledging. In essence they demanded that practitioners should learn from, mimic, and work with – rather than against – nature. This implies that we should design complex, integrated, even multi-storied, systems within which all organisms – primarily perennial rather than annual crops – perform not single and competitive, but multiple and mutualistic functions. Permaculture also emphasises working paradoxically (even seeing so-called enemies as potential allies) and catalytically (making the smallest intervention, using the minimal amount of resources, for the greatest benefit). It emphasises specificity in regard to both time (when to do things), and place (where to do things), and it aims to produce no 'waste'. All systems are designed to be as self-sustaining as possible (Mollison, 1988).

Over the past 20 years, Holmgren has been further elaborating on these ideas and has now summarised them into twelve key design principles – or imperatives – and these will form the basis of separate chapters in his forthcoming book, provisionally entitled *Permaculture Principles and Other Ideas*.

Soon after the publication of *Permaculture One*, Holmgren met Haikai Tané (1995) at the second 'Down to Earth' festival at Bredbo, near Canberra. Mollison was also there and, during a presentation on Permaculture, he made what for conventional farmers would be regarded as an extremely heretical statement. He argued, for example, that rather than spraying the common weed Briar Rose to get rid of it, we should rather recognise its capacity to heal the land. When Tané commented to Holmgren that he didn't quite agree with this, he had assumed that Tané was going to elaborate on the dangers of not controlling such weeds. Instead he then went to explain why another weed, Gorse, is even more effective in building soil and contributing to the health of the local landscapes in question. This 'deeper' knowledge considerably impressed Holmgren. Tané, a 'Renaissance man' with degrees in Law, Geography and Ecology and Planning, and a 'student' of Richard St Barbe Baker (the Man of The Trees (1970; rpt 1979) who willed his library to Tané), became Holmgren's next extraordinary mentor.

Tané was also impressed with Holmgren and invited him to New Zealand, where he was working as a planner. Here he introduced him to his land systems approach to the integration of diverse sources of information into comprehensible patterns which could then inform site selection and subsequent design processes. Together they ran workshops in New Zealand over a three-month period in 1979 (and again in 1984) on a version of Permaculture made more comprehensive by Tané's input. They were also instrumental with others at that time in establishing 'Permaculture New Zealand'.

At Bredbo, Holmgren had also reconnected with his mother, Venie Holmgren, who had left their home in Western Australia after the death of her husband, and after the last child had left home, for a two-year campervan journey around Australia to rediscover both herself and forgotten landscapes. When they met at Bredbo, Venie told David that she intended to sell the family home in WA in order to buy a bush property in the eastern states and build an ecological home. It would be an opportunity to implement some of her son's Permaculture design principles. David agreed to help her realise this dream, and soon after she purchased a bushland property in the Towamba Valley in southern New South Wales. Here, for

David Holmgren with author Stuart Hill [left] in his garden at Hepburn,
Victoria, in 1998 (Stuart Hill)

18 months (1980–81), they worked tirelessly together to establish systems
that would subsequently require minimal maintenance. This experience
put Holmgren on a steep learning curve because the scale of the task was
much greater than any he had previously tackled. The intensity of designing and building his first ecological homestead left him feeling drained and
exhausted, but it was also a valuable test for his design ideas, and he documented the experience in his first solo publication – *Permaculture in the Bush*
(1985; 1993).

After taking some time to recover from this experience, Holmgren
moved to Hepburn in Central Victoria and set about building a
Permaculture home and garden for himself and his new partner Sue
Dennett and her two children. He also documented this experience – his
second significant case study – in the book *Hepburn Permaculture Gardens:
10 Years of Sustainable Living*, which was not published until 1995 even
though the draft manuscript was completed in 1991. Since 1990, Holmgren
has used his Hepburn home and garden as the site for teaching his
'advanced' Permaculture design courses. In addition to this work,
Holmgren has also worked as a consultant on broad acre properties.
Examples of this work are included in his report *Trees on Treeless Plains:
Revegetation Manual for the Volcanic Landscapes of Central Victoria* (1994).

Recently, Holmgren and Dennett have teamed up with another
couple and embarked on their next major project, 'Fryers Forest', which

aims to apply Permaculture principles to both forest management and community development. As Holmgren puts it:

> The vision is of an equitable community [of 11 dwellings] . . . whose simple resolve is to enjoy the place, co-operate for mutual benefit . . . and live [on one-acre lots] in sustainable accord with and within a superb forest location [of 300 acres, 12 kilometres from Castlemaine].[9]

At the time of his interview in 1999, Holmgren was not only working on the manuscript of his latest version of Permaculture design principles, but also on the manuscript for a book on the controversial topic of introduced species, provisionally entitled *Migrant Plants and Animals: Ecological Imperialism or Co-Evolution*. Public debate on this topic was stimulated in 1998 by the release of the book *Feral Future* by Tim Low and, as already indicated, Holmgren is much less inclined than many environmentalists to regard all introduced species as 'enemies'. Now that he has had more time to test and refine the principles of Permaculture design, it is to be hoped that, with these new publications, Holmgren's work will attract the attention it deserves. Like Yeomans, he has been a pioneer thinker and practitioner and the work of both men should be required reading for all people involved with landscape design.

THE BRADLEY SISTERS AND TED TRAINER

Although Yeomans' design principles had been developed within rural, farm landscapes, his later work on cities (1971), like the contributions of Holmgren and Mollison, had attempted to breach the segregation of thinking about rural and urban landscapes. Although the most important pioneers of the bush regeneration movement in Australia, Joan and Eileen Bradley, began their work in neglected urban landscapes, the movement they fostered has also helped to breach the urban/rural divide, especially in the way it has been integrated into many government-subsidised rural Landcare projects (Campbell, 1994). The Bradleys were among those who regarded introduced 'weeds' as enemies, because they had watched some fast-growing introduced species overwhelming less competitive native plant species. The choking out of the indigenous flora resulted in the destruction of the habitats of native birds and mammals. However, as we shall see, the Bradleys discovered that a simplistic 'war on weeds' is not only futile but frequently counter-productive, and that a more subtle intervention in the ongoing battles between communities of plants is required to be effective in the long term. Like both Yeomans and Holmgren, the Bradleys were guided by the principle of learning from nature through persistent, deep observation and trial-and-error.

Joan and Eileen Bradley first became involved in bush regeneration work almost accidentally. From their home in the harbourside Sydney suburb of Clifton Gardens, they enjoyed walking in nearby Ashton Park (now part of Sydney Harbour National Park). Here they found themselves instinctively pulling out the seedlings of the increasing number of invasive, non-indigenous, weeds – such as lantana, camphor laurel, privet and morning glory. However, what may have begun as a casual activity rapidly evolved into a daily habit, and the sisters were pleased to see that the natives gradually began to recolonise areas that they were tending in this way. In the early stages their work focused on an ability to identify and eradicate the invasive plants, and in 1967 they published a booklet entitled *Weeds and Their Control*. Joan Bradley later wrote (1971) that it was only in the writing of this book that they realised that what they were essentially doing was to tip the balance in favour of the natives in their competition with the invaders. She noted that, in the final analysis, it was the natives themselves that had to do the real work of re-establishing their complex communities. Furthermore, they learned from experience that it is best to start, using hand-weeding and careful cutting with easily carried hand-tools like secateurs, where the natives are strongest and the weeds least well established, and work back from there to where the weeds become more dense. Because the roots of natives tend to be sensitive to disturbance, it is also best to work slowly and carefully so as to avoid any such disturbance, and to keep the cleared ground covered with as much mulch as is available. Their third imperative, also learned through experience, was to avoid over-clearing, because natives can only slowly reclaim bare ground and such extensively cleared areas are much more likely to be re-invaded by weed species. The key is to allow the regeneration process to dictate both the rate and extent of clearing, and this requires extraordinary patience. As Joan Bradley has stressed:

> However bad it looks, don't start to clear a block of solid weeds until you have brought good bush right up to it. (1971, p. 8)

To avoid mistakenly pulling out young natives, it is necessary to become competent at identifying the different plant species that are present, especially in their seedling stages. Their most surprising discovery was that healthy native plant communities could be regenerated without doing any replanting. There is apparently enough native seed stock in the soil and the remnant vegetation to recolonise an area, provided that the invasive weeds are kept in check. Their insights and experiences were in sharp contrast to the common assumption that what is required is an all-out eradication of the weeds, followed by an extensive replanting of natives. Their more effective approach involved working slowly and methodically over a longer period of time. In fact, most of the bush regeneration work that the Bradleys did was accomplished while taking their dog for a walk, Eileen in the morning and Joan in the afternoon, with about three-quarters of an hour of pulling weeds for each of them each day. They also stressed the importance of seasonal awareness, noting that early in the growing season is the best time to weed, and to avoid weeding later in the year when most weeds are shedding their seeds.

The Bradleys' method flew in the face of conventional 'wisdom' about weed control and revegetation of degraded areas. Dominant practices relied on the use of toxic chemicals and powerful machinery. Patience and incremental change is rarely valued in a society committed to quick-fix 'solutions'. So it's not surprising that the Bradleys were frequently ridiculed by the 'experts' as being unscientific 'ratbags'. Such critics ignored the fact that Joan Bradley had a degree in chemistry, and that the sisters had been very systematic in their experimentation, relying on reproducibility before making any claims. More importantly, their approach was consistent with a deep understanding of plant biology and ecological processes, whereas the overkill, technology-intensive approaches were not. Indeed, the Bradleys had closely observed and documented what happened on the many sites around Sydney that had been subjected to over-clearing. As Joan wrote:

> The result of all this labour was either re-growth of the same weeds. Or more usually the appearance of a bewildering variety of quick-growing, quick-seeding weeds which have proved infinitely harder to control than those which were removed . . . and indeed in many places weeds have spread into what was previously quite good bush. (1971, p. 1)

Joan Bradley's second book, *Bush Regeneration*, was published in 1971. However, it was not until 1975, when Joan, with colleague Toni May and their small team of regenerators, was commissioned by the National Trust to demonstrate their approach in Blackwood Reserve near Beecroft, in Sydney's north-west, that the wisdom and effectiveness of the 'Bradley method' was finally acknowledged. This led to numerous demands for Joan to teach the method to other bush regeneration groups throughout the country. Sadly, Eileen died in 1976, before this more extensive recognition had been achieved, and Joan died six years later, with her final manuscript still incomplete and unpublished. Fortunately, three of her colleagues, Joan Larkin, Audrey Lenning and Jean Walker, took on the challenging task of completing and editing the manuscript, which was published in 1988 as *Bringing Back the Bush: The Bradley Method of Bush Regeneration*. This work has subsequently been reprinted and it remains a valued text for bush regenerators throughout the country.

Although the Bradley sisters, especially Eileen, did not receive the recognition they deserved while they were still alive, the wisdom of their patient and systematic approach is now familiar to most small-scale bush regenerators in Australia. Happily, the counter-productive methods that they critiqued have now been largely discredited, and there are many thriving plant communities – providing valuable habitats for native birds and mammals – that owe their existence to the innovative approach of the Bradleys.

Another Australian ecological pioneer who has been busy making his mark on a particular block of land near Sydney is Ted Trainer, who has been working for decades to make his 20-hectare property at Pigface Point on the Georges River in Sydney's south-west a demonstration site for sustainable living practices. Trainer's message, however, is more global, and his challenge is much more radical than that emanating from the rather genteel sisters of Clifton Gardens. In fact, he is openly critical of the fragmented, save-the-threatened-species-and-spaces, green marketing, waste management and pollution control focus of most environmental groups. Instead, he calls for much more demanding changes in our values, lifestyles and organisational structures. Trainer argues that we must abandon our:

> . . . obsession with high rates of production and consumption, affluent living standards, market forces, the profit motive and economic growth. [He insists that a] sustainable and just world order cannot be achieved until we undertake radical change in our lifestyles, values and systems, especially our economic system. (Trainer, 2001, p. 1)

Trainer is encouraged that an increasing number of people now acknowledge that we live in an unjust, ecologically unsustainable society in which quality of life is now getting worse, not better. He is frustrated, however, that these same people seem to be immovably addicted to the two driving forces that have brought us to this sorry state – 'growth' and 'affluence'. He also notes that even the most vocal environmentalists continue to act as if they believe sustainability can be achieved without a radical change of lifestyle.

Trainer has used his academic training as a lecturer in the School of Social Work at the University of New South Wales to systematically build his argument against preservation of the *status quo*. So, for example, in his balance sheet of the 'achievements' of our growth-oriented global economy he draws attention to the following:

- Over a billion hungry people in the Third World, another two billion malnourished and 30,000 children dying each day, largely as a result of this deprivation.
- An ever-widening gap between rich and poor, with the wealthiest 20% of the world population receiving over 85% of all income, and the poorest 20% receiving only 1.5%.

- Rising unemployment and social problems (drug abuse, homelessness, depression, violence and suicide).
- Increasing world debt and associated interest payments.
- Widespread rural decline.
- Increasing foreign ownership in most countries.
- Loss of public services and small enterprises.
- Deteriorating working conditions (longer hours, more stress, less job security and falling real wages).
- Increasing economic vulnerability of small- to medium-sized businesses and nations.
- Decline in community spirit and civic culture.
- Widespread loss of biodiversity and environmental degradation.

Of course, the people most directly affected by all of this are in the poorest countries, or in the poorest communities within the wealthier nations, and it is too easy for the affluent to feel that the problem is not theirs. As Trainer stresses, however, all of us are already being affected, at least indirectly, and the direct effects will continue to escalate until we change our lifestyles.

Trainer identifies 'unregulated competition within the market' and our 'obsession with affluent living standards and economic growth' as the two factors most in need of fundamental rethinking and change. The market, he argues, will always 'lead to the development of the most profitable industries, as distinct from those that are most necessary or appropriate' (Trainer, 2001, p. 4). Because of its emphasis on maximising sales and profits, the market always allocates most wealth and resources to the rich, and it provides 'no incentive to think and behave cooperatively or to focus on what is good for society' (Trainer, 2001, p. 6). And now, with increasing globalisation of the economy, all of these negative effects are being amplified manyfold.

With the rapidly expanding power of resource-hungry and market-driven transnational corporations, governments are being pressured through international trade agreements to remove all forms of market protection, tariffs and controls, and to lower environmental and work safety standards, all of which are regarded as impediments to global competition and free trade. Countries that resist are punished by the withdrawal of investment. As a result, all countries now find themselves in fierce competition to secure and maintain a place in the global export market. This is why looking to our present politicians for leadership is likely to lead to disappointment, as they are all, whether they like it or not, locked into this treadmill of unsustainable development. What is particularly distressing to Trainer is that this is the model that underpins nearly all conventional thinking about development, and also most of our aid programs to the Third World (Norgaard, 1994).

For Trainer it is obvious that the development of a caring, collaborative and responsible society is unlikely to be achieved when the main organisational forces are competition, self-interest and limitless growth in output, consumption and affluence.

To support his arguments, Trainer has calculated the resource needs and environmental impacts of a global population of 10 billion, which is the level at which it is expected to peak some time after 2060. Even if there was no further growth in per capita consumption of resources beyond present levels in Australia, annual production would have to increase eight to 10 times to meet the needs of this population. To try to meet even this goal would result in the exhaustion of all 'probably recoverable' fossil fuels by 2045, with petroleum becoming economically unavailable long before then. To enable these 10 billion people to live the way we live in Sydney today would require 50 billion hectares of productive land, which is seven times all the productive land on the planet! Trainer argues that because this level of consumption cannot be sustained by either nuclear or solar technologies, and because the environmental impacts of such consumption would be devastating, we have no other choice than to reduce consumption. The disturbing implication is that, for such a population to survive, each member would need to reduce their per capita consumption of energy to 1/18th of what we currently consume. Trainer notes that such projections are clearly unachievable, yet development policies continue to be based on the assumption that we can maintain our current levels of growth in living standards and GDP. Trainer argues that the only rational way to address this situation is to adopt a 'limits to growth' policy framework. Furthermore, he stresses that we must start the process of social transformation now, especially given that world petroleum supply is predicted to peak within 10 years. He finds the fact that so few people seem able to clearly recognise the implications of the situation we are in both astounding and disturbing.

Trainer argues that this 'limits to growth' analysis of the global situation leads to clear conclusions as to the form that a sustainable and just society must take. This must involve simpler lifestyles that are compatible with ecological processes and reduced consumption. He also advocates a move towards a more decentred society, organised around local economies that are largely self-sufficient. Revitalised local communities would encourage more participation and collaboration, and political systems would start to exhibit the features of what Dale Hunter and her colleagues (1997) have called 'co-operacy' (the next step along from autocracy and democracy). Instead of growth, the aim would be the achievement of steady-state economies (Daly and Cobb, 1994).

Although an academic himself, Trainer is critical of many of those who have made similar points about prevailing economic and social policies. With undisguised cynicism he told the author that he has found very few academic colleagues willing to tackle the issues of affluence personally, noting that:

> . . . even the very left ones tend to like their affluence, their jet travel to conferences, and they tend to like to keep their wine racks in fairly good shape. But when it comes to living simply, they tend to be not so forthcoming.[10]

Clearly, Trainer sets very difficult standards for himself and those who might agree with his basic analysis of global trends. Such an uncompromising message has made it difficult for him to build any long-term working relationships, and his greatest supporters have been his own family and occasional students who have worked on some of his projects. When asked to nominate a team of Australians whom he would trust to lead us in a process of necessary social change, he sadly replied:

> I'd have to scrape the barrel very hard. There's a lot of concern and green activism, but hardly any radical green activism.

Trainer has developed his vision of a 'conserver society' gradually, encouraged first by his socialist father, and later by Paul Ehrlich's (1968) insistence that our species must recognise and live within ecological limits. At his home at Pigface Point, where he has spent his whole life, he receives a steady stream of visitors and explains his ideas to anyone who will listen. He effectively uses graphs, models and examples of 'soft' technologies to show such audiences how existing suburbs could be transformed into eco-villages (1995; 1996; 1998). In such village communities, life would be simpler, more meaningful and ecologically sustainable. Most food could be produced in community and market gardens, with much of the neighbourhood being redesigned as 'edible' and aesthetically nourishing 'Permaculture' landscapes. Most essential products could be produced locally. The economy would integrate a range of economic and exchange systems, with barter and LETS (Local Employment Trading System; (Linton and Greco in van Andrauss et al. (eds.), 1990) playing key roles. Also included would be a small locally regulated market economy and banking system, with loans only being given for initiatives that would improve community wellbeing. Most adults would probably only need to devote one or two days per week to paid work, as most needs would be met outside the money economy; and most would also be able to walk to work. With more leisure time, such communities would be more supportive of participation in arts and crafts, and other creative activities. Entertainment would be more based on participation than on consumption. Tools and equipment could be shared and stored in communal workshops. Much of the work within the community would be accomplished through voluntary rosters and working

Ted Trainer addressing a group of visitors to his property at Pigface Point
in Sydney in 1999 (Stuart Hill)

bees, rather than through centrally controlled government agencies (Biehl
and Bookchin, 1998). People would have more time and energy to participate in neighbourhood meetings, and to follow through on decisions that
were made.

Trainer calls on those who read his books, visit his property and visit
his website (http://www.arts.unsw.edu.au/socialwork/trainer.html) to
put their energy into circulating such ideas and visions rather than just
focusing on changing their own lifestyle. He especially looks to those living
in the growing number of urban and rural eco-villages to act as ambas-
sadors in demonstrating the viability of ecological community living
(Hagmaier et al., 2000). Trainer hopes that such initiatives will achieve
greater maturity over the next 20 years and so build the critical mass
needed to trigger a much broader social change that he believes is essential.

Ted Trainer's uncompromising message and style is not designed to
win friends, and many who agree with the gist of his arguments still
dismiss his ideas as being impractically utopian. From all accounts,
P. A. Yeomans evoked a similar response in his day and, although David
Holmgren was more cautious than his collaborator Bill Mollison, he too has
been criticised for being too radical. The Bradley sisters, ironically, were
also criticised for being utopian in their cautious and patient approach to
their work. What these particular pioneers had in common was that they
all recognised the need to adopt some humility by being willing to

genuinely learn from nature and abandon prevailing practices that were, from their perspectives, based on short-term greed and/or ecological ignorance. They all sought to apply ecological principles in redesigning our relationships with natural systems, and they all saw direct benefits for people in learning how to think and act both ecologically and mutualistically (Boucher, 1982; Hill, in press).

Yeomans' story – with its elements of long observation of nature's processes, chance events that led to a partly subconscious yet driven pursuit of a goal, profound paradoxical insights and the development of theory as a product of creative practice and persistent experimentation – is perhaps the most inspiring of all. His introduction of design theory and practice into farming marked a quantum leap from the old style of husbandry that still, unfortunately, dominates agriculture. Holmgren has expanded on these concepts of design and developed them into a set of principles that can be applied much more widely. The Bradley sisters, though more narrowly focused on the bush, provide a striking model of working paradoxically (starting in the opposite place to where one might be inclined to start). They also demonstrated the wisdom of letting the rate of an intervention be determined by nature's time scale and not by our impatient tendencies. Trainer injects a note of urgency into all of this by reminding us that we simply cannot continue to ignore the devastating global implications of our irresponsible personal actions. In politicising this 'redesign' work, he argues that to be effective in its broader goals it must become linked to the necessary transformation of our institutional structures and processes, particularly our economic policies. More than the others, he urges us to work not only at the local level – using ecological redesign principles that others have pioneered – but also simultaneously on the global issues that continue to undermine such local sustainability work. In this way, he has extended Holmgren's principles of ecological design into a vision for radically redesigning our social and economic systems.

The basic message of ecological redesign is radical because it challenges many of the prevailing assumptions of western culture. It also requires long-term planning and a strong commitment to both personal and political change. The communicators of this message, including those profiled here, have frequently been dismissed by the rest of society as overly critical voices that offer only discomfiting advice. However, through their suggestions for better ways to act in the world, they have also offered the hope of both personal and ecological renewal. They are town criers with important messages that we should heed. To avoid disadvantaging future populations, it is imperative that we no longer postpone the need to fundamentally redesign our numerous unsustainable practices. The time for action is now.

CHAPTER 9

CHALLENGING *TERRA NULLIUS* VIEWS: *The Aboriginal Land Rights Movement*

INTRODUCTION

For a very long time after the colonisation of Australia began in 1788, the colonisers had little or no interest in learning about the beliefs and practices of the indigenous people. The first serious attempt to record cultural practices of the Aboriginal people in a relatively unbiased way came when the explorer Edward John Eyre attached a 360-page description of practices he had observed to the journal of his 1845 expedition. Anthropology as a study only reached Australia in 1894 when the biology professor at the University of Melbourne, Baldwin Spencer, and Frank Gillen of the Alice Springs telegraph office set out to record cultural practices of the Arunda people west of Alice Springs. Tom Griffiths (1996) has said that early Australian anthropology was 'driven by the expectation of Aboriginal extinction' (p. 26). The first book by Spencer and Gillen was greeted by a reviewer at the Melbourne *Argus* as being a valuable study of 'the lowest in the scale of civilization' (Haynes 1998, p. 146). The first chair in anthropology was established at the University of Sydney in 1927. Like early naturalists who collected exotic specimens for 'scientific purposes', the anthropologists believed that they should document the passing of a 'primitive people'. Surely there was nothing that such 'stone-age people' could teach 'civilised' people about contemporary issues.

By the middle of the twentieth century white Australians had come to believe that Aborigines, as a 'dying race', wanted nothing more than a chance to assimilate themselves into contemporary society. They were taken aback, therefore, in 1966 when a 'bunch' of Aboriginal stockmen and their families at a place called Wattie Creek in the Northern Territory announced that they were reclaiming their tribal lands because they were no longer prepared to tolerate conditions of life on the cattle stations of white pastoralists. In staging their much-publicised walk-off, the Gurindji people of Wattie Creek (Daguragu to them) were following in the footsteps of a less publicised earlier campaign by the Yolngu people of north-eastern Arnhem Land to reassert their rights over their tribal lands. Between them the Yolngu and Gurindji effectively launched the 'modern' Aboriginal land rights movement.[1] As well as challenging the widespread belief that Aboriginal people desired assimilation, the Yolngu and Gurindji were also challenging the fictitious notion of *terra nullius* (empty land) used to justify the forced expropriation of land.

A legal challenge to *terra nullius*, mounted by the Yolngu in the Supreme Court in Darwin in the late 1960s, ended in failure in 1971. But this very failure prompted the setting up of the now-famous Aboriginal Tent Embassy outside the old Parliament House in Canberra in January 1972 and the spreading of the campaign for legal recognition of traditional Aboriginal rights to land. In 1982, a group of Torres Strait Islanders – led by Eddie Koiki Mabo – asked the Queensland Supreme Court to rule on their traditional rights to land on Murray Island and, although they lost that case, the Supreme Court ruling was subsequently overturned by the High Court in Canberra which ruled, in 1992, that a form of native title should have been recognised by the white settlers. The historic 'Mabo decision' finally overturned the myth of *terra nullius* in respect to rights to land.

As mentioned in chapter 6, some prominent indigenous Australians have pointed out that white environmentalists have maintained a *terra nullius* attitude in their promotion of the idea that 'wilderness' areas should be pre-served because they have never been altered by human intervention. It is now widely accepted that Australian landscapes have been changed by the presence of people for over 60,000 years and that the indigenous people developed a high sense of responsibility in their role as caretakers of 'country' that could sustain them and other forms of life. The 'modern' Aboriginal land rights movement emerged at a time when the nature conservation movement in white Australia was starting to be transformed into a broader environ-mental movement (see chapter 6). As a result there was a larger constituency of people who could see that indigenous Australian philosophies regarding 'right relations' between people and the land stood in sharp contrast to the exploitative attitudes of the white settler society that were held responsible for widespread degradation of the environment. There was obvious poten-tial for an alliance of environmentalists and the indigenous people. The alliance has been problematic because deep philosophical differences divide the two 'camps', but the dialogue has moved along. Furthermore, non-indigenous people who have responded to the cause of the indigenous people on the grounds of 'social justice' have also been influenced by their ecological worldviews, becoming more sympathetic to ecological critiques of western 'civilisation' and its 'development' practices.

The 'modern' land rights movement has had a major impact on a wide range of discourses relating to Australian identity and the land. It has thrown up many 'heroes' and leaders, but indigenous people prefer to point to the achievements of communities rather than individuals, so this chapter focuses on pioneering communities – the Gurindji and Yolngu – who helped bring some ancient ecological insights into contemporary ecological discourses.

THE GURINDJI STOCKMEN'S STRIKE

On Easter Sunday 1966 Lupgna Giari, a tall Gurindji stockman with a proud disposition, returned to Newcastle Waters station in the Victoria River district of the Northern Territory after a week of mustering cattle. Having worked as a stockman on various properties in the Territory for 30 years, Giari was head stockman at Newcastle Waters and was being paid $10 a week compared to the standard rate for Aboriginal workers of $6 a week. Despite his long experience with horses, Giari had had an accident the day before, Easter Saturday, as he tried to get control of a particularly troublesome horse. He was knocked out after hitting his head against a tree, injuring his neck.

'You'll be alright Major', the white cattle station manager Roy Edwards had told him. (Giari was widely known as Captain Major in accordance with the tradition of adopting colourful or comical nicknames that white people could remember.) 'I'll get the missus to put some liniment on it.' So Giari wrapped a piece of leather around his aching neck and went on working. By the time he returned to the station he was feeling sore and disgruntled. His own version of what happened next is recorded in Frank Hardy's *The Unlucky Australians*:

> That eebning I finished breakin' horses and I went down to the top of the yard and I was thinkin' to mesel': I was reckon I only get ten dollar and all these other men only get six dollar. And them women might book a few things down at the store, lucky if we get thirty bob left after two months. That not right. And I bin thinkin' agen: Wish we had someone behind us somewhere.
>
> And there was this letter for me. A man called Prentice had brought it. First letter I ever got. I never been learn to read so I asked Prentice to hand the letter over Arty, young fella who had been to school. He tol' me what that letter said. That letter was from Dexter Daniels. I bin hear about that young Dexter, an aboriginal who work for that Union mob in Darwin. That letter said: 'All right, all you stockmen, you must wait for me. I will come back. Don't go away. Finish that job, then I want you to stop work and wait for me.

Dexter Daniels had visited Newcastle Waters after going to a horse race meeting at Renner Springs to talk to Aboriginal stockmen about the recent decision of the Federal Arbitration Commission to delay the implementation of equal pay for Aboriginal workers for a further three years. At Renner Springs he was told about Lupgna Giari's reputation for standing up to the cattle station managers. But when he got to Newcastle Waters, Giari was out on the muster so he left him a note. Daniels later told Hardy that he had not really expected a reply to the note. So he was surprised to receive a telegram saying the strike had already begun. By remarkable coincidence the Newcastle Waters stockmen began their strike on May 1, 1966 – exactly 20 years to the day after a similar major strike by Aboriginal stockmen in the Pilbara district of Western Australia had been launched.

It was also a happy coincidence that Frank Hardy, renowned writer and story-teller, happened to be in Darwin at the time when the strike started. He heard about the strike by talking to Dexter Daniels, who was agitated by the fact that his boss Paddy Carroll, the secretary of the North Australian Workers Union, had reacted negatively to the news of the strike. He had asked Daniels to return to Newcastle Waters and advise the stockmen to resume work and wait for the ACTU to appeal against the decision of the Arbitration Commission. Daniels was not prepared to do that and he told Hardy he was trying to think of ways to support the strikers without the assistance of the union. When Hardy asked him if he thought the stockmen would have the determination to continue their strike, Daniels replied: 'They know how to wait, those people.' This proved to be prophetic because most of those stockmen had begun what would turn out to be a strike without end. Many of them never returned to work for Newcastle Waters again.

In conversation, Daniels and Hardy decided that they would try to revive the defunct Northern Territory Council for Aboriginal Rights to back the strikers. Daniels went away to talk to Darwin Aboriginal leaders like his brother Davis Daniels and Philip Roberts. Hardy spoke to Communist Party and trade union activists Brian Manning and George Gibbs. The NTCAR was reformed at a meeting held at Rapid Creek on July 29, 1966 with Dexter Daniels elected as president, brother Davis as secretary, and Robert Tudawali – star of the popular film *Jedda* – as vice president. The organisation started raising money to buy food and provisions to keep the strikers going. As support started coming in from Darwin and from people in the trade union movement 'down south', Dexter Daniels went on another trip to cattle stations in the vicinity of Newcastle Waters. When he got to Wave Hill station to meet with Gurindji tribal leader (*kadijera* man) Vincent Lingiari, Lingiari told him: 'We have waited a long time for you.'

Wave Hill station is in the heart of Gurindji country. When Hardy later visited the Wave Hill 'mob' Lingiari told a similar tale to that of Lupgna Giari about the wages and conditions of the stockmen. However, not only did they have little left from their meagre wages after they had

Famous image: Prime Minister Gough Whitlam pours a handful of Daguragu soil into the hands of Vincent Lingiari at a hand-back ceremony in 1975. (Newspix)

bought food from the station store, they were sometimes not paid at all for many weeks. To compound matters, Gurindji women were regularly harassed by white men living and working at the station. Lingiari was quite prepared to take on the station owner, British food mogul Lord Vestey, and he had no sympathy for the station manager Tom Fisher (pronounced Pisher by the Gurindji). His main concern about going on strike was the welfare of the cattle, so he arranged for someone to visit the water pumps every day to make sure they had enough water. He told Hardy: 'We not bin let them cattle die of thirst. Them big Besty mob not hear them cattle die. But I bin hear them cattle die.'

Soon after he heard that the Wave Hill mob, led by Lingiari, had gone on strike, Lupgna Giari led his mob in a walk-out from Newcastle Waters so they could join their tribespeople at Wave Hill. With the strike still stalemated in mid-August, the Gurindji decided they would also abandon Wave Hill and set up camp at a place they call Daguragu (known as Wattie Creek to white people). This place was selected because it had once been a Gurindji camp, adjacent to Bungaroang (Seal Gorge) where the bones of the dead had been laid to rest for thousands of years. Many of the older people at Wave Hill, including Lingiari, had been born at Daguragu and the gorge not only contained the bones of the ancestors but cave paintings that signified a long history of association.

The older members of the community could remember back to 1928, when some survivors of the infamous Coniston massacre (occurring well to the south) had fled as far as Daguragu. However, the survivors – mostly Walbiri from the desert – were pursued by a party of police and vigilantes who attacked the Daguragu camp and shot a number of people at random. After the massacre was over, the Gurindji took the bodies into the Bungaroang gorge for burial and a new painting was made to honour the victims.[2] For older Gurindji, Daguragu was the place where they had come into this world and they hoped it would be the place where their bones would return to the soil. It was a place that could not be erased from the memory or identity of the Gurindji leaders and they were determined to take their community home. There was a deep sense that the land was rightfully theirs.

So a strike that began over pay and working conditions had turned into a struggle for land rights. The Gurindji squatted 'illegally' at Daguragu from 1966 until 1975 when the Whitlam government excised an area from the Wave Hill lease and leased it to the Gurindji, pending the adoption of land rights laws that were then being developed. Whitlam had a good sense for the dramatic and so he had a flag-pole (together with electric lights) erected at Daguragu for a lease-signing ceremony, which would include the lowering of the Australian flag. The tall white leader flew into the desert by plane to acknowledge the accomplishment of the Gurindji. As hoped, this led to a good photo opportunity as the towering Whitlam poured a handful of sand into the hands of the frail-looking, but immensely strong Gurindji leader, Vincent Lingiari.

In 1996 – thirty years after the Gurindji started their struggle – August 14 (the day they walked off Wave Hill station) was declared 'Gurindji Freedom Day'. In the same month, the Governor-General William Deane delivered the inaugural Vincent Lingiari Lecture in Darwin. To quote from a song written by Kev Carmody and Paul Kelly about the Gurindji struggle 'From little things big things grow.'

THE YOLNGU LANDS RIGHTS BATTLE

The Gurindji strike was not the first big Aboriginal strike. As already mentioned, Aboriginal stockmen in the Pilbara had done something similar in 1946 and many of them never returned to work for the white cattle stations either. The Gurindji land claim was not the first Aboriginal land rights struggle in Australia. In her book, *Invasion to Embassy: Land in Aboriginal*

Politics in New South Wales, 1770–1972, Heather Goodall has pointed out that access to land was at the heart of Aboriginal resistance to colonisation from the time white people arrived. But the Gurindji strike and the campaign also begun in the 1960s by the Yolngu at Yirrkala on the Gove Peninsula of Arnhem Land (to stop the construction of a bauxite mine and aluminium smelter on their land) are acknowledged as the efforts that gave birth to what might be called the 'modern' Aboriginal land rights movement. Both the Gurindji and the Yolngu got enormous support from white organisations 'down south'. In the case of the Gurindji, the Communist Party played a key role in mobilising the support of the trade unions and student organisations. The Yolngu got support from the Federal Council for the Advancement of Aborigines and Torres Strait Islanders (which headed the successful campaign for the 1967 referendum on citizenship rights for indigenous Australians). Once the Yolngu took their case to the Supreme Court in Darwin – in a court challenge that was like a dress rehearsal for the later Mabo case in the High Court – there was enormous interest in the outcome.

The Yolngu struggle for land rights first gained national attention in 1963 when the community at Yirrkala sent a petition to the Federal Parliament drawn up on a piece of bark and bearing the names of all the clan leaders from the Yolngu areas. The Yolngu were protesting about the fact that the Methodist Church, which ran the mission, had failed to consult them in negotiating a mining lease with the Swiss-Australian mining consortium Nabalco on Yolngu land. The petition, and those who were leading the campaign at Yirrkala, got support from the missionary of the day, the Rev. Edgar Wells (who later wrote an excellent account of events). But, for his efforts, Wells was sacked by the church, and the agreement with Nabalco was signed.

The Yolngu were not able to prevent the mining at Gove but they certainly gave it their best shot. When Justice Blackburn was hearing evidence for the case in the Supreme Court in Darwin in 1971, ten Yolngu leaders stayed in Darwin for the two weeks of the hearings and they took with them two young men, Galarrwuy Yunupingu and Wulanybuma Wunungmurra, who had learnt English by taking up church scholarships to attend Brisbane Bible College to act as interpreters (Williams, 1986). Supporting evidence was given by the anthropologists W. E. H. Stanner and Ronald Berndt. Blackburn could not deny that the evidence clearly showed that the Yolngu had a long and close association with their tribal territory. But, in a decision that stunned the Yolngu and their many supporters around the nation, he ruled that they failed to show that they had a system of land ownership. In his judgment, Blackburn said:

> I think that property, in its many forms, generally implies the right to use or enjoy, the right to exclude others, and the right to alienate. I do not say that

all these rights must co-exist before there can be a proprietary interest, or deny that each of them may be the subject to qualifications. But by this standard I can not characterize the relationship of the clan to the land as proprietary.

It makes little sense to say that a clan has the right to use or enjoy the land. Its members have a right, and so do members of other clans, to use and enjoy the land of their own clan and other land also . . . That the clan has a duty to the land – to care for it – is another matter. This is not without parallels in our law, which sometimes imposes duties of such kind on a proprietor. But this resemblance is not, or at any rate is only in a very slight degree, an indication of a proprietary interest . . .

In my opinion, therefore, there is so little resemblance between property as our law, or what I know of any other law, understands that term, and the claims of the plaintiffs for their clans, that I must hold that these claims are not in the nature of proprietary interests. (in Howitt, 1992, pp. 16–17)

So, by using a narrow, European, interpretation of what land 'ownership' means, Blackburn found a way of upholding the doctrine of *terra nullius* (i.e. the fiction that Australia was essentially unoccupied when the colonists arrived). He failed to acknowledge that the Yolngu had their own system of law and his criteria made it impossible for their claim to succeed.

Blackburn's decision upheld a tradition established by Lord Watson of the British Privy Council in 1889, when he ruled against an Aboriginal land claim on the grounds that Australia had been 'practically uninhabited' in 1788 (Reynolds, 1996). Watson, in turn, had echoed the conjecture of Joseph Banks, who initiated the concept of Australia as *terra nullius* when he wrote in his journal of the voyage of the *Endeavour*:

We saw only the sea coast: what the immense tract of inland country may produce is totally unknown: we may have liberty to conjecture however that they are totally uninhabited. (in Reynolds, 1996, p. 18)

As Henry Reynolds has pointed out, the colonists quickly came to realise that Banks' conjecture was erroneous. Yet the fiction of an empty land remained as the basis of the system of land tenure. Unlike Watson, Blackburn was presented with detailed evidence as to the Aboriginal system of land tenure. Yet he found a way to sustain the myth of *terra nullius*. It is interesting to note that Blackburn's central argument about proprietorship versus occupation had been considered by another figure in authority – South Australian Governor Gawler – 130 years earlier. But, as the following quote from a letter written by Gawler in 1840 indicates, he reached the opposite conclusion:

The natives here . . . [have] very distinct and well defined proprietary rights. These rights afford them protection from other tribes and bodily support – they hunt the game upon, catch the fish in and eat the roots of their own districts just as much as the English gentleman kills the deer and sheep upon or the fish in his private park. The property is equally positive and well defined. (in Reynolds, 1996, p. 27)

When the Yirrkala case was being heard in May 1970, the young anthropologist Nancy Williams travelled to Darwin to observe the proceedings. She later wrote of her sense of frustration as she watched old Yolngu men and women patiently explaining, through interpreters, their systems of land management, while lawyers tried to reinterpret it all into the language of the foreign, white, law. In 1986 she wrote:

The Yolngu leaders perceived the court situation less in adversary terms than as a setting where their role was to assist the court to learn about their ownership of land. They saw an opportunity 'to explain', and explanation in their terms involved 'to demonstrate': they revealed to the judge some of their *rangga*, sacred objects which their counsel described as title deeds to land. They attempted to explain their ownership of land in terms they thought the English-speaking court could understand. For the Yolngu the situation was marked by explanation that would result in understanding . . . Having seen the court situation as analogous to traditional meetings where they expected explanation and persuasion to lead to the expression of consensus, the Yolngu leaders were unprepared for a situation in which Europeans explain only enough to 'win'. (in Howitt, 1992, p. 8)

Williams was very impressed with the work of the two translators. One of them, Galarrwuy Yunupingu, would go on to become the long-time chairman of the powerful Northern Land Council. While fluent in English, he was also fluent in the Yolngu understandings of the land. When he became the NLC chairman in 1976 he explained that he had learnt much about the land from his father Mungurrawuy Yunupingu, who had been one of the initiators of the 1963 bark petition. In a letter, he wrote:

When I was 16 years old my father taught me to sing some of the songs that talk about the land. He told me . . . [they] are the history of the Gumatj [a Yolngu clan] people, which talks about us being one with nature . . .

One day, I went fishing with Dad. As I walked along behind him I was dragging my spear on the beach which was leaving a long line behind me. He told me to stop doing that.

He continued telling me that if I made a mark, or dig, with no reason at all, I've been hurting the bones of the traditional people of that land. We must only dig and make marks on the ground when we perform or gather food.

> The land is my backbone ... I only stand straight, happy, proud and not
> ashamed about my colour because I still have land. I can dance, paint, create
> and sing as my ancestors did before me. (in Roberts, 1978)

By sitting in the Supreme Court, Williams got the feeling that she was
observing an important historical moment. She described the scene in
courtroom in the following way:

> Court 1 in the Supreme Court building in Darwin was almost always filled
> with spectators or observers having various interests in the case during the
> two weeks of the hearing. Public servants, with responsibility for some
> aspect of Aboriginal affairs or just on their lunch breaks, missionaries, and
> anthropologists, were among those I identified in court. Press coverage
> implied interest throughout a wide spectrum of the Australian population
> and suggested that the outcome of the case would be regarded as important.

The outcome of the case was certainly important but it was not what
Williams hoped for. She emerged from the experience with intense frustra-
tions and decided to dedicate herself to the task of educating white
Australia about the Aboriginal understanding of land and the resulting
book – *The Yolngu and their Land: A System of Land Tenure and the Fight for its
Recognition*, published in 1986, is a classic in its field.

THE CANBERRA TENT EMBASSY

The outcome of the Yirrkala court case was not only important to Williams
and her career. It became the subject of public debate around the country
over the following year. In January 1972, on the eve of Australia Day, Prime
Minister William McMahon responded to the public debate by announcing
that the Yolngu would receive some royalty payments from the Nabalco
mine, but he argued that the government was bound to accept Blackburn's
ruling that there is no such thing as Aboriginal land rights (see Goodall,
1996, p. 338). That evening, three young Sydney-based koori men – Billy
Craigie, Tony Koorie and Michael Anderson – decided to act on a suggestion that had been made by the writer Kevin Gilbert. They packed up a tent
and some blankets and, in a car driven by the *Tribune* (the newspaper of the
Communist Party of Australia) photographer Noel Hazard, they travelled
to Canberra and set up an Aboriginal tent 'embassy' on the lawns outside
Parliament House. When the press arrived in the morning they said they
had established the embassy because McMahon's speech the day before
had made them feel like foreigners in their own land.

The tent embassy captured the imagination of the nation. For six
months it stood as a dramatic symbol of Aboriginal alienation along with
their growing sense of a separate national identity. When the tents were

finally torn down by a big party of police (following the adoption of special legislation to legalise this act), hundreds of people gathered in protest. Already there had been thousands of visitors to the embassy and the now-popular Aboriginal flag had been launched. A film, *Ningla-a-na* (translated as 'hungry for our land') captured the experience of the imaginative protest.

The tent embassy was also chosen as the site where the then Opposition leader in Canberra, Gough Whitlam, announced that the ALP would go into the coming federal election with a policy in favour of creating legislation to legalise Aboriginal land rights. Whitlam and his team was duly elected in December of that year and they established a special commission (headed by Justice Woodward) to develop proposals for land rights legislation in areas that came directly under Commonwealth jurisdiction (most notably the Northern Territory). Although the 1967 referendum on citizenship rights for indigenous Australians had clearly given the Commonwealth the power to override the states in regard to Aboriginal affairs, the Whitlam government, like subsequent federal governments, was unwilling to override the states on matters of land ownership.

The Whitlam government was brought down before the Woodward Commission had completed its work. The incoming Fraser government received the report and, despite watering down some recommendations in regard to mining and the power of government veto over Aboriginal land claims, enacted the appropriate legislation. A commission, headed by Justice Toohey, was set up to hear Aboriginal land claims and two land councils – Northern and Central – were set up to provide necessary resources to Aboriginal communities wanting to make a claim under the act.

The 1976 land rights legislation was a bittersweet victory for the Yolngu. They could now claim ownership of their traditional lands, but they had not been able to stop the mine and the smelter. The old people also lost their struggle to ban the sale of alcohol in the new mining town of Nhulunbuy (built on the Nabalco lease), which they felt would become a corrupting influence on the young. When they lost their case in the Supreme Court, the Yolngu got locked into an agreement on the payment of royalties which was far inferior to mining royalty agreements reached elsewhere in the Territory after the adoption of the 1976 legislation. Yolngu who lived through the struggles of the 1960s and early 1970s remember with sadness their defeats at the hands of the church, the Supreme Court and the mining company.

THE MABO DECISION

But there would not have been land rights legislation in 1976 if not for the efforts of the Yolngu and Gurindji. When Eddie Koiki Mabo began his long and lonely legal struggle for recognition of native title in the early 1980s he was undoubtedly helped by the controversy that surrounded Blackburn's decision. Mabo pursued his dream from Townsville to Brisbane to Canberra over a period of 10 years and he died before the decision was handed down on June 3, 1992. Mabo would also have lost his case if the decision had been in the hands of only one of the High Court judges, Justice Dawson, who ruled that any native claim to land rights had been properly extinguished by the act of colonisation (or 'annexation') (Reynolds, 1996). Fortunately, the other six judges on the bench disagreed with Dawson and upheld Mabo's claim. Reflecting a degree of discomfort with earlier judgments (such as that of Blackburn), one of the majority, Justice Brennan, observed that:

> The fiction by which the rights and interests of indigenous inhabitants in land were treated as non-existent was justified by a policy which has no place in the contemporary law of this country. (Reynolds, 1996)

Speaking at the inaugural Yarramundi Lecture at the University of Western Sydney – Hawkesbury in August 1997, Cape York Aboriginal Land Council leader Noel Pearson suggested that the anniversary of the Mabo decision should be regularly celebrated by all Australians. Arguing that Mabo's stunning triumph was much more than a moral victory of indigenous Australians, Pearson said:

> Mabo was the opportunity Australia needed to put our colonial past behind us. [It] was the start of Australia's redemption.

Of course, it takes more than a single decision of the High Court – no matter how momentous – to overcome the legacy of 200 years of injustice. Both the Mabo decision and the consequent and related Wik decision of the High Court became the subject of intense political manoeuvring in the wake of the election of a Liberal–National Party coalition government in 1996. Although the High Court found in the Wik case that native title and other forms of title could coexist, the Howard government moved to extinguish native title in areas where pastoral and mining leases were operating. Consequently, indigenous Australian spokespeople said an historic opportunity had been squandered for short-term political expediency. Nevertheless, the historic legacy of the Mabo decision could not be overturned completely and the Howard government's attack on the spirit of Mabo was widely condemned. As a former executive director of the National Farmers' Federation said during the debate on the Howard government's amendments to the Native Title Act, the issues surrounding native title issue will never go away because as long as there is indigenous Australian culture there will be a land agenda. Aboriginal people conceive of the land as mother, Rick Farley said, and a 'mother's love is big enough for everyone'.[3]

THE GURINDJI AND THE YOLNGU

In different ways, and for different reasons, the Gurindji at Daguragu and the Yolngu at Yirrkala emerged in the 1960s as the leaders of the renewed struggle for land rights for indigenous Australians. Both communities went through long and bitter struggles for their rights and, in the process, developed strong leadership and a strong sense of identity. Both communities live in the northern half of the Northern Territory. They are linked by both their Aboriginality and their recent history. But they are also as different as chalk and cheese. It is important to recognise that Aboriginality embraces a wealth of diversity.

Long before the white colonists arrived in their homeland, the Yolngu were accustomed to having visitors from afar. A Dutch ship, the *Arnhem*, first visited the area in 1572 and when the English explorer Matthew Flinders arrived at Elcho Island, off the coast of Arnhem Land, in 1805 (during his trip around Australia in the *Investigator*) he heard white people referred to as 'balanda' – an Indonesian term corrupted from the word 'Hollander'. The Dutch and English discovered that Macassan traders, from the island now known as Sulawesi, had been visiting Arnhem Land on an annual basis for hundreds of years. The Macassans were particularly interested in gathering trepang – sea slug – which was a delicacy in their homeland but of little interest to the Aborigines. At Gove, the Macassans left behind some artefacts, tamarind trees and a few words. They took back trepang, artefacts and, perhaps, some wives. Recently some Yolngu students studying at Bachelor College (near Darwin) visited Sulawesi and they came across a very old Yolngu woman who had gone back with the Macassans when she was young. Some of the students learnt, to their amazement, that the woman was related to them.

However, the English colonists did not make serious inroads into Arnhem Land until well into the twentieth century. There were attempts to establish cattle stations in south-east Arnhem Land in the 1920s but these were not successful and in 1931 the Commonwealth government declared Arnhem Land a reserve 'for the exclusive use and benefit of the Aborigines'. On several expeditions into Arnhem Land in the 1920s and 1930s, Donald Thomson was able to compile a brilliant photographic record of people living their traditional lifestyle. Thomson was followed by a virtual caravan of anthropologists and missionaries and the Methodist Mission at Yirrkala was established in 1934. When the miners arrived in early 1960s, the Yolngu were not used to dealing with foreigners who had a commercial interest in their land.

By contrast, the Gurindji did not sight white people until the middle of the nineteenth century. In 1839, the *HMS Beagle* – the ship made famous for carrying Charles Darwin to the Galapagos Islands in 1835 – sailed into the mouth of the river that runs through the heart of Gurindji country and named it after the newly crowned queen of England: Victoria. But it was not until 1855–56 that a party of men, led by the Gregory brothers, ventured into the area on foot, following the Victoria River upstream. They were the first white men to walk into Gurindji territory. The Gregory expedition excited interest in the Victoria River area as a site for cattle stations but it was not until 1879, after a couple of false starts, that the first cattle station, Victoria River Downs, was established. The following year, Nathaniel Buchanan took up the lease on the first Wave Hill property. The first cattle arrived in the area in 1883.

The Gurindji initially resisted the cattle influx and cattle killings and reprisals took place from around 1886 until the 1920s (McConvell and Hagen, 1981). However, the massacre that began at Coniston in 1928 probably broke the back of this resistance and thereafter more Gurindji sought work in the cattle stations. By 1966, when the Gurindji strike began, they had become the backbone of the regional industry. The Gurindji came to identify themselves primarily as stockmen. When a leader of the Daguragu community, Pincher Numiari, stayed with the author during a visit to Sydney in 1976 he could not be parted from his bush hat. On a visit to the beaches on Sydney's northern peninsula, he laughed loudly when we passed a Riding School where people would pay money to get on a horse and he wandered out into the foreign territory of a seaside beach in his hat and riding boots.

When they began their struggle in 1966 the Gurindji had to extract themselves from the cattle industry to claim their lost rights, whereas the Yolngu were trying to stop the miners making new encroachments onto their land.

The Yolngu word 'dhimurru' refers to the 'cheeky' wind that sometimes blows up unexpectedly during the Dry Season to create a change of weather. For this reason, it is the name they used when they established a new organisation in 1993 that would train Aboriginal rangers to drop in on Aboriginal sites of significance in the Yolngu lands to make sure that white people from the mining town Nhulunbuy were not visiting these sites without permission of the Aboriginal Land Council. Sitting in the Dhimmurru Office, Djawa Yunupingu, younger brother of Galarrwuy and Mandawuy, and Mandaka (Sam) Marika, son of Yirrakala leader Roy Markika, talked about how proud they were of the achievements of the older men who were now mostly dead. 'We have to carry on what those old men started', Djawa said firmly. Behind the two young men was a huge photo of Roy Marika – bearing an uncanny resemblance to his son. The presence of this photo broke a tribal tradition to never show the image or mention the name of the dead. But it is clear that the leaders of the campaign to stop the mine are held in particular reverence by the younger generation of Yolngu.

Roy Marika was the younger brother of Mawalan Marika who, together with Mungurrawuy Yunupingu, first raised the alarm about the activities of miners on Yolngu land in 1963. Unfortunately, Mawalan died before the case went to court and the Marika clan came to be represented by Roy Marika and his brother Wandjuk Marika (whom Edgar Wells remembered as the one with a typewriter and a determination to document their claim). The Marikas are from the Rirratjingu clan, who are traditional owners of the site where Yirrkala mission was built. Their land was badly affected by bauxite mining. Mungurrawuy Yunupingu was a leader of the Gumatj clan, whose traditional area included the beautiful Melville Bay where the aluminium smelter was built. Gumatj and Rirratjingu, now separated by the town of Nhulunbuy, were neighbours and, by tradition, commonly intermarried.

Speaking at her home at Yirrkala in July 1997, Banduk Marika, a daughter of Mawalan Marika, explained that women also played leading roles in the struggle against Nabalco and the betrayal of the church:

> There came to be a lot of focus on the old man [Roy Marika] because the church always wanted to deal with a male leader. But women also signed the petition. Women kept their power even though the church only dealt with men. In our culture men and women are partners. There are separate things to do with men's business and women's business. For public ceremony, women always have a role, always decide part of the ceremony. I know for a fact, from my father's aunties, and from what I saw with my own eyes, that my father always negotiated with his aunties before doing anything. Women are the strong voice in the community.

Marika was talking while sitting on the top step of her house overlooking the aqua-blue sea and a long, white beach stretching beyond the 'old men's camp' to an even more distant headland. Her young daughter Ruby came in and out of the conversation. A pet lorikeet nibbled at the shining jewellery of the visitors and lay on its back to have its tummy scratched. The high house with walls that could be opened up to allow cooling sea breezes to pass through was designed by famous Australian 'ecoarchitect' Glenn Murcutt and Marika hoped that it would be the first of a number of Murcutt-designed houses for Yirrkala.

'My people have always lived near this beach', Marika explained with a sweeping gesture. 'Before the mission, Yirrkala was a two-clan camp – Rirratjingu and Gumatj. After the mission, this grew to 30 clans and the mixing up of clans caused a lot of problems.'

Pointing towards the old mission house that lay behind the beach on the other side of a beautiful freshwater lagoon, Marika said angrily that 'they' used to use a public address system to denounce people for participating in singing and dancing whenever they heard it going on.

They stripped everything from the old people. We had healing people who were made to stand in front of the others while the things they used for healing were taken from them. Ladies were stripped of the things they used for ceremonies and all of this was burnt while people watched.

We haven't got old people now. They've died of leprosy, diet problems or alcohol. We don't have 80-year-olds in our community. My father died in 1967 before the court case began. It makes you feel sad. Like Eddie Mabo, he died before the court made its decision. The courts take too long and people die without knowing what happened.

But we are proud that our old people took action against the government at that time to have Aboriginal people recognised as land owners.

Pride in what the 'old people' did is something you hear often at Yirrkala. It was the same at Daguragu in 1975 when the author visited with a party of land rights supporters. On that occasion, our party was met at the closed gate of the settlement by a young man who collected our details. He went away to consult the elders and returned with Pincher Numiari, a leader of the strike. Having established that we were supporters from 'down south' Numiari made us feel very welcome and asked his son, Charlie Pincher, to be our guide for the next week.

Banduk Marika with daughter Ruby and their pet lorikeet on the steps of her Glenn Murcutt-designed house at Yirrkala in 1997 (Martin Mulligan)

As well as pride in the old people, the Gurindji and Yolngu share a cheeky sense of humour and a ready laugh. A good example of the Gurindji sense of humour is recorded in Frank Hardy's book. Visiting the 'new' camp at Daguragu for the first time, Hardy was taken on a tour by Vincent Lingiari, during which the latter pointed out new buildings and even a brand new sign reading 'Gurindji Mining Lease and Cattle Station'. While they were doing the rounds Hardy and Lingiari came across a man who Lingiari introduced as Rook:

> 'This bin Rook,' Lingiari said. 'He bin Jesus man. Got 'em Bible and grama-phone record of Jesus songs. I got ebrything new. I bin reckon. I got 'em fencing contractor, got 'em own teachers, got 'em own sign, got own Jesus man.'

> 'But I thought you said you didn't want Jesus men any more,' Hardy replied.

> 'Well,' said Lingiari with a grin, 'If Welfare come there and say: "Why you got no Jesus man?" I say: "We got we own Gurindji Jesus man."' (Hardy, 1968)

The Gurindji sense of humour even got turned into a popular song by Northern Territory singer/songwriter Ted Egan. Egan picked up on a conversation between Hardy and Pincher Numiari about the possibility of the Gurindji buying out the lease on their tribal lands. During this conversation, Numiari had noted wryly that 'We work for Bestey for plour, chugar and tea. Maybe that proper fee [to pay out the lease].' Out of this came the song *Gurindji Blues*, which went:

> Poor bugga me, Gurindji
> My name Vincent Lingiari
> Me talk allabout Gurindji
> Daguragu place for we
> Home for we Gurindji.
> But poor bugga blackfella this country
> Gov'ment boss him talk long we
> Build you house with electricity
> But at Wave Hill for can't you see
> Wattie Creek belong to Lord Vestey.
> O poor bugga me.
> Poor bugga me, Gurindji
> Me bin sit down this country
> Long time before Lord Vestey
> Allabout land belongin' to me
> O poor bugga me.

Poor bugga blackfella this country
Long time work no wages we
Work for good old Lord Vestey
Little bit plour, chugar and tea
For the Gurindji from Lord Vestey
O poor bugga me.

Poor bugga blackfella, Gurindji
S'pose we buyin' back country
What you reckon proper fee
Might be plour, chugar and tea
From the Gurindji to Lord Vestey
O poor bugga me.

This song, recorded by Egan at Daguragu with a full chorus of Gurindji singers, became a national hit.

Gurindji humour is often self-deprecating, but it can also be used to ridicule the 'strange' ideas of white people. In their 1981 report supporting the Gurindji land claim, anthropologists McConvell and Hagen report that the history of the Wave Hill name is 'a source of some amusement' to the Gurindji. The name was first used by Nathaniel Buchanan when he built the first Wave Hill homestead near the banks of the Victoria River in 1883. The house was endangered in every wet season and, eventually, the 1924 floods washed it away. A new house was built on higher ground, but the homestead was moved twice more – the last time in 1967 after the Gurindji set up camp at Daguragu. What amuses the Gurindji is that every time they moved the homestead the white people took the name Wave Hill with them. As far as the Gurindji are concerned names belong to particular places and cannot be applied to any other. The ignorance that led Buchanan to build on the banks of a flooding river is seen as being repeated in the effort to carry a place from one location to another. It confirms the impression that the whites have little connection with the land.

Although this story might be told with a grin it highlights the enormous difference in the way that Aborigines and white Australians relate to specific places. For Aborigines who have lived in their 'country' for countless generations, each place has a set of stories that relate to its natural and human history. They see white people as having shallow and transitory relationships with such places. David Abram (1997) has made the point that indigenous people around the world have stories that are entirely site-specific. He develops the theory that it was the emergence of written languages, making use of phonetic alphabets, which created the possibility of separating stories from particular places and landscapes. The result, he argues, has been the abstraction of humans from the natural world; the emergence of abstracted and homogeneous perspectives of space and linear concepts of time. Literate societies have gained a capacity to generate far-reaching discourses, but they have lost a capacity to have strong, sensuous relationships with 'more-than-human' nature. To Aboriginal people, western concepts of land ownership appear to be arbitrary and reckless. It is easy to imagine how affronted the Yolngu people felt when Justice Blackburn ruled that they had not demonstrated an adequate claim to land ownership.

As mentioned earlier, Nancy Williams began research for her book *The Yolngu and their Land: A System of Land Tenure and the Fight for its Recognition* after watching the proceedings in the Darwin Supreme Court. She wanted more white Australians to understand that not only do the Yolngu have a clear system of land tenure but that it embraces some important principles of land and resource management.

All Yolngu, Williams explained, have a special relationship with the places where they are born and where they expect to die. Because they live in the places where their ancestors have lived for many thousands of years, the recent history of those places soon merges into ancient stories that incorporate creation myths. The past merges into the realm of an eternal spirit world (sometimes translated as 'dreamworld') from which the future also emerges. Living in close contact with natural cycles – day/night, seasons, etc. – people like the Yolngu have a cyclical concept of time instead of the linear segregation of past, present and future in western worldviews.

In Yolngu cosmology, spirits from the other world choose to animate new human life in the body of a woman at a particular place. That same spirit will return to the spirit world through the soil at the place where that person's bones are buried. This means that Yolngu have a strong sense of belonging to their mother's country. But they also have patrilineal connections with particular sites as well. Traditionally, a wife would join the community of her husband and so the 'baby spirit' that enters the world through the mother's body does so in the father's country. The role of the father then is to care for this spirit in its human form.

According to an Aboriginal sense of place, every place has its stories. Williams reports that she once travelled with a Yirrkala man to a place he had never visited before and the man's comment was: 'We don't know who owns this land; we haven't heard the story. But we know that somebody owns it.'

Flying across Arnhem Land on a visit to Yirrkala in 1997, the author (Martin Mulligan) was struck by the same thought. From the air, the landscapes appeared to be uninhabited (except for an occasional road or airstrip). The human story was not as obvious as it is in south-east Australia where the landscapes are heavily modified (even as seen from the air). However, on the ground, you meet the people who can tell you about the landscapes you are in and the humanscapes start to become more visible. It helps to be given a map of Arnhem Land with the names of clan groups scattered across the territory because that looks more like a system of land tenure we white people are accustomed to. For Aborigines the stories of particular sites are more important. We all look at landscapes with different eyes and see different stories (natural and human) embedded there. But we are all looking at storied landscapes.

Speaking at a conference of the New South Wales Association for Environmental Education in September 1997, Allan Fox told an interesting story of his experience in working with 'Big' Bill Neidjie (of the Gagadju people) in the establishment of Kakadu National Park. Working for the National Parks and Wildlife Service on establishing the case for the new national park, one of Fox's duties was to arrange for visiting delegations of politicians and Canberra bureaucrats to be taken to sites of Aboriginal significance by Neidjie. After a while Fox noticed that the old man was telling different stories to different parties even when they visited the same location and so, concerned that this inconsistency would damage the case for the national park, he confronted Neidjie about it. Calmly, the old man replied that every place has many stories. For some sites he might know 15 or 16 different stories and he always has to decide which one is appropriate on a particular day for a particular audience. Each story has its purpose and the existence of multiple, and multi-layered, stories reflects the fact that Aborigines have different connections with the sites concerned (e.g. matrilineal and patrilineal). The story-keepers have responsibilities for the sites and the most detailed stories belong to those with the greatest authority and responsibility.

Stories related to particular sites communicate a system of law that has its roots in antiquity. Ancient landscapes and ancient stories give this law a sense of permanence that Aboriginal people find missing in white Australian law. An old man of the Yarralin people in the Victoria River district made this point to anthropologist Deborah Bird Rose when he said:

> You see that hill over there? Blackfellow Law like that hill. It never changes. Whitefellow law goes this way, that way, all the time changing. Blackfellow Law different. It never changes . . . The Law is in the ground.[4]

The sense of permanence can be an illusion. Stories change and rules of behaviour change. But this perspective helps white Australians realise that they were initially seen by indigenous Australians as flighty and disconnected; not the sort of people with whom to share the secrets of the land.

Multi-layered stories of landscapes reflect the fact that Aboriginal people have a range of overlapping identities. They belong to their father's clan, mother's clan and spouse's clan. People have connections with places where they live and with other places related to other identities. In her study of the Yolngu, Williams found that a patrilineal clan living in a particular area has a sense of owning the land. But they cannot make important decisions regarding land use or the holding of important ceremonies without consulting those related to the land-owning clan matrilineally, but living elsewhere. At the same time a man might be allocated a portion of his mother's land to 'look after' provided others who may claim the same right approve.

Williams also found that many Yolngu clans have 'ownership' of two non-contiguous areas of land; one on the coast and one inland. They would move between these areas depending on the seasons and different senior men would have particular responsibility for each area. Boundaries between clan areas are designated by reference to natural features, such as hills or rivers, and anyone wishing to visit an area would seek (and expect to get) the approval of the land owners. But as well as the boundaries that separate there are other lines – what Williams calls 'myth tracks' but which have also been called 'songlines' elsewhere – that connect sites through creation stories about the journeys of spirit-beings. These sites – known to the Yolngu as *ringgitj* – 'belong' to all the Yolngu although there are rules regarding access and care for the sites. The 'myth tracks' connect and unify rather than separate.

In developing his theory (mentioned earlier) about how literate societies have abstracted themselves from sensuous relationships with the 'more-than-human' world, David Abram discussed the way that Aboriginal people in Australia have a capacity to 'sing up' stories embedded in the land (Abram, 1997, pp. 163–177). He cites an amusing anecdote told by the American environmental poet Gary Snyder after a visit to Central Australia in 1981 (p. 173). On one occasion during this visit Snyder found himself on the back of a table-top truck with a Pintupi elder named Jimmy Tjungurrayi as they drove into Pintupi land. At one moment, Tjungurrayi began to rattle off, at great pace, a story of an encounter between some Wallaby Men and Lizard Girls at a mountain they could see in the distance. As soon as he had finished that story he launched into another one involving another landmark and then another, and another. After about half-an-hour of this, Snyder, his head spinning, suddenly realised that what he was getting was a speeded-up version of what was designed to be told over several days of walking at leisurely pace through the same landscapes. Abram points out that stories that could be told as people walked through the landscapes of their ancestors were partly designed to guide them to hidden waterholes or sources of food – a kind of oral survival map. However, they also included strong messages about how those places should be cared for. Such stories were often retained in the form of rhythmical chants or songs that made them easier to remember and pass on – hence the idea of 'singing up' the stories. Citing evidence from Central Australia, Abram points out that myth, or dreaming, tracks – i.e. the string of sites marking the creation journeys of spirit-beings from the dreamtime – often traversed vast distances and intersected with each other. People belonging to particular clans had the responsibility for keeping the stories of sites within their area, but other people had access to those sites and could exchange their stories of sites further along the same track. Songs and chants were also easier to share with people from other language groups and this is why the word 'songlines' is sometimes used to refer to dreaming tracks.

Overlapping and shared connections to places makes a concept of exclusive land ownership irrelevant. Aboriginal systems of land tenure start with a concept of belonging rather than ownership, although their system of responsibility *can* be related to a European concept of ownership. Their complex relationships to land are difficult to translate into English concepts of land tenure because, by comparison, the latter is rather one-dimensional and exclusive.

Rose (1992) has made the point that the English language is the product of a particular cultural history and, as such, it cannot articulate Aboriginal concepts of land tenure. Simultaneously, she explained, '(d)reaming strings [the term Rose uses for 'myth tracks'] differentiate, connect and cross-cut'. The Ngarinmen, she explained, have one word for 'country' – *ngurra* – but in different contexts it can mean a person's camp, a family area, a clan area, a language region, a geographic region, or an ecological zone. Ecologically, the Ngarinmen distinguish areas such as 'saltwater country' (estuaries) from the area of 'too much fresh water' (flood plains). But they have other terms to describe separate rivers and their catchments. Perhaps the concept of ecosystems has enabled white Australians to develop a similar understanding of overlapping landscapes and what they support but the language of ecologists has only recently moved into popular use. Rose has suggested that Aborigines pioneered a 'dreaming ecology'.

In a later work Rose (1996) makes the point that Aboriginal people prefer the English word 'country' to 'landscape' because it denotes a sense of belonging rather than a separation or distance. She writes:

> Country in Aboriginal English is not only a common noun but also a proper noun. People talk about country in the same way they would talk about a person: they speak to country, sing to country, visit country, worry about country, feel sorry for country, and long for country. People say that country knows, hears, smells, takes notice, takes care, is sorry or happy. Country is not a generalised or undifferentiated type of place, such as one might indicate with terms like 'spending a day in the country' or 'going up the country'. Rather, country is a living entity with a yesterday, today and tomorrow, with a consciousness, and a will toward life. Because of this richness, country is home, and peace; nourishment for body, mind, and spirit; heart's ease. (p. 7)

This captures well a relationship to landscapes that is affective and caring. It is a two-way relationship; the land will provide physical and spiritual nourishment, but it also requires care and attention. People have well-defined roles to play in caring for country. However, this is far from the western concept of 'proprietorship' because the role of humans is not privileged. Rose goes on to explain:

> A 'healthy' or 'good' country is one in which all the elements do their work. They all nourish each other because there is no site, no position, from which the interest of one can be disengaged from the interests of others *in the long term*. Self-interest and the interest of all of the other living components of country (the self-interest of kangaroos, barramundi, eels and so on), cannot exist independently of each other *in the long term*.

> The interdependence of all life within country constitutes a hard but essential lesson – those who destroy their country ultimately destroy themselves. (p. 10)

Another difference is that sea-oriented people, like the Yolngu, did not draw a sharp distinction between responsibility for land and sea. As Banduk Marika told the author:

> We have just as many stories about the sea as we have about the land – stories about the winds or the way the clouds come. We know about particular places in the sea and we think about those places a lot.

As a result, the Yolngu system of 'land' tenure extended into care for areas in the adjacent seas. The stories and ceremonies of the coastal people in Arnhem Land were gathered as evidence to support a 'sea claim' launched by the Northern Land Council in 1997, amid howls of protest from the Northern Territory government and the commercial fishing industry.

In Aboriginal law the right to own an area of land or sea is accompanied by clear responsibilities for conservation or 'resource management'. For example, when Yolngu women dig for yams they leave a portion of the tuber in the ground so that it will regenerate. Banduk Marika explained the ways in which conservation principles and rules are passed on:

> We have to observe different rules when we catch mud crabs, long-necked turtles, turtle from the sea, etc. There are rules about how you prepare each of these for cooking because we have to pay our respect for their life so we can continue our life. When we take something out of the soil we always put the soil back and we teach the young people to respect these rules. If you look after nature, it will support you.

Aboriginal systems of land care and nature conservation have evolved with traditional ways of living in particular places over many thousands of years. While this has generated very deep connections between people and particular places, it has not created understandings that can be easily translated into different places and different cultures. The Gurindji and Yolngu showed that profound disruptions – from massacres to missions – could not extinguish their sense of belonging. But what will be the impact on new generations of 'global culture' in the form of television and the arrival of more and more tourists? While Aboriginal communities, even in the most remote areas, are going through profound cultural changes, it is hardly likely that white Australian communities will take up traditional Aboriginal lifestyles. But it is not too late for white Australians to come to the realisation that, compared to indigenous Australians, our roots in this land are shallow and unreliable. What we can learn from the Aboriginal land rights movement is that we need to find a new love for the land and all its stories.

There can be little doubt that over the last 20 years white Australians have begun to wake to this realisation. A much stronger interest in Aboriginal stories, music and art has emerged. While the classic 'dot paintings' of Central Australia and 'x-ray' paintings and sculptures of Arnhem Land have been expropriated as new Australian icons, there is finally a recognition that such images can help us to visualise the stories embedded in the landscapes and in other living things (see chapter 3).

It is interesting to speculate about why the Yirrkala 'bark petition' of 1963 has gained something of a mythic status. According to Galarrwuy Yunupingu,[5] the painted images surrounding the names on the petition are decorative only. But the form of the petition evoked the image of Aboriginal Australia choosing to communicate with white Australia in its own way. The image of the old people painting their message to the government in Canberra caught the attention of the media and the petition is now on display in a glass case in the national Parliament House. Even those who know little about the Yolngu and their story have heard mention of the 'bark petition'. According to Yunupingu, the Yolngu have been surprised by the symbolic success of the petition. It is a legend worth preserving. Yunupingu told the authors that when students at the Nhulunbuy school recently decided to complain about some rather trivial matters of school policy, they decided to present their demands to the school principal in the form of . . . a bark petition!

Yirrkala is a beautiful and seemingly peaceful place. But it is certainly not without its problems. Ever since Nabalco started paying royalties, there have been new divisions between clans over the payments. These divisions came to a head in 1995 when Gatjil Djerkura – then head of an Aboriginal and Torres Strait Islander Commission sub-committee fostering business enterprises in indigenous Australian communities – announced his intention to set up a breakaway Ringgitj Land Council for East Arnhem Land, independent of the Northern Land Council headed by Gumatj leader Galarrwuy Yunupingu. The federal minister for Aboriginal affairs of the time, Robert Tickner, commissioned anthropologist David Martin to assess the level of support for the breakaway council. Martin's report found that the proposal for a new land council had little serious support but it did reflect some tensions between clans about royalty payments.[6]

Furthermore, while Yolngu leaders have pointed out that their claim to land rights is based on a long-standing association with special places, the people of Yirrkala have not always respected their immediate environment. On his visit to Yirrkala in 1997, the author was surprised to see old car tyres and a bicycle dumped in the lagoon and a large pile of litter on one street corner. Banduk Marika suggested two main reasons for the apparent lack of respect. In the first place, people from 30 different clans were forced or cajoled into moving to Yirrkala once the mission was set up in 1934. For most of them it was not their own country and the missionaries had usurped the authority of those who were the traditional owners. Secondly, there has been a growing problem with alcohol abuse since nearby Nhulunbuy was built in the late 1960s. According to Marika, this has robbed people of their ability to care for themselves, let alone their environment. However, Banduk and her sister Raymattja, have established a landcare program that is starting to clean up the litter and plant more trees. Systems are being developed for dealing with rubbish and for replacing weeds with indigenous plants. Marika stressed that her aim is to involve young people in this project so that this might develop an ethic of care within the next generation. When we visited the landcare office, she encouraged us to interview some of the young workers and volunteers. 'Come on, tell 'em what you think about this place in your own words', she goaded the shy youngsters, who gradually warmed to the task.

'We have learnt respect for the soil,' one of them said, 'because we know it contains the bones of our ancestors. When we dig up the soil we feel we are digging up our ancestors, so we always treat the soil with respect.' Another volunteer explained that she had learnt many ways of telling the seasons. 'When the wattle flowers, we know it is a good time to eat seafood', she said.

Marika pointed to one young man who had been to Batchelor College near Darwin to learn horticulture:

> A lot of young people leave here at some time but they all come back eventually. Some of us might travel and live outside for a while. But we don't really like to live anywhere else.

Some of the tensions at Yirrkala have diminished with the growth of the 'homelands movement'. With the adoption of land rights legislation in

1976, Aboriginal people all over the Northern Territory realised they were no longer obliged to live in the missions and settlements set up by churches and governments in the early part of the century. From Maningrida in Arnhem Land to Hermannsburg near Alice Springs, clans started moving out of the settlements to return to their 'homelands' – or 'outstations' as they were called in many places. In the case of Yirrkala, 26 clan groups made the decision to quit the mission and move back to their homelands. They were supported in this move by the Yolngu Homelands Association, which had the resources to construct necessary infrastructure and provide education services for the children. The education program focuses on supporting local teachers who work in the language of the community and it ensures that those who want to, get the chance to move into schools and colleges outside the community. According to Banduk Marika, the homelands movement has not only reduced tensions at Yirrkala, it has also given young Yolngu a stronger sense of their real identity and, consequently, a better chance of dealing confidently with white Australian cultures.

Yolngu identity grows in the soil where the bones of the ancestors are buried. This point was expressed well in a speech to the National Press Club by Galarrwuy Yunupingu in February 1997. Explaining that he decided against bringing a written speech because he wanted to speak from the heart, Yunupingu went on to say:

> What is the strength of our claim to our rights? It's nothing to do with property ownership. It is nothing to do with running cattle and sheep. When it dries up and it doesn't rain any more, cattle are dying, sheep is dying, there's no life on the land . . . I'm going to pack up my family and I'm going to go and I'll sell that land or give up that land because there's nothing happening on that soil?

> You tell Aboriginal person that a horse has died and the cattle have died and it's dry as a bone. Aboriginal person will not leave that land, ever. That is the true entitlement to that land. Because Aboriginal person treats that land as not only the surface, it is the very bone of the land . . .

> Same thing as mining companies. You dig my country and you make your money and you go laughing all the way to the Swiss bank, but you leave me a hole and the pollution. No thanks, but you take all the goodness out of my land and you leave me nothing but I'm not disappointed. I might be a bit angry with you, the way you treated me and being unfair. But I'm still there nursing the hole and the pollution that you left behind.

> I will not go away from that hole and that pollution because my right is to die in that land. Because the very soil that you have taken, you have taken my bone and you took it to the bank in the Swiss bank. That's my bones and

you ought to pay me some time, because some of my bones are being banked. That's the truth of it.[7]

As Noel Pearson said in the Yarramundi Lecture delivered in August 1997 (referred to earlier), Mabo – the culmination of a 30-year struggle for land rights – represents a new opportunity for 'Australia's redemption'. It offers redemption from the lie of *terra nullius* and the possibility of a reconciliation with our brutal past. It can also offer reconciliation with a 'hostile' land because we can learn how to love this strange and unpredictable country. We are encouraged to confront our own fears and prejudices and develop a new sense of responsibility through a deep sense of belonging to the land that nurtures our existence. It has taken white Australians a very long time to realise that we have much to learn from the indigenous people of this country, and even now that sentiment is still repudiated by a very significant minority. Struggles about our future are still being fought on the basis of our knowledge of the past.

This story began with the Gurindji so it seems appropriate that the last word should go to a Yolngu elder, a signatory of the 'bark petition', Wirilma Mununggurr, who told the author in 1997:

'We still fight – not only for us but for history.'

CHAPTER 10	GREEN POLITICS IN THE WIDE BROWN LAND: *Wilderness Politics and Social Justice Agendas*

INTRODUCTION

As discussed in chapter 6, a 'second wave' of environmentalism took shape in Australia in the late 1960s as the conservation movement began to tackle a broader agenda and adopted more radical campaigning methods. The movement was also enriched by the arrival of organisations that began life in other parts of the world, such as Friends of the Earth and Greenpeace, who brought with them a concern for global issues like the environmental threats of the nuclear industry and the loss of biodiversity on a global scale. However, even as it changed direction, the environmental movement in Australia still carried the legacy of the frontier mentality that fostered the hyperseparation of people and 'the bush'. For much of the movement the focus continued to be on the preservation of 'pristine' wilderness and the pinnacle of this preservationist movement probably came with the campaign to prevent the flooding of the Franklin River in the south-west corner of Tasmania in the late 1970s and early 1980s. While the Franklin campaign can be seen as a continuation of the preservationist movement of old, it also represents a turning point in the way that wilderness issues were taken directly into the political arena and it led to the birth of a green political party, firstly in Tasmania and then nationally.

Ironically, the pinnacle of ecocentric, wilderness politics – the Franklin campaign – also marked a turning point towards a broader green political agenda because the Tasmanian Greens, once elected to Parliament, saw a need to develop policies on a wide range of issues. Although the Franklin campaign helped to get Greens elected to the Tasmanian parliament, the political organisation was built on the legacy created by an earlier formation – the United Tasmania Group (UTG) – which had quite a broad agenda because it was influenced by a number of social movements (e.g. the women's movement and the peace movement) that were prominent in the early 1970s when the UTG was formed. Just as the 'proto-greens' (in the

243

shape of the UTG) were emerging in Tasmania, the New South Wales branch of the Builders Labourers' Federation (BLF), led by the prominent Communist Party member Jack Mundey, was using 'green bans' to prevent urban developments that would eliminate open space, heritage buildings and vibrant working class communities. Although the green bans had emerged rather accidentally out of an alliance between community groups and a trade union promoting a social justice agenda, a conceptual link was forged around the defence of quality of life in the face of mindless developments driven by a greed for quick profits. Although Mundey and his colleagues in the leadership of the New South Wales BLF were overthrown by a 'coup' engineered by a political opponent of theirs in the national leadership of the BLF, Mundey's concept of a red–green alliance had significant influence inside and outside Australia. German Greens leader Petra Kelly, for example, specifically credited Mundey as a major inspiration for the formation of green parties in Europe.

The leader of the ecocentric 'wing' of green politics in Australia, Dr Bob Brown, was also influenced by Mundey's ideas and so the political movement that began in the wilderness tried to make itself relevant to people living in cities, towns and farming communities. A conception of green politics that was shaped by people as different in their origins and experiences as Mundey and Brown has been a significant Australian contribution to second-wave environmentalism on a global scale. The Greens have been able to build a certain base of electoral support without ever threatening to become a major party. In many ways they remain condemned to a life on the periphery, yet, as we saw in chapter 7, people forced to operate in 'the margins' can often have a major influence by promoting innovative and 'unsettling' ideas.

LAKE PEDDER, THE FRANKLIN RIVER AND THE TASMANIAN GREENS

On March 22, 1972 more than 400 people crammed into the Hobart Town Hall for a public meeting called by the Lake Pedder Action Committee (Law, 1997; Walker, 1989). Interest in the meeting was so great that people who could not fit in the hall waited in the stairway or in the street outside to find out what was happening. It was already late in the day for Lake Pedder. Construction was underway on a dam that would flood the remote inland lake with its whisky-coloured water and surprising pink quartz beach. However, the state's minority Liberal government had been forced to call an early election when it lost the support of the sole representative of the Centre Party, Kevin Lyons, who was forced to resign his seat following a personal scandal. Perhaps the elections would offer Tasmanians a last opportunity to express their opposition to the flooding of the lake. The trouble was that the Labor Opposition was an enthusiastic supporter of the dam and of the powerful Hydro-Electricity Commission (HEC) that was behind the development. So how could it become an election issue?

Many Tasmanians were feeling a great anguish about the likely fate of the lake. A year earlier, in March 1971, more than 1,000 people had participated in a 'Pedder Pilgrimage' (Walker, 1986), camping together for a weekend on the now-famous beach, from where light planes were able to take off for sight-seeing tours of the surrounding Western Arthurs ranges. In the same month, the concert pianist and convenor of the Hobart Arts Club, Brenda Hean, and the Legislative Council member Louis Shoobridge called a public meeting in the Hobart Town Hall that drew a capacity crowd. This meeting called for a referendum on the fate of Lake Pedder and Shoobridge took this proposal to the Parliament, where it was duly rejected (Kiernan, 1990). Soon after the Pedder Pilgrimage, the Hobart Bushwalking Club called a public meeting, at which the Lake Pedder Action Committee (LPAC) was formed. The new committee campaigned hard on the issue over the following year – organising a major symposium with some high-profile speakers. A petition signed by 184 scientists and another signed by 17,500 Tasmanians were presented to Parliament pleading for a reprieve for the lake (Kiernan, 1990, p. 26). But the government and the HEC would not even acknowledge the LPAC's existence, failing to respond to invitations to attend the symposium. As far as it was concerned, the fate of the lake had been sealed in 1967 when a Labor government headed by Premier Eric Reece had approved the HEC's development proposal. Premier Reece had earned the nickname 'Electric Eric' because of his renowned enthusiasm for the economic strategy of hydro-industrialisation[1] and it filled conservationists with dread that he could again become premier if Labor won the 1972 poll.

LAKE PEDDER

Those attending the March 1972 meeting agreed that the prospects for Lake Pedder were bleak and that it would not even get a mention in the forthcoming election campaign unless they put up candidates who would make it an issue. So, the decision was taken to stand a number of candidates under the name United Tasmania Group (UTG) in order to ask voters to give the lake a last-minute reprieve. The chairman of the LPAC, the Tasmania University botany lecturer Dr Richard Jones, was asked to co-ordinate the campaign (Walker, 1986, pp. 11–13).

The UTG was able to put together a ticket of 12 candidates to stand in four of the state's five multi-member electorates,[2] with most effort being put into the southern electorates of Denison and Franklin (Walker, 1989, pp. 162–163). The experienced, Sydney-based conservation campaigner

Lake Pedder before it was flooded by the construction of a dam in 1972.

Milo Dunphy volunteered his services and came to Hobart to help Jones co-ordinate the effort. A donation also came from the Australian Union of Students and students from Tasmania University did a lot of the hard work in putting up posters, letterboxing and door-knocking. The result vindicated the effort. In Denison, the ticket won 9.6% of the vote, just a few hundred short of what was needed to win a seat in Parliament. The statewide vote was 3.9%; a significant protest vote. However, this vote was not enough to turn the tide on Lake Pedder; several months later the water started backing up behind the completed dam wall. But it did bring the issue to the attention of many more Tasmanians, many of whom were horrified by the callous disregard shown for such a natural treasure. Grief over the loss of the lake was deep and enduring; more than 20 years later a high-profile committee was formed to look at the feasibility of restoring the lake by draining away the dam waters.[3] The good showing for the UTG also meant that Tasmanian conservationists had found a new way to put their concerns on the political agenda. The decision was taken to maintain the UTG for future election campaigns.

Between 1972 and 1977, the UTG stood candidates in ten state, federal and Tasmanian Legislative Council elections, without having any of the candidates elected (Walker, 1986, p. 44). The high water mark was achieved in the first campaign and it was more difficult to attract attention without the critical focus issue that Lake Pedder provided. Nevertheless, the party did manage to make a distinct contribution to public discourse about the future of the state. It argued that hydro-industrialisation was not only environmentally disastrous but also economically short-sighted because the end result would be an expensive over-supply of electricity.[4] It argued instead for a broader range of smaller economic projects that would generate a diversity of jobs. It suggested that progress could not only be judged by a net growth in the state's economy but by less tangible gains in quality of life. And it suggested that Liberals and Labor had become so similar in their ideas that the term 'Laborials' could be used to refer to both.[5] The media tended to ignore the UTG but an editorial in the Launceston *Examiner* in 1974 suggested that 'Dr Jones' little party has produced more teasing relevant ideas for Tasmanians than all the other party policy writers put together' (Walker, 1986, p. 166). When the Tasmanian Wilderness Society was formed in 1976, most of those who had been active in the UTG rechannelled their energies into the new organisation. The UTG was never formally wound up[6] but it effectively ran out of puff in 1977. However, as we shall see, it planted the seed that later grew into the Tasmanian Greens.

The UTG has been described as the world's first green political party, closely followed by New Zealand's Values Party formed several months later (Walker, 1986, p. 22). These were followed by Britain's Ecology Party, formed in 1976, and the German Greens (*Die Grünen*), formed in 1978. Some commentators have suggested that the UTG was not really a political party but rather a collection of individuals concerned about Lake Pedder (Walker, 1989, p. 163). But many political parties begin as similarly loose formations. There is a clear line of development from the UTG to the Tasmanian Greens and the Australian Greens. So the claim about it being the first green party is a reasonable one. The UTG began as the political wing of the LPAC with Dr Richard Jones as chairman of both organisations. But it outlived the LPAC and the Lake Pedder issue.

Richard Jones was undoubtedly the intellectual leader of the UTG. He wrote most of the party documents[7] and was principal author of a small booklet called *Damania*, which used this clever pun on the name of the state to critique the strategy of hydro-industrialisation. Together with Hugh Dell he drafted and promoted the party's most famous document *A New Ethic*, adopted in 1972, which urged a return to non-material values that might offer more in the way of 'spiritual refreshment'. As a botanist, Jones had a strong understanding of ecological science and argued that humans must again begin to see themselves as part of the 'web of life' and not separate from it. This is reflected in *A New Ethic*, which said, in part:

> We, citizens of Tasmania and members of the United Tasmania Group,
> UNITED in a global movement for survival . . .
> UNDERTAKE to husband and cherish Tasmania's living resources so that we do minimum damage to the web of life of which we are part while preventing the extinction or serious depletion of any form of life by our individual, group, or communal actions;
> And we shall . . .
> CHANGE our society and our culture to prevent a tyranny of rationality, at the expense of values, by which we may lose the unique adaptability of our species for meeting cultural and environmental change.

Such sentiments might be seen as pretty standard for environmentalists of today, but in 1972 they were radically new, certainly as the basis of a political platform. Jones very quickly pushed the UTG beyond the single issue of Lake Pedder. He wanted to promote ideas and values that had been missing from traditional political discourse and to dramatise this he promoted the slogan: 'Neither left nor right, just out in front' (Friend, 1997, p. 4). The UTG was formed because both the major parties in Tasmania were committed to a state development agenda that saw wilderness as nothing more than an untapped economic resource. Because it put wilderness first the UTG can be called an ecocentric formation (see, for example, Crowley, 2000), however it also reflected the social agenda of a number of social movements that were prominent in the early 1970s, such as the women's movement and the peace movement. This meant that it presented a radical challenge to prevailing traditions in Tasmanian politics.

Ironically, Jones himself came from a very conservative political background in provincial Queensland.[8] He was born in Mackay with a father who was a supporter of the Labor Party. However, his mother had come from a family of sugar cane-growers who were solid supporters of the Country Party, the forerunner of the National Party. As a young man he became president of the metropolitan branch of the Country Party in Brisbane (where he met his wife-to-be Patsy in 1959). In the mid-1960s he took up a position in CSIRO and was based at Deniliquin in south-west New South Wales. Although he was considered to be a 'suitcase resident' (a term used for people temporarily based in the town) he decided to stand for the shire council and campaigned for the support of fellow 'suitcase residents'. Against the odds, he won a position on the council and this gave him a taste of political influence. He went to Melbourne to complete his postgraduate studies in botany before arriving in Hobart to take up a position at Tasmania University in 1970. Soon after he arrived there he heard colleagues mumbling in the corridors about the terrible decision to flood Lake Pedder. He suggested that it was not too late to change the minds of the decision-makers and when he attended the meeting called by the Hobart Bushwalking Club in March 1971, he found himself elected as chairman of the new Lake Pedder Action Committee.

Jones began an important collaboration with Dr Bob Brown in 1976, after the latter had staged a one-person protest against a visit to Hobart of the nuclear-armed US aircraft carrier, the USS *Enterprise*. The two men stood on a joint UTG ticket for the Legislative Council elections late in 1976 and Jones was largely responsible for convincing Brown to stand again in state elections held in 1982. Tragically, Jones died after falling from a ladder at his home in 1986, but he had left a mark in his adopted state. Not only did he play the leading role in both the LPAC and the UTG, he was also largely responsible for the establishment of the highly acclaimed Centre for Environmental Studies at Tasmania University. After his death this centre and the Tasmanian Environment Centre collected money for a trust that would sponsor an annual Richard Jones Memorial Lecture. The first of these public lectures was delivered in 1987 by a founder of Friends of the Earth and the Ecology Party in England, Jonathan Porritt. Later lectures in the series were given by Paul Ehrlich and David Suzuki.

Bob Brown was not among those who founded the UTG, because he arrived to work in Launceston two months after its foundation. Yet he would become the most prominent of all those who were associated with the fledgling party. He first came to the attention of UTG organisers in 1972 when he used his own money to buy space in the national newspaper the *Australian* and all three Tasmanian dailies for a notice that likened the destruction of Lake Pedder to the extinction of the Thylacine (Tasmanian Tiger) and the genocide of Tasmanian Aborigines. In the following years he built a reputation in the north of the state for his radical views on conservation and politics, often having letters published in the Launceston *Examiner*. So the party kept an eye on him. But it was not until he came to Hobart to make his protest against the visit of the USS *Enterprise* that Richard Jones had the chance to talk to him at some length and suggest that he stand as a UTG candidate. Brown certainly demonstrated his mettle during his one-man protest against the US warship, which took the form of a hunger strike staged at the top of Mt Wellington. A cold snap brought snow and ice to the mountain but Brown stayed on in his tent and was interviewed by Norm Sanders for ABC television's *This Day Tonight* program.[9] The protest also rated a mention in Singapore's *Straits Times* (Thompson, 1984, p. 37) but was studiously ignored by the Hobart *Mercury*. When it was over, and Brown was back in Launceston, he thought about the proposal that had been put to him by Richard Jones and he told the UTG he would stand provided he would have the right to say publicly if there were any party policies that he disagreed with. After some negotiation, an agreement was reached and he stood as the northern representative on a joint ticket with Jones for the Legislative Council. Brown only polled 414 votes in the election but he found that he agreed wholeheartedly with the party's ideas and he developed a healthy respect for Jones and the 'young Pedder activists'.[10]

On June 26 1976, several months before his protest against the visit of the *Enterprise*, Brown had hosted a special weekend gathering at his home at Liffey, near Launceston, organised by UTG activists Kevin Kiernan and Geoff Holloway. Like Brown, Kiernan was very fond of the south-west corner of the state but he felt the South West Tasmania Action Committee had fallen into a rut after the heady days of the Lake Pedder campaign and needed a major revamp. Sixteen people attended the gathering and they all agreed with Kiernan that the best approach would be to launch a fresh new organisation with a more up-market image (Thompson, 1984, p. 54). They even agreed with his suggested name – the Tasmanian Wilderness Society. When the new organisation was launched in Hobart in August 1976, 19 of the 23 people in attendance were also members of the UTG (Walker, 1986, p. 41). There was a growing sentiment that the UTG had just about run its course and that it was time to build an organisation that could influence all the political parties. Brown became actively involved in the TWS – eventually becoming its third director after Kiernan and Norm Sanders. But he agreed with Richard Jones that the UTG still had a role to play, so he agreed to stand as a candidate.

Bob Brown had missed out on seeing Lake Pedder up close. He only arrived in the state, to take up a locum position in a medical practice in Launceston, in May 1972.[11] He met the 'Pedder activists' when they were touring the state in a caravan and was impressed by their commitment. Then he got a chance to see the lake from the air when he took a day to visit a surgeons' conference in Hobart where a friend from Sydney had said he had an afternoon off and asked Brown if there was anything he would like to do in Hobart. Brown suggested that they hire a light plane to see the doomed lake. It was a stormy day with sunbeams occasionally bursting through the low cloud while sheets of rain swept across the button-grass plains where the Serpentine River drained the lake (Thompson, 1984, p. 19). The view made such an impression on Brown that he knew he must return to the area on foot. However, the lake started going under water in July 1972.

Brown had opportunities to visit the south-west and other wilderness areas when he joined the Launceston Walking Club not long after coming to Tasmania. However, his most important opportunity came out of the blue when a bearded young man stopped him on the steps of the Launceston Library one day in November 1975 and said 'You're Bob Brown aren't you?' 'Yes.' 'I've asked lots of people to come rafting with me down the Franklin River and they won't come, will you?' Brown had never met Paul Smith, a local forester, before and he did not even know where the Franklin was. He had no experience in rafting. Smith told him the Franklin had a reputation as an awesome place and that he could provide all the equipment and experience they would need. At the time Brown was organising a walk to the Western Arthurs and so he told Smith that if he came on that walk he would go with him to the Franklin. He would want to do some research about the river before they went there and he had a month off work in February. After a rather short and business-like discussion they had reached an agreement.

What Brown found out in his research on the Franklin did not fill him with confidence about the forthcoming trip. He learnt that the river had been named by a surveyor, James Calder, who had peered down into the gorge from the top of a ravine and noticed that flood-logs were caught in trees some 30 feet above the river (Thompson, 1984, pp. 79–80). He learnt that the river then attracted the interest of white settlers when they noticed that it sometimes flushed large Huon Pine trees into the Gordon River that, in turn, flowed into Macquarie Harbour. This suggested that the Franklin developed a powerful flow through steep and heavily wooded country after winter rains and this assumption was confirmed by an exploration by two brothers from the timber town of Strahan, who spent 17 days pushing a three-metre punt up the river from its junction with the Gordon in 1940 (Thompson, 1984, p. 48). The first successful canoe trip down the wild river came in 1958 when five men completed a hair-raising journey down waterfalls and very steep rapids. No one tried it again until 1970. Only a handful of white people had ever laid eyes on the places where Smith and Brown were about to venture. Furthermore, his mind was not set at ease when, soon after they began the journey, they passed some local fishermen who asked them where they were headed. When told, the reply came 'Yer bloody mad!' (Thompson, 1984, p. 50).

But after 11 days without seeing another person, they finally reached their destination feeling exhilarated. Smith had selected the equipment well and guided them safely through a treacherous passage. Many years later, Brown told his biographer Peter Thompson that the trip had been, without doubt, the highlight of his life. His love of wild places had been dramatically enhanced. But something else had happened on the trip that would also change his life. After travelling through country that seemed a million miles from any human settlement, they were jolted back to reality when they came across the signs of a construction camp being built at the end of a new road that had been carved through the bush to reach the mid-section of the river. Several days later they came across another such camp further downstream. Smith had told Brown that the Hydro-Electricity Commission had plans to build a series of dams on the river. That was why Smith was so keen to make the journey, before the damage was done. But both men were shocked and saddened to see that the preliminary work had already begun. It was another Lake Pedder in the making and they returned from the trip determined to do what they could to let the public know what was happening. Brown and Smith both attended the gathering at Liffey that discussed the establishment of the Tasmanian Wilderness Society. Another participant was Helen Gee, who had been liloing down the Jane River, a tributary of the Franklin, at much the same time that Brown and Smith were rafting down the river. The Liffey gathering decided that the Franklin could well become the 'next Lake Pedder' and that the new organisation should make it a priority issue.

The campaign against the flooding of the Franklin River made little progress in 1977 and 1978, but neither did the work by the HEC. Sensing that time was running short, however, Brown made the momentous decision to quit the medical practice in Launceston and move to Hobart to campaign full-time. He arrived in Hobart at the beginning of 1979 and started working out of the tiny office of the TWS on the second floor of the Tasmanian Environment Centre. He took over from Norm Sanders as TWS director and convinced the organisation that the Franklin had to be its sole priority in the immediate future. He took a big gamble by using most of the money in the TWS account to pay for a company to conduct a public opinion poll on the issue but was delighted when the result was a surprising 53.5% in favour of preserving the river rather than allowing the construction of any dam that would flood it. In his time in Launceston he had developed a good relationship with the editor of the *Examiner* and the paper reported the results of the survey in a small article on its front page. Next Brown took a young and inexperienced Sydney film-maker down the river to make the film *Franklin: Wild River* and he raised the money to have 10,000 copies of a glossy brochure with dramatic photos printed. He talked the National Parks and Wildlife Service into commissioning an independent study on a proposal for a Wild Rivers National Park centring on the Franklin and Gordon – a move that proved to be a masterstroke later when this proposal was sharply counterposed to the plans of the HEC.

Brown developed a good working relationship with Kevin Kiernan who managed to convince the world-renowned violinist Yehudi Menuhin to become a patron of the TWS. This was announced at a press conference in Melbourne in 1979, attended by Menuhin and Brown. Kiernan also told Brown about a hidden cave he had found on a trip down the river and when the two men and Bob Burton went back to investigate in 1981 they were overwhelmed to find unmistakable evidence of ancient Aboriginal occupation – information that was used extensively in the campaign. When the NPWS decided it would publicly promote the idea of a new national park, it commissioned an experienced film-maker, Sydney-based Bob Connelly, to make a 35 mm film and asked Brown to organise the trip. Making this film proved to be difficult and dangerous but when it was released – as *Franklin River Journey* – Premier Doug Lowe was in the audience for the first showing. At the end of the film, there was a spontaneous burst of applause and someone shouted out 'Save the Franklin!' (Thompson, 1984, pp. 117–118). All eyes turned to Lowe, who sat white-faced and visibly moved.

Doug Lowe became Tasmania's Labor premier in December 1977 following the retirement, in 1975, of the ageing 'Electric Eric' Reece and the rather lacklustre premiership of his successor Bill Neilson. Whereas Neilson had been a member of the party's old guard, Lowe was a bit of a gamble – a 'young Turk' who could perform well on television (Thompson, 1984, p. 95). He was much more accessible than previous Labor leaders and Brown had been quite impressed when given the chance to put his case to him concerning the fate of the Franklin. But Lowe was still the leader of a party that was strongly influenced by the powerful HEC, and he was reluctant to risk his fragile hold on leadership. In May 1980, he summoned Brown to his office to suggest a compromise proposal: to move the dam further upstream on the Gordon River, above its junction with the Olga River, so that the Franklin would not be affected. Brown said he would obviously welcome any move to save the Franklin but the alternative would still cause the loss of precious wilderness in the south-west heartland and the Wilderness Society would campaign against the construction of any dam in the region. Lowe continued to push the compromise proposal and, when he announced it publicly, Brown told the press that although this was the first time that a Labor premier had been prepared to stand up to the HEC, the Wilderness Society could not accept any dam in the area and would take its campaign into the national arena. In May 1980 he travelled to Melbourne to address a public meeting on the issue and was overwhelmed to find that an estimated 1,200 people crammed into the Storey Hall to hear him.

Lowe was able to get his compromise proposal through the lower house of the Tasmanian Parliament, but it was rejected by a very conservative Legislative Council. The only way to get around this difficulty was to propose a referendum in which Tasmanians would be asked to make their choice between the two dam proposals. Again Lowe summoned Brown to ask for his support, saying that he would support the declaration of the Wild Rivers National Park if Brown supported the Gordon above Olga option. However, Brown's response was to say that the referendum should include a third choice of no dams at all, and no agreement was reached. On November 11 1981, Lowe was dumped as Labor leader, to be replaced by the more conservative Harry Holgate. It was clear the Labor government was in trouble and would soon be replaced by a Liberal government led by an aggressive, pro-development politician, Robin Gray. Holgate refused to accept the proposal for putting a 'no dams' option in the referendum, so the Wilderness Society campaigned hard to get people to simply write the words 'No Dams' on their ballots, thus casting an informal vote. Clearly, this was a high-risk strategy because defeat for Lowe's compromise could simply entrench the original dam proposal. The result of the ballot was narrow majority for the original Gordon below Franklin proposal (53%), but a massive 38% had followed the instructions of the Wilderness Society. The Gordon-above-Olga option was now off the agenda and no one could say what might happen if voters were asked to choose again between the remaining options of Gordon below Franklin or no dams at all. Clearly the issue had not been resolved, but in early 1982 the Holgate government fell on a no confidence motion – caused by the resignation from the ALP of Lowe and another parliamentarian Mary Willey, who was angered by the government's failure to save the Franklin – and the incoming Gray government acted quickly to legislate for the Gordon-below-Franklin project. Saying that there was no need for any further discussion on the issue, Gray made the memorable public comment that 'For eleven months of the year the Franklin River is nothing but a brown ditch, leech ridden, unattractive to the majority of people' (Thompson, 1984, p. 160).

This was a dark moment for Brown and his colleagues. With a hostile government in power in Tasmania, the only hope seemed to be federal intervention but when Bob Brown went to see Prime Minister Malcolm Fraser on the issue soon after the Tasmanian elections his response had been to say that people with lost causes in Tasmania should not come to Canberra seeking respite (Thompson, 1984, p. 154). However, the dark mood of the Tasmanian campaigners was countered by a much more buoyant mood among those campaigning on the issue on the mainland. Interest in the issue was manifested by the fact that 'No Dams' posters and stickers were now spreading across the nation and that 8,000 people attended a rally in Sydney on the issue in March 1982 and 15,000 people a similar rally in Melbourne. Karen Alexander, an organiser of the 'No Dams' campaign in Victoria, told the authors[12] that wilderness supporters on the mainland had already reached the conclusion that the battle had to be won on a national level and they knew they could exert considerable political influence by campaigning in marginal electorates. Alexander said that work by Franklin River supporters in marginal seats in Victoria in early 1982 bore fruit when the prominent Victorian Labor senator John Button proposed a surprisingly strong motion on the question at the ALP national conference in July 1982. Bob Brown has said that he was encouraged to attend that Labor Party conference in Canberra by the party's spokesperson on the environment, Stewart West, who was hopeful of getting the conference to support the call for a public inquiry into the construction of the proposed dam in order to buy time for further work on the issue. However, the signs were not good when debate on the Franklin dam followed hard on the heels of a very emotional debate on uranium mining, in which the party had committed itself to a policy of allowing the development of three uranium mines. Energy was very low in the conference room when West's motion for a public inquiry was put and, sitting in the public gallery, Brown was dismayed when it was lost comprehensively. However, to his complete surprise, Button then put his motion opposing the construction of any dams in the region and this was passed with a strong majority.

Now Malcolm Fraser was realising that his stance of leaving the issue to the Tasmanian government could alienate voters on the mainland. At a by-election held in the Melbourne seat of Flinders in December 1982, a stunning 42% of voters wrote the words 'No dams' on their ballot papers. Fraser could see that the issue could influence the results in a number of seats in federal elections due in 1983, so, in early 1983, he announced that his government would give Tasmania a $500 million grant to build a thermal power station instead of proceeding with the dam. When he made this offer public he already knew that Premier Gray was going to reject it, arguing that the question of the dam had long been settled. But Fraser hoped to neutralise a swing of support to the Labor Party on the issue. It was a forlorn hope. Emotions were now running very high and the issue of the dam was on the front pages of the nation's newspapers almost every day. Construction work had begun and protesters had resorted to an intense campaign of civil disobedience to hamper the operation. On December 14, 1982 the 'Franklin Blockade' had begun when 53 people were arrested at the site of the dam and at a loading wharf, Warners Landing, further down the Gordon River when they tried to obstruct the construction workers. Brown himself was arrested on December 16, along with 48 others and was only released on Christmas Eve, in time to organise a visit to the river by federal Labor leader Bill Hayden. In an amazing six-week period, 2,613 people registered their participation in the blockade with the organisers (Thompson, 1984, p. 174), with 900 of them coming from Tasmania, 650 from Victoria, 600 from New South Wales, 145 from Canberra, 142 from South Australia, 73 from Queensland, 56 from Western Australia, three from the Northern Territory, and 67 from overseas. There were 1,272 arrests, with 450 people being remanded in Risdon Prison. The issue was starting to make news outside the country. To cap off a highly effective campaign, Bob Brown was selected as Australian of the Year in January 1983, with nominations coming from people all over Australia.

In February 1983, Bill Hayden was dumped as Labor leader on the same day that Malcolm Fraser announced an early election. The high-profile new Labor leader Bob Hawke went on to win the election comfortably and he had promised that his government would stop construction of the dam. The Wilderness Society, the Australian Conservation Foundation and 17 other conservation organisations had run a united, high-energy campaign urging voters in 13 targeted marginal electorates to vote Labor.[13] Hawke thanked the environmentalists for this support and he put the issue high on the agenda of his incoming government. The new government made use of the fact that south-west Tasmania had recently been declared a world heritage area to justify the use of 'external powers' in overriding the jurisdiction of the Tasmanian government. Predictably, the Tasmanian government challenged this move in the High Court but on the narrow margin of 4:3 the court upheld the Commonwealth's right to take the action it had. The river was saved and the hard line pursued by Brown as Wilderness Society director had finally been vindicated. In the process, the Wilderness Society had been transformed into a national organisation of similar significance to the other big two – the Australian Conservation Foundation and Greenpeace.

In the 1982 elections that had brought Robin Gray to power in Tasmania, Bob Brown had stood as an independent in the electorate of Denison in order to inject the 'No Dams' option into the election debate. He polled well but lost, narrowly, to Australian Democrats candidate Norm Sanders, who had first entered parliament in a by-election in 1980. Just after Christmas 1982 Sanders announced that he was resigning his seat in protest over the way the government was handling the protests against the building of the dam. He said it was impossible to do anything worthwhile in a parliament dominated by Robin Gray. According to Tasmania's unique electoral system, his seat would be filled on the basis of a recount of votes for the other candidates in the previous election and, as Bob Brown had come in just behind Sanders, it was highly likely that he would be given the seat if he agreed to participate in the exercise. This unexpected development came while Brown was on remand in Risdon Prison and he agonised over his decision. Like Sanders, he had real doubts as to what could be achieved in parliament with Robin Gray in power and he knew that he would be personally vilified. Back in 1976 he had taken the brave decision, as a doctor in Launceston, to announce publicly that he was homosexual and he knew that this would also be used to smear his reputation. He already had plenty to do in the Franklin campaign so the responsibilities of being a parliamentarian could be an unwanted burden. But, against all the negatives, he knew that a position in Parliament would give him and his concerns an even higher profile. So, on the day he emerged from prison, carrying a box containing the clothes he had been arrested in and some gifts from supporters, he was formally elected to Parliament. With this he began his 10-year career in the Tasmanian Parliament.

On the day that Bob Brown was born, in 1944, none of those present could have imagined that they were witnessing the birth of a future Australian of the Year and high-profile national figure. They certainly could not have guessed that he would achieve all that as a prominent Tasmanian because he was born, shortly after twin sister Jan, in Oberon, on the western edge of the Blue Mountains in New South Wales. His father was a humble country policeman who accepted a posting to a small town called Trunkey Creek, just south of Bathurst, when the twins were two years old. The family stayed there for seven years (before moving on to Armidale in New England) and it proved to be a good place for young children. The police station was adjacent to a five-acre paddock that, in turn, backed onto the bush. In talking about influential childhood experiences, Brown told the authors that he could definitely trace his love of the bush back to his child-hood at Trunkey Creek:

Bob Brown at the centre of media attention (William W. Smith)

I didn't know this at the time, but the bush was still recovering from bad bush-fires in 1939 so it was going through a process of regeneration that we were able to watch. My sister and I stuck pretty much together in those days and we used to play a lot in the paddock and through the fence to the bush behind. There was a creek running down there and I became aware of everything from frill-necked lizards to spitfires on the tips of the gumleaves. My earliest memory is finding myself standing on top of an ants' nest with my mouth open screaming until my mother came through the gate, took my clothes off and put me in a pail of water. You learn that you've got to be careful in the bush but I could always find something fascinating to look at.[14]

After starting school at Trunkey Creek, where there were 23 children and one teacher, Brown got a shock when he then found himself at the highly competitive Armidale Demonstration School, where he was once caned for daydreaming. He started to become more competitive and did well academically, but he always looked forward to the weekends. His mother had come from a dairy farm near Glen Innes, where some of her family were still living, and on some weekends they would go there.

I climbed the granite rocks and boulders and knew every one of them, knew exactly how high they were. I was pretty obsessive and competitive at school by then, but I loved the freedom I felt in that place. I felt relaxed in an endlessly fascinating and non-competitive bush environment.

From Armidale the family moved on to Bellingen, in beautiful rainforest country, and from there to Windsor on the Hawkesbury River west of Sydney, where they experienced the huge floods of 1963–64. While they were living at Windsor, Brown took the old motor rail to Blacktown where he attended the state's most densely populated high school. He was a good student in a school unused to much academic success, and the headmaster wanted him to become a role model for other students. He was made school captain, against his will, in his final year. Despite feeling very uncomfortable in the leadership role he had been thrust into, Brown did well enough in the final exams to win a place in the medical school at Sydney University. However, he remembers the six years spent at Sydney University as the most lonely and turbulent of his life (Thompson, 1984, p. 10). He hated living in the city and found that he had become disaffected from the church that played such a big role in his parents' life. He still kept the faith in Prime Minister Robert Menzies, whom he openly admired. But he found that he had little in common with his fellow medical students. When the chance came he moved to Canberra in 1968 to work as a junior doctor at Canberra Hospital, but he continued to feel so lonely there that he remembers a time when he sat on the shore of Lake Burley Griffin contemplating suicide (Thompson, 1984, p. 13). A rare highlight of this period was a three-month residency in hospitals in Darwin and Alice Springs. While he was in Alice he met John Hawkins, a surgeon who, unbeknown to Brown, had participated in the first trip down the Franklin River in 1958.

Brown fled Canberra and went to London for some work experience in 1970. The highlight of his time in Britain was a walking trip in the Scottish Highlands and he returned to Australia still feeling very unsettled. He took a job as a ship's doctor working on Pacific cruise ships so that he could see some more of the world. But he soon tired of this and, during a stopover in Fiji, noticed the advertisement for a locum in Launceston that brought him to Tasmania. After feeling so much at sea – both on land and on ships – Brown was surprised to find how comfortable he felt from the moment he set foot in Tasmania. He landed at Devonport in May 1972 and decided to take a long route to Launceston to see what the place looked like.

> That first night I drove across the central plateau from Devonport and stayed in a pub at Ouse where I would probably get lynched today. Then I drove to Bridgewater and back up the highway to Launceston. I was just knocked out by what I saw.

He also settled quickly into his role in the Launceston medical practice. His direct and friendly manner made him popular with patients and, as a young and apparently eligible bachelor, he became the subject of much speculation (and attention from young ladies). When his locumship was over, the practice's partners offered him a permanent job, which he quickly accepted. From childhood days he had been fascinated by what he heard about the Thylacine (Tasmanian Tiger) and he spent much of his spare time researching the animal and its fate. Within a year, Brown had decided that he would settle in Tasmania and he bought a little cottage set on 11 hectares of land outside Launceston on the Liffey River, with the spectacular, 1,340-metre high Drys Bluff as a backdrop. Brown did not know it at the time but the previous owner, John Dean, had been on the first trip down the Franklin River with John Hawkins and three others in 1958. The only trouble with the house at Liffey was that Brown enjoyed being there so much that he began to resent having to go to work in the medical practice. He negotiated a reduction in his hours so that he could spend more time at home. Apart from getting a garden established and getting to know his neighbours, he loved to take the arduous walk to the top of Drys Bluff where he could sit and meditate in peace. He also loved to sit in or outside his house listening to the sound of the river passing by, and he got into the habit of keeping a journal of what he experienced and how it made him feel. 1974 and 1975 were quiet and peaceful years for him; the calm before the storm.

As we have seen, Brown's chance encounter with Paul Smith on the steps of the Launceston Library in November 1975 triggered a chain of events that shattered the calm of his country life. The year 1976 was that in which he staged his protest against the Hobart visit of the USS *Enterprise* and committed himself to the cause of saving the Franklin River. It was also the year in which he decided to go public with his admission that he was homosexual, a very brave thing to do in conservative Tasmania where he could have been imprisoned for the 'crime'. His revelation came in an interview for the ABC television program *This Day Tonight* and it was followed up with a front-page article about him in the Launceston *Examiner*. Being able to admit this to the world was a great relief for a man who had often been plagued by self-loathing, leading to loneliness and depression. Fortunately, the newspaper article described his admission as courageous and that was the way it was seen by most people who knew him. There were a few patients who now refused to be treated by him, but many more who respected him even more for his honesty and integrity. The experience had been a positive one and he told the authors that his ability to come to terms with his own sexuality also gave him a stronger empathy for other people who faced discrimination:

> That has given me the jump, if you like, on understanding people in minority situations; people who suffer personal travail because of a prevailing 'wisdom' that is actually incomprehensible and destructive. During the Franklin campaign I was often asked how I could put up with the abuse that was being hurled at me and I used to say it was nothing compared to what was being done to dissidents in Argentina who were being dropped alive out of helicopters. I had come from a place where I had been locked up in my own little world feeling that society didn't approve of me and I had to struggle to approve of myself before I could do anything else. But that probably made me stronger in the end.

Brown's personal strength is something that strikes most people who meet him. When we went to his Senate office, in the bunker-like Parliament House in Canberra, to interview him for this book, he asked us to excuse him for ducking out from time to time to cast his vote in the Senate. Missing the debate did not worry him because he knew what would be said but he wanted to make sure his vote was counted. So each time the division bell rang, he calmly finished what he was saying before excusing himself. When he returned, he picked up the thread of the conversation as if there had been no interruption. This same capacity to stay calm and focused is what really came to the fore during the difficult days of the Franklin campaign. Working in a small office with zealous volunteers, who were mostly much younger than him, he had to use all his personal skills to keep the campaign focused and moving, especially when the prospects for stopping the dam seemed very bleak (Thompson, 1984, p. 109).

Brown also had to use his personal strength to survive the experience of being the sole Green independent in the Tasmanian Parliament after he took his seat in 1983. He told the authors that very soon after he arrived he volunteered for a parliamentary delegation to inspect the construction site for the Gordon below Franklin dam. Robin Gray refused to endorse his participation, which meant he could not travel on the plane provided for the delegation, but he drove to the site overnight and the presiding HEC official felt obliged to give him an entry permit. The government members had inspected the site the day before and now was the turn for the Labor members and independents. The construction workers had been tipped off that Brown would be coming so they prepared a very hostile reception for him. They carried abusive placards and chanted 'Kill the bastard'. According to Brown:

> One of them came up to where we were standing and drew a line in the dirt around me and said: 'If you step outside that circle we'll knock your brains out.' The Labor people, bless their brave hearts, all went and stood at a distance. That was the strength of their commitment to democratic representation. I moved out of the circle and two policemen said: 'Look, this is really dangerous, we can't guarantee your safety here.' So I went with them down to the boat we had come on while the Labor people went and had scones and tea with the dam builders.

Being a parliamentarian was not easy but Brown discovered that he could get access to a lot of information not available to the public and when the 1986 elections came around he convinced Dr Gerry Bates, a lecturer in environmental law at Tasmania University, to stand with him. Both were elected and they started using the term Green Independents to refer to themselves and to encourage others to stand in future elections. Now that he had a parliamentary colleague, Brown found he could be much more effective and both of them got heavily involved in a new campaign to stop logging in the Picton Valley, south of Hobart, at a place called Farmhouse Creek. They campaigned for the declaration of World Heritage areas in Tasmania and called for comprehensive legislation to protect remnant native forests across the state. However, the next really big confrontation flared up in the north of the state in 1987 when the Australian resource company, North Broken Hill, announced that it had formed a consortium with the Canadian company Noranda to build a 'world class' paper pulp mill on farmland at a place called Wesley Vale. Farmers were outraged and local residents, who knew how dirty the existing paper mills at Burnie were, refused to believe the claim that the new mill would hardly be noticed. On investigation, they learnt that the new mill would be dumping 13 tonnes of toxic organochlorides off the coast in Bass Strait every day and its fumes would include carcinogenic dioxins. More than 1,000 people attended a protest rally in Devonport.

People opposed to the development formed an organisation called Concerned Residents Opposing Pulp Mill Siting (CROPS), for which a highly articulate schoolteacher and daughter of a Wesley Vale farmer, Christine Milne, became chief spokesperson. In a 1997 interview, Milne told the authors that she stepped into an organising role with CROPS because she had gained some experience during the campaign against the flooding of the Franklin River. She assumed that an organisation such as the Wilderness Society, the Australian Conservation Foundation, or Greenpeace, would want to take the leading role in the campaign with CROPS providing the active volunteers. However, when she rang their offices they all said they were already over-committed with their staff and resources and could only play a supporting role to CROPS. With support and advice from Brown and Bates, she had little difficulty in handling the media work in Tasmania, but it was pretty obvious that the political work would have to be done in Canberra. If the mill was going to pollute Bass Strait and require an export licence for its products, then there were opportunities for the federal Labor government to intervene once again to override the enthusiastic support for the project by the Gray government in Tasmania. But Milne would have to go to Canberra. She took leave from her job and became a full-time conservationist.

Milne proved to be a very capable campaigner and the press began to anticipate 'another Franklin'. Having done the groundwork in Tasmania to demonstrate the extent of local opposition to the plan she got a good hearing in Canberra. The federal government decided to impose very tough environmental guidelines on the granting of an export licence. The Canadian company, Noranda, started to be concerned about the bad publicity. It was even more concerned that if it met very tough environmental standards in Australia, people living near its existing mills in Canada would demand the same. So it pulled the plug and the project gurgled down the drainpipe. Brown and Bates were so impressed with what Milne had achieved that they asked her to stand with them on a joint ticket for the May 1989 state elections. She agreed and encouraged her friend and fellow Wesley Vale campaigner, Diane Hollister, to stand as well. Brown, Bates, Milne and Hollister were joined by Hobart Uniting Church minister Rev. Lance Armstrong to make five Green Independents and, in a stunning result, all five were elected. They held the balance of power in the new parliament and negotiated an accord with the Labor Party to enable it to form government. The Tasmanian Greens had arrived as a significant political force.

Since the time of the UTG, Tasmania had been the centre of green politics in Australia, but it did not form a political party containing the word 'Green' until well after that name had been used in other parts of the country. The first organisation to register the name was the Sydney Greens, formed in 1983. This 'party' restricted its activities to a relatively small, inner-urban, area. They were followed by a group in Queensland, led by an experienced political activist Drew Hutton, which adopted the name Queensland Greens in 1985. As early as 1984, the German Greens leader, Petra Kelly, on a visit to Australia, urged green political activists to come together to form a single national Greens party. However, a number of things made this difficult. One was the fact that a Nuclear Disarmament Party, formed by the Canberra doctor Michael Dernborough in 1984, suddenly took off when a national conference of the Labor Party that year reaffirmed that party's commitment to uranium mining. In federal elections that were held late that year the new party managed to get one of its candidates – the peace activist Jo Vallentine – elected to the Senate from Western Australia and missed out narrowly on having the rock music star Peter Garrett join her from New South Wales. The following year the NDP suffered an acrimonious public split and, when Jo Vallentine stood for re-election in 1990, she did so as a representative of a group calling itself the Greens (WA). However, a remnant of the NDP continued to exist long after the 1985 split. A second problem has been the existence in federal politics of the Australian Democrats, who have often campaigned on conservation issues. In 1986 Bob Brown approached the Australian Democrats about the idea of a merger between that party and Tasmanian Greens and a proposal to form the 'Green Democrats' was discussed at high level in 1991 (Brown and Singer, 1996, p. 80). But nothing came of it and the Democrats continued to compete fiercely with the Greens in Tasmania. The other thing that made the formation of a national Green party difficult was the sheer size of the country and the fact that different groups moved to claim the name on a state basis, without any agreement to form a national organisation.

THE 'RED AND THE GREEN': JACK MUNDEY AND THE AUSTRALIAN GREENS

The idea of a single national green party was put to a 'Getting Together' conference at Sydney University in April 1986, attended by around 500 people, but no consensus could be reached in the diverse gathering. After further failed attempts to reach a broad consensus the decision to launch the Australian Greens was finally taken at a gathering held in Sydney in 1992, attended by just 50 people.[15] Although the Greens (WA), which had two members in the Australian Senate in the period 1993–96, continued to operate as an autonomous organisation, the Australian Greens finally became the recognised green party at a national level. Bob Brown was elected to the Australian Senate as a representative of the Australian Greens in 1996.

Bob Brown has credited Sydney trade unionist Jack Mundey as being the initiator of the term 'green politics' internationally (Brown and Singer, 1996, p. 65). It became well established as a political term after 23 members of the German Greens (*Die Grünen*) were elected to that country's national parliament in 1983. But Brown points out that a founder of the German Greens, Petra Kelly, picked up the term after talking to Jack Mundey during a visit to Australia in 1977. By the time Mundey met Kelly he had already spent six months in Britain in 1975 on a lecture tour organised by the Centre for Environmental Studies in London.[16] During this visit he had spoken at the national conference of the Labour Party and had been instrumental in a campaign to stop the demolition of the Birmingham Post Office – an action that led Birmingham unionists to form a Green Bans Committee to promote the 'social responsibility of labour'. In 1976 he was one of '24 world thinkers' invited to address the first United Nations conference on the built environment in Vancouver. The following year he was invited to the World Wild Life Congress in San Francisco. However, his contact with Kelly put him in touch with young environmental and anti-nuclear activists across Europe. Kelly initiated a number of invitations for him to visit several countries and the term 'green politics' started being used in Switzerland, Belgium and Germany. According to Brown (Brown and Singer, 1996), Swiss environmentalist Daniel Brelaz was the first green to be elected to a national parliament when he won a seat in 1979.

In Australia, Mundey was known in the early 1970s as the key leader of the New South Wales Branch of the Builders Labourers' Federation, which imposed 'green bans' on building or demolition work that would destroy open space or heritage buildings in Sydney and throughout the state. What impressed the Europeans was the idea of a coalition between a working class organisation – a trade union – and 'middle-class' groups concerned with environmental conservation. The idea that Mundey promoted in his visits to Europe was that concerns for social justice and conservation could be merged in the fight for liveable environments and the green political movement in Europe soon had this idea as a central tenet.

According to Mundey,[17] he first used the term 'green ban' in a conversation with a journalist from the *Australian* newspaper in 1973. He had suggested to the journalist, Malcolm Colless, that the traditional union term 'black ban' was too narrow and negative to characterise the sort of bans the BLF was then imposing on demolition and building work in various parts of Sydney. The term 'green ban' might have more positive connotations; suggesting that the bans were of interest to the wider community and not just the union members. When Colless used the term in an article it was quickly picked up by other journalists and by members of various residents action groups that had invited the BLF to impose such bans in their areas. In 1973 there was intense interest in the bans because they were tying up millions of dollars worth of property developments and had made Mundey and other leaders of the BLF public figures. Although the term was not in use at the time, the first green ban was imposed in 1970 to prevent the destruction of an area known as Kelly's Bush on the Parramatta River at Hunters Hill to make way for a housing estate. Between then and 1974 they were used to prevent the demolition of historic buildings and houses in The Rocks, Kings Cross, Woolloomooloo, and the city; to prevent the destruction of trees in the Botanic Gardens to allow the construction of an underground carpark for the Sydney Opera House and to prevent another carpark being built on part of Centennial Park and Moore Park; and to prevent the demolition of houses in inner urban suburbs like Waterloo, Glebe and Ultimo to make way for high-rise Housing Commission apartments or freeways. Outside the inner-city area of Sydney, green bans were used to protect open space in the Sydney suburbs of Eastlakes and Merrylands, as well as in Gosford and at Port Stevens, north of Newcastle. They were also invoked to protect a heritage area in Newcastle and to prevent the demolition of historic buildings in Newcastle, Wollongong and Inverell.[18] At the height of a building boom in Sydney, Mundey, in particular, became a *bête noir* for wealthy property developers[19] and the pro-development state government led by Premier Robert Askin. But he became a folk hero for conservationists and inner-city residents. In 1974, he and his close colleagues Bob Pringle and Joe Owens were forced out of office in an internal 'coup' organised by BLF national secretary Norm Gallagher. But they had introduced a brand of trade union politics never before seen in Australia and one that would have its reverberations around the world.

What added to the intrigue regarding Jack Mundey was that he was also a prominent member of the Communist Party of Australia. A 'red' trade union leader prepared to risk a possible loss of jobs for his union members in order to defend a patch of bushland in the prosperous harbour-side suburb of Hunters Hill seemed to be an enigma. Mundey argued that there was no shortage of useful work that his members could be employed to do and that his members had as much interest in the preservation of public space and the city's heritage as anyone. But his personal passion for the preservation of green space also stemmed from the fact that he grew up in a particularly beautiful part of Queensland, the Atherton Tableland, where he was able to run free (without shoes) for the first part of his life.

When Mundey was born his parents owned a small dairy farm outside a small town called Malanda, located on the Johnstone River. Upstream from the town was an Aboriginal settlement; downstream were the Malanda Falls and a patch of undisturbed rainforest. His mother came from a small mining town in Queensland; his father – also Jack – from a line of Irish Catholics who first started farming at Bega on the New South Wales

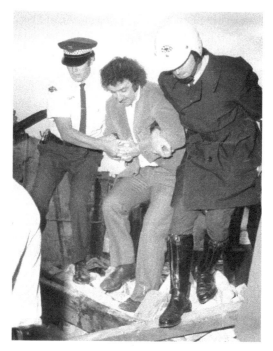

Jack Mundey being arrested during a 'green ban' protest in Sydney
in 1974 (Newspix)

south coast. Mundey has described his father as quick-witted and gregarious. He had been encouraged to head for the good soil and climate of north
Queensland by his father. But in the Depression era he could not survive as
an independent farmer so, soon after Jack was born in 1932, he sold his
property and became a share farmer. Young Jack's mother died when he
was only six, with the family still struggling financially. Later, Jack senior
would have some success as a cattle auctioneer and his oldest daughter,
Josie, took on the role of mother to her younger siblings. The family moved
around a lot; knew most people in the district and knew of the rest. Young
Jack remembered donning his first pair of shoes for a trip to Cairns when he
was 12 years old. It was also the first time he saw the sea. Before that, his
world had been the Tableland; swimming in the river, riding a horse to
school, exploring the rainforest.

After completing his early schooling in the Tableland, Mundey
went to a Catholic high school in Cairns. However, he left at 16 to return
home and start an apprenticeship as a plumber. What he had learnt in
Cairns was that he had an aptitude for sport and this became the consuming interest in his life; a bit of cricket and boxing, but mostly Rugby League.
When representing Atherton Tableland in Rugby League he has said he

trained harder and longer than his team-mates and soon caught the eye of some of the player-coaches in north Queensland who had earlier played in Brisbane or Sydney. Someone mentioned him to Vic Hey – a legendary Australian player who had taken on the job as inaugural coach of the new club Parramatta in the Sydney first grade competiton. An invitation to play for Parramatta arrived in the mail. By now Mundey was ready to try his luck in the big smoke, especially as it would give him the chance to make his name as a footballer. He arrived in Sydney in 1951 thinking he would try it out for one football season. He was still living in Sydney 47 years later.

Mundey had three seasons with the Parramatta side before becoming player-coach for developing teams in the Parramatta district. When he first arrived in Sydney the football club had found him a job as an ironworker, where he could make use of his training as a plumber. He shifted from that into sheetmetal work; moving at the same time from a union dominated by conservative people associated with the Catholic Church into one run by a coalition of people from the Communist Party and Labor Party. When he attended union meetings he bought copies of the Communist Party's newspaper, *Tribune*, and admired the stand the paper took in opposing the development of nuclear bombs internationally. Soon after attending a party forum during the 1953 state elections he joined the Communist Party, thus completing his disaffection from the Catholic Church.

As a party member and union activist, Mundey got jobs on building sites, firstly at St Marys and then in Granville, where he joined the BLF. At that time the BLF was under the leadership of people who were extremely hostile to the Communist Party and Mundey was horrified to discover that he was put on a kind of employment blacklist when people in his own union told employers he was a 'trouble-maker'. He could not get a job with one of the big employers but he managed to survive in the industry by working at a multitude of small sites. A change in the union's leadership at elections in 1961 meant that he was no longer a pariah and he was able to get a job on a large site – building the University of New South Wales – where he became a union delegate. In 1964, the BLF encouraged its members to participate in protests against any Australian involvement in the Vietnam war and at his first protest Mundey was one of three people to be arrested for disobeying a police order to disperse. This did nothing to dim his enthusiasm for left-wing politics and, in 1966, he became the Sydney district president of the Communist Party. In this position he was heavily involved in the campaign to support the land rights claim of the Gurindji people in the Northern Territory (see chapter 9). In 1968 the BLF secretary, Mick McNamara, resigned suddenly and Mundey was asked to take on the job until the next elections. He was then elected to the position in his own right.

During the first two years of Mundey's term as BLF secretary, the union's membership rose from 4,000 to 10,000 (Mundey, 1981, p. 44) on the back of a big boom in the building industry. The surge in membership, coinciding with a surge in demand for labourers, put the union in a strong position to campaign for improvements to job conditions, with the big issues being job safety and security of employment. However, a series of skirmishes with particular employers in 1970 quickly escalated into a statewide industry strike that lasted five weeks. Facing huge financial losses, the employers finally settled in favour of the union demands, but, according to Mundey, the union only just managed to outlast its opponents with members and leaders alike going without pay for an extended period. It was a close-run thing that could have ended in disaster for the union but the result was a battle-hardened organisation with a network of experienced organisers who had formed a highly effective strike committee. It was a high point for the union but also a worrying time for Mundey and his leadership team. They were concerned that the building boom had been so sudden and frenetic that it could easily end in an equally spectacular bust. They became increasingly concerned about new projects that seemed ill-conceived and possibly unsustainable. A rogue developer could do enormous damage to the whole industry, maybe even triggering an industry collapse. As early as 1966, Mundey wrote an article in the union journal that criticised the boom in the construction of office blocks, calling on builders to develop more socially useful projects (Burgmann and Burgmann, 1999, p. 40).

It was in this climate of concern about the short-sighted behaviour of Sydney property developers, that Mundey and his colleagues started to look more critically at the broader social and environmental impact of the industry. As a member of the Communist Party, Mundey was well aware of the international impact of Rachel Carson's 1962 book *Silent Spring*. He was predisposed to the view that capitalist development was entirely profit-driven and consequently oblivious to social and environmental concerns. However, it was the Sydney building boom that placed these concerns on the agenda of the union. As Mundey told journalist Denis Minogue in 1972 'it's not much use getting great wages and conditions if the world we build chokes us to death' (quoted in Burgmann and Burgmann, 1999, p. 36). So, in 1970 Mundey got the union to release a public statement to the effect that 'as the workers who raised the buildings we had a right to express an opinion on the social questions relating to the building industry' (Burgmann and Burgmann, 1999, p. 81). It was this statement that came to the attention of three Hunters Hill women who had started a committee calling itself 'Battlers for Kelly's Bush' to stop the housing development that would destroy the bush of that name. They came to Mundey to ask for his help and he agreed to put their request to the union executive. According to

Mundey, the idea of imposing a ban to save bushland in an affluent suburb like Hunters Hill gave rise to a 'spirited debate' on the union executive. However, the decision taken was to inform the Battlers that the union would be willing to impose the ban provided they could show there would be broad public support for it in the area. The Battlers organised a public meeting attended by 450 people, at which the sentiment for saving the bushland was clear and strong. A union ban on the site was enough to convince the developer, A. V. Jennings, to abandon the project. Thus, the green bans movement was born.

When Mundey and his colleagues were eventually excluded from the BLF in 1974 by the hostile national secretary based in Melbourne, Norm Gallagher, they moved out of the public spotlight. An article by Helen Pitt in the *Bulletin* in 1992 suggested that after the heady days of the early 1970s Mundey's achievements as an urban environmentalist were probably better appreciated in Europe and North America than in Australia and she quoted Patrick White as saying that he was 'the great wasted Australian'.[20] However, Mundey himself has pointed out[21] that he maintained a very active involvement with the preservation of heritage buildings and areas that green bans had saved. He served a term as an independent councillor on the Sydney City Council and had been consulted about urban development issues in every capital city of Australia. He was regularly invited to speak at union meetings and at universities and by 1999 he estimated he had visited every major campus in the country at least once. He was elected to the national council of the Australian Conservation Foundation in 1973, while still in a leadership position in the BLF, and stayed on that governing body until 1993. When he declined to stand for re-election in 1993 he was made a life member of the ACF. In 1998 both the University of New South Wales and the University of Western Sydney presented him with honorary doctorates for his outstanding contributions to heritage preservation.

Mundey has pointed with pride to the fact that the internationally acclaimed ecologist Paul Ehrlich once said that the green bans movement in New South Wales heralded the international birth of urban environmentalism as distinct from nature conservation.[22] During a visit to Australia at the time of the bans, Ehrlich is reported to have said that this form of direct action to preserve the environment was 'the most exciting ecological happening, not only for Australia but overseas as well'. In 1976, the ACF published a tribute to the New South Wales BLF, describing it as 'a group that has achieved more for urban conservation in Australia than many a government'.[23]

Norm Gallagher's campaign against Mundey and his close associates Bob Pringle and Joe Owens was partly fuelled by divisive, sectarian labour movement politics. Whereas Mundey and Owens were members of the Communist Party of Australia (CPA), Gallagher was a prominent member of the Communist Party of Australia–Marxist/Leninist (CPA–ML), which had split from the CPA at the time of the Sino-Soviet split in the 1960s. While the CPA became increasingly independent from the international communist movement in the 1960s, the CPA–ML aligned itself with China and the 'Maoist' Gallagher was determined to undermine the influence of the CPA in 'his' union. According to Mundey,[24] property developers and building companies in New South Wales supported Gallagher's move against the New South Wales leadership by offering 'bribes' to their employees to change their allegiances. Mundey insists that he and his colleagues never lost the support of the union rank and file. But Gallagher eventually used a bogus interpretation of the union rules to remove them from their positions and have them expelled from the union in 1974. Although the Federal Court eventually ruled, in 1978, that the expulsions had been illegal, this ruling came too late. Gallagher supporters had been duly elected to leadership positions in the New South Wales branch and Gallagher had succeeded in having his rivals blacklisted out of employment in the industry.

Bitter political divisions within the trade union movement that led to the demise of Mundey and his colleagues left a bad taste in the mouths of those who had been inspired by the green bans. This helps to explain why those who formed new green political organisations were determined to create a new kind of political culture. As early as 1972, Richard Jones had argued that the UTG should see itself as 'neither left nor right but way out in front' and when the Ecology Party was formed in Britain in 1976, it said it was seeking a 'third way' in politics. When Brown and the internationally acclaimed ethicist and animal rights campaigner Peter Singer stood on a joint ticket for the Greens in federal elections of 1996 they collaborated to produce a booklet *The Greens*,[25] that was much more comprehensive than any election manifesto could be. It included a chapter on 'Green Ethics' and one titled 'A New Politics', in which the authors suggested that it is highly unusual to make personal ethics a political campaign issue but they were campaigning for a shift from an ethic of self-interest to a communitarian ethic that included a sense of responsibility for future generations of people and for non-human life on Earth. Anticipating the criticism that such pious statements have no place in the pragmatic world of politics, they wrote:

> When we think of ethics, we have to stop thinking of bishops thundering against promiscuity, or conservative professional organisations telling doctors and lawyers how large their brass nameplates can be. An approach to ethics that is based on our ability to think rationally and critically about our values, combined with empathy and concern for others, could become the most powerful force for change the world has yet seen. (*The Greens*, pp. 54–55)

During the same campaign Brown also said that Greens should feel no shame when they are accused of being political idealists. 'The Greens emerge from a location in the heart, not on a map', he wrote (Law, 1997).

By conventional standards, the Greens have had limited and patchy success in Australian politics. Jo Vallentine was elected to the Australian senate from Western Australia as a Green Independent in 1987, having earlier represented the Nuclear Disarmament Party. She stood down for Christabel Chamarette, who was joined by her colleague Dee Margetts from WA in 1993, and they held the balance of power for three years. Chamarette lost her seat to the Australian Democrats in the 1996 elections, but Bob Brown won a seat in Tasmania and joined Margetts. Margetts lost her seat to the Australian Democrats in 1998. Greens have been elected to state parliaments in Tasmania, New South Wales and Western Australia and the legislative assembly in the ACT. They have tasted considerable success at local government level in many parts of the country and in Queensland elections in 1995, Green preferences effectively decided the outcome of the election, as they did in Western Australia in 2001.

The Greens have had their most sustained success in Tasmania – partly because the state's unique Hare–Clark voting system has allowed representation by minority parties and independents in the lower house of Parliament and partly also because the Labor Party in Tasmania has been more hostile to environmental concerns than its counterparts in other states (Crowley, 2000). The Greens, led by Bob Brown, kept a minority Labor government in power from 1989, but the experience made Labor even more hostile to the Greens and their agenda. After 1996 the Greens again held the balance of power and this time, led by Christine Milne, they entered an agreement to support a minority Liberal government. Milne was much less adversarial in her style than Brown and she told the authors in a 1997 interview that she was proud of the successes she and her colleagues had achieved in having more transparent consultations around the shaping of the state's budget and in getting the Parliament to adopt a convention against the use of violent language in parliamentary speeches in the wake of the tragic Port Arthur massacre of 1997, when a gunman shot dead 35 people during a random shooting spree. As parliamentary leader of the Tasmanian Greens, Brown probably followed a more ecocoentric agenda that that pursued by Milne as leader; however in the decade following the Franklin River campaign there was a general shift towards a broader agenda, reflecting a combination of wilderness politics and social justice politics. Of course, it was not easy to be sandwiched between two political parties that were both hostile to the Greens even when they wanted their support. When the authors asked Brown why he had decided to leave the Tasmanian Parliament he said that he got to a point when he thought he might do something 'untoward' if another journalist asked him about political machinations rather than issues. Yet, when asked if he regretted his decision to have a career in the state parliament, he replied:

> Oh, I wouldn't trade it for anything. An increase in the World Heritage in the state from 769,000 to 1.384 million hectares would take decades of campaigning. We got freedom of information legislation through the parliament; local employment initiatives which now employ 1000 people; the Douglas–Apsley National Park and reserved areas dotted across the state.[26] I knew what we were in for and we drove ourselves to the limit to get what we could. Even in the last week before the government [led by Premier Michael Field] fell on its own sword we managed to get The Friendly Beaches added to the Freycinet National Park.

When Milne was asked for an assessment of her experiences in supporting two minority governments – one Labor and one Liberal – she listed some achievements in environmental law on issues such as land clearance and planning processes, but she also highlighted the achievement of being able to get some funding put back into education after the Greens had earlier felt obliged to support cuts to the education budget when they were convinced the state's budgetary position was very fragile. She felt that the Greens had changed the culture in Parliament by being willing to listen and talk to people on both sides. To an extent, Brown and Milne probably reflect two ends of the spectrum of green politics in Tasmania – from hard-nosed advocacy for the environment through to efforts made to humanise the parliamentary processes. Yet, no change of leadership style was going to convince the major parties that the Greens should be welcomed into the club because a 1998 proposal by the Liberals to reduce the numbers of parliamentarians was made into a more radical reform by a Labor amendment. One result of this was to make it much harder for the Greens to achieve the quota of votes needed to get people into the new Parliament and in the 1999 elections only Peg Putt (who had earlier replaced Brown in the electorate of Denison) was returned. The major parties probably both hoped that the reform would eliminate the Greens from Parliament but Putt continued to be a thorn in their sides (with polling in 2000 showing an improvement in the level of support for the Greens). The Greens may have been pushed further into the margins, yet, as Kate Crowley (2000) has said, there are some advantages in being 'peripheralised' by both major parties because this adds weight to the critique made of those parties.

Critics say that green parties can promote an eclectic mix of policies and ideas because they do not have to assume the responsibility of governing. The Tasmanian Greens might take exception to this by pointing to the responsibilities they assumed in negotiating the terms for allowing two minority governments to take power. Nevertheless, the dismissal of the voices in the margins is a facile one because, as we saw with the career of some 'marginal' scientists in chapter 7, those who are in the margins are often the ones able to contribute innovative ideas that can regenerate systems in decline. Clearly, marginal voices often are irrelevant, but a society that is not receptive to marginal influences can suffer stagnation. Marginal voices need to be destabilising in order to be regenerative as well.

For Jack Mundey green politics was an extension of his concern for quality of life. For Bob Brown, quality of life should extend to an empathy with all living things. He has frequently said that he finds the energy to sustain a frustrating and tiring political career by going to inspirational places, like Drys Bluff, which rises majestically behind his home on the Liffey River in northern Tasmania. In his contribution to a book published in 1990 (Pybus and Flanagan (eds.), 1990) he explained why he thought it was so important to have a retreat like Drys Bluff and he went on to say:

> One thing is clear. Alienation begets alienation. It is vital for everyone's future that the bond with nature is restored, nurtured and made our guide in planetary affairs. It is important that we all have a bluff or a river, a tree or a vacant patch where other things are busily in control, a place where we can not only remember the Earth, but find ourselves and meet the Universe which, awesome as it is, has nurtured us through billions of years into this amazing existence. The Earth is us, and as we diminish it, so we diminish ourselves . . . (Pybus and Flanagan (eds.), 1990, p. 249)

> Each of us is here for just the blink of history's eyelid. Rather than fearing our mortality we should cherish our continuity with all life before us, and with life in the future. It is our role to offset the disaster and make life as secure as we can – not just human life but that of all the species, the whole splendid fabric of life on the planet with our millions of interdependent fellow species. (p. 252)

Brown had been a restless soul until he found a sense of home in Tasmania; and the experience changed his life and career. The young doctor who found inspiration in peaceful and remote parts of Tasmania ended up having much in common with the farmer's son from the Atherton Tableland who found his home in the hurly-burly of Sydney's trade union politics. A surprising pair of soulmates, united by an emerging conception of green politics. Starting from different directions, Mundey and Brown have helped to build a bridge between the environmentalism of the city and the bush.

CHAPTER 11

TOWARDS A
COMMUNICATIVE ETHIC:
*Australian Contributions to
Ecophilosophy*

INTRODUCTION

This book has argued that Australian settler society has produced pioneer-ing ecological thinkers and practitioners for well over a century. As suggested in chapter 1, the paradox is that *because* the settler society began so inauspiciously from the perspective of relationships between people and landscapes the tensions that arose in these relationships forced some people to think more deeply about their beliefs and practices. Rather ironically, the nation at the edge of an empire has been able to produce ecological pioneers who have had a significant international influence; certainly out of propor-tion to the size of population or economic importance of that nation in global terms. Yet, at home in Australia, such pioneers have often been in the margins and their influence has been limited in a society that has remained addicted to economic growth and consumerism. When 'second-wave' envi-ronmentalism reached a peak in its political influence in the 1980s many environmentalists believed that it was enough to convince people that humans are in the process of damaging the planet's life support systems to turn around public opinion and produce behavioural change. It has not been so easy and, in fact, there is some evidence to suggest that messages of global catastrophe began to fall, increasingly, on deaf ears.

Of course, there are plenty of people who say that the decline in overt public concern reflects the fact that environmental concerns have been 'mainstreamed' to the effect that new and clever solutions to the problems are being devised. In a seminal work in 1972, the Norwegian philosopher Arne Naess described this as a 'shallow' response to the crisis, when we need a much deeper exploration of our beliefs, attitudes and practices. Naess went on to outline the principles for what he dubbed a 'deep ecology' movement. Others joined him in exploring the cultural roots of our alienation from the non-human world without necessarily accepting his principles. Interest grew in exploring the roots of anthropocentrism and in trying to outline some principles for more ecocentric philosophies and worldviews that would be able to sustain a more radical change in attitudes and behaviours. Once again, in this work, Australians have been able to play a disproportionate role internationally.

Around the time that Arne Naess was articulating his ideas, some Australian philosophers were working independently on an exploration of western anthropocentrism. Prominent among them were Richard and Val Routley who worked both independently and together on this project during the 1970s. When their partnership broke down, Val changed her name to Val Plumwood – in honour of the plumwood trees that grew in pockets on her bushland property outside Braidwood, New South Wales – and she began to bring together her work as a feminist philosopher and ecophilosopher to help forge an international ecofeminist movement, which, in some important ways, became a rival of the deep ecology movement led by Naess. In the 1980s Plumwood was joined by another prominent Australian ecofeminist in Ariel Salleh, while Hobart-based Warwick Fox became a prominent voice in Naess' deep ecology movement. Meanwhile, a founder of the world-famous Rainforest Information Centre in Lismore, New South Wales, John Seed, went to North America to build a bigger base of support for rainforest protection globally and there he linked up with deep ecology philosopher and activist Joanna Macy. Macy and Seed collaborated on the development of ecocentric workshops – called the Council of All Beings – and together they collaborated with Naess and Pat Fleming to write an influential book called *Thinking Like a Mountain.*

In their efforts to understand the cultural roots of western anthropocentrism, the ecophilosophers were trying to understand what Plumwood has called the 'hyperseparation' of humans from nature. While most of these philosophers were also environmental activists they often wrote for academic audiences or for people who were already sympathetic to their general critique of western culture. In a very different way, the Australian cartoonist Michael Leunig also took up themes related to alienation and tried to convey ideas in simple pictorial form to very broad newspaper audiences. The self-effacing Leunig has been a reluctant philosopher but there are many critics who say that he has been able to go further and deeper than any other cartoonist, into issues that newspapers rarely touch. Encouraged by the response to his more philosophical works, Leunig went on to openly explore issues related to spirituality and the loss of 'soulful' relationships with the non-human world.

Those who have started with an interest in the root causes of our alienation from the 'more than human' world have often gone on to suggest ways of building more 'ethical' forms of engagement. In Plumwood's terms we need to work on a 'communicative ethic' that involves us in an interactive ('dialogical') kind of relationship. If the familiar messages of looming global crisis are losing their motivational power we desperately need to find more inspirational forms of engagement. So say the philosopher, cartoonist and 'hippie drop-out' whose stories are featured here.

THE PHILOSOPHER: VAL PLUMWOOD

In February 1985 Val Plumwood was paddling a canoe on peaceful waters in Kakadu National Park, on a break from her work as a lecturer in philosophy and feminism at Sydney's Macquarie University. Early in the wet season, these wetlands offer spectacular scenery; a profusion of water lilies at their peak, majestic waterbirds in large congregations, still waters reflecting the image of massing thunderclouds on the horizon, and backdrops of paperbark trees and the more distant craggy cliffs of the Arnhem Land Escarpment. Plumwood had borrowed the canoe from a park ranger who had advised her to stick to the backwaters because the current in the East Alligator River was dangerously strong and there were plenty of crocodiles waiting for anyone who might get into trouble there.[1] He provided Plumwood with a sketch map to help her make her way to an Aboriginal rock art site via the backwaters.

This year the rains were late and the water levels were still low. A slight drizzle had begun when Plumwood pushed off from the canoe launch site but, having enjoyed a glorious outing in the canoe on the lagoon the day before, she was determined to make the most of her opportunity. By the time she had made her way across the lagoon and into the maze of shallow channels on the other side, the drizzle had turned to squalling rains with strong winds and she found herself having to stop a number of times to tip the rainwater out of the canoe. The sketch map proved inadequate for negotiating the maze and the going was very slow.

Plumwood moved into a channel that seemed to take her in the direction she wanted to travel, but with the water level low the banks were high and she found it very difficult to get her bearings. She stopped the canoe on a sandbar and, having checked carefully to establish that no crocodiles were in sight, she got out and walked to the top of a sand dune to establish her position. She was horrified to discover that the fast-flowing main body of the river was only a hundred yards in front of her. A sense of unease that had been with her for much of the journey now grew into outright alarm as she realised that 'as a solitary specimen of a major prey species of saltwater crocodiles, I was standing in one of the most dangerous places on the face of the earth' (Plumwood, 1996). Before hurrying back to the relative safety of the canoe, her gaze was captured by the tumbling rocks of the escarpment cliffs on the other side of the river, where the sight of a large rock balanced precariously on a much smaller one made her think of her own vulnerability and the precariousness of human life. It also made her realise that she had not sought the advice of the local Gagadgu people about visiting this place.

Plumwood was relieved to make it safely back to the canoe. Crocodiles will not attack you if you stay in the canoe, the ranger had assured her. But when she looked at the fibreglass walls of the canoe standing just inches above the water, she felt only partially reassured. Having abandoned all thought of getting to the rock art site, she was now entirely focused on the task of getting back to the camping ground as quickly as possible. Now travelling with the current, she began to make good progress and it seemed that the moment of greatest danger had passed. Rounding a bend in the high-banked channel, Plumwood noticed what appeared to be a log floating in midstream. As she tried to use the paddle to steer well clear of the object she noticed that it was a crocodile and she found herself powerless to prevent the current dragging her in its direction. She could see that it was not a large crocodile – smaller, in fact, than her canoe – so she was sure it would swim away to avoid a collision. She reminded herself of what the ranger had said. But the crocodile not only maintained its course, it launched itself against the side of the canoe with great force. Fearing that she would be tipped into the water, Plumwood turned the canoe and began paddling furiously towards the bank as the crocodile returned to the attack again and again. Now fully aware that she had become prey, she was caught in a deadly dilemma. The canoe offered little protection but if she got onto the steep, muddy bank she would be quickly hauled in. She decided instead to try to leap into the branches of an overhanging paperbark tree. As she prepared to jump, she looked directly into the golden eyes of the crocodile and tried, desperately, to shoo it away with waving arms. She leapt for the nearest branch but was intercepted by the crocodile bursting out of the water and clenching her between the legs.

The crocodile took her into a death roll – 'a centrifuge of whirling, boiling blackness, which seemed about to tear my limbs from my body, driving water into my bursting lungs' (Plumwood, 1996). After what seemed an eternity, the fury ended and Plumwood was amazed to find herself alive with her head above water, coughing and sucking in air. She was taken into a second death roll of similar intensity but again survived the ordeal, this time grasping the stout branch of a sandpaper fig growing in the water. As she held tightly to the fig tree, the crocodile relaxed its grip and she managed to break free, pulling herself behind the tree but not onto the bank. Again the only avenue of escape appeared to be the paperbark tree but 'in the repetition of a nightmare, when the dreamer is stuck fast in some monstrous pattern of destruction', she was again intercepted by the crocodile and taken into a third death roll. The third death roll was of less intensity and again Plumwood found herself adjacent to the sandpaper fig when it was over. This time, when the crocodile – apparently tiring – relaxed its grip she dragged herself onto the slippery bank and threw herself into the task of clawing her way to the top with all the strength she could muster. Slipping and sliding, as tufts of grass failed to hold her weight, she discovered that she could get just enough purchase by digging her fingers directly into the mud. When she got to the top of the bank she hauled herself to her feet and began to think, for the first time, that escape was possible. She later wrote:

> The course and intensity of terminal thought patterns in a near-death experience can tell us much about our framework of subjectivity. A subjectively centred framework capable of distinguishing action and purpose must, I think, view the world 'from the inside' . . . investing it with meaning, reconceiving it as sane, survivable, amenable to hope and resolution . . . This desperate delusion split apart as I hit the water [in the first death roll]. In that flash, when my consciousness had to know the bitter certainty of its end, I glimpsed the world for the first time 'from the outside', as a world no longer my own, an unrecognisably bleak landscape composed of raw necessity, that would go on without me, indifferent to my will and struggle, to my life or death. (Plumwood, 1996)

Viewing her life 'from the outside' as death was upon her, some rather surprising thoughts popped into her head. At one point, she told the authors, she found herself imagining how her friends might receive the news of her death, feeling annoyed that she would not be there to tell them what really happened. At another moment she found herself thinking that she now had a good excuse for being late with an article she was writing for the *Australian Journal of Philosophy*. In her seemingly hopeless struggle for survival, she was able to blank out the pain from her horrific injuries. It was not until after she had scrambled to the top of the bank and started walking in the direction of the camping ground and ranger station that she discovered that her left thigh had been badly shredded and she was bleeding profusely. (She would later learn that she narrowly missed having a major artery severed.) Somehow, she managed to walk about three kilometres before collapsing; realising, with a sinking heart, that she was on the opposite side of the water to the canoe launch site with no prospect of crossing over. She then managed to crawl the rest of the way to an open spot opposite the launch site. Daylight was fading fast and again she began to despair about her chances of surviving. With the onset of darkness, the heavy rain and wind stopped, but then a cloud of mosquitoes gathered around her and dingoes howled in the distance. Loud, swirling noises in the water nearby reminded her that she remained 'easy meat' for another crocodile. If she was not found she might die from loss of blood. After all that had happened, the apparent certainty of death had returned. She hoped for unconsciousness but 'consciousness persisted'.

Fortunately for Plumwood, the ranger who had provided the canoe began to wonder about her return and noticed that the light had not come on in her caravan. He rode to the canoe launch site on a motor-trike and noticed that the canoe had not been returned. On the opposite bank, Plumwood managed to raise herself to her elbows and call out for help. The ranger heard enough to turn off his engine and listen more carefully, and Plumwood established voice contact with him. He hurried away for help and returned about an hour later with some assistants. Plumwood was rescued and taken by helicopter to Darwin Hospital. The next day the spot where she had lain while waiting for help was under four metres of water. The late rains had arrived with unusual intensity.

> Obviously, I think about what happened that day a lot. Clearly, I under-estimated the danger in what I was doing and I think that a stronger appreciation of danger has given me more respect for nature. But I also think about why I was so naïve about the danger. I mean, my rational mind told me that what I was doing was safe. Plenty of other people had been on these waters in canoes. The day before I had been paddling in brilliant sunshine, surrounded by the extraordinary beauty of Kakadu. The danger was hidden by all the beauty.
>
> I was incredibly lucky to survive and my luck extended to the fact that I can still walk well and have lost few of my capacities. Despite the injuries and the pain, the wonder of surviving the attack cast a golden glow over my life for about a year. After that I continued to feel grateful for the gift of life. But the real impact of the experience on my ideas and work as a philosopher took much longer to mature. It was maybe ten years before I felt I had really absorbed the lessons of the experience. Before the encounter with the croco-dile I had been actively campaigning for about 10 years against the destruction of native forests. As a philosopher, I had argued for the inclu-sion of nature in our ethical framework. But the crocodile event changed my emphasis to one of better understanding our dependency on nature and our over-estimation of our powers. We need to understand danger because we have a sense of security that is illusory. The dominance of rationalism in our culture creates a degree of over-confidence. (Plumwood, 1996)

At the time of the crocodile attack, Plumwood had been working for more than a decade as a philosopher with a particular interest in the ways in which western philosophical traditions have conceived the relationship between humans and non-human nature. In a review of a book by John Passmore published in the *Australian Journal of Philosophy* in 1975,[2] for example, she criticised the assumption that nature has no value except as something for humans to use; a resource.

> However, the event with the crocodile began to shift my sense of nature from the passive to the active voice. That is, I became much more aware of nature as a powerful agent, and also aware that the dominant culture devalues and

denies this agency as part of its human-centred dynamic. We see ourselves as independent users of passive nature that we are not part of. We live in 'culture' and other things – wild ones – live in 'nature'. We think we can manipulate nature from outside, from a distance. This is dangerous both because we lose a sense of kinship with the non-human elements of our world and also because we become less aware of our vulnerability. As crocodile prey I was brought face-to-face with my own inclusion in nature, as an embodied and edible part of the food chain. This disrupted the idea that as a human I couldn't be food because I belong in the realm of culture. This was shocking, even for someone who had thought a lot about human-centredness. The experience went very deep and made me realise that we have cultural blindspots that are dangerous to us because we lose sight of our dependency and vulnerability.

So, rather than dimming her enthusiasm for nature, the crocodile attack gave Plumwood a stronger incentive to continue her work on trying to understand how and why people in western societies have developed such distorted and illusory views about nature. She was able to return to her house deep in the forest near Braidwood, New South Wales, where her surroundings are very conducive to this kind of work. She returned to Plumwood Mountain, which, like Plumwood herself, takes its name from the extraordinary *Eucryphia* trees from the Gondwanan rainforests that survive in pockets on this high plateau above the coast.

Plumwood has lived at her home at Plumwood Mountain since 1974. Her house – located in a small clearing in the forest – is octagonal in design and constructed from local timber and stone. To one side is a well-tended garden ('my family'). A short walk through the forest in one direction brings you to the edge of the escarpment with a magnificent view towards the sea. A slightly longer walk in another direction down into a creek-bed brings you to a fairyland grove of ferns and plumwood trees. But even here danger lurks in the form of the abundant burrows of funnel-web spiders. 'You have to keep your eyes open when you walk around this country', Plumwood warned us, 'there are lots of tiger snakes.' At the time she was interviewed for this book, Plumwood was receiving frequent visits to her house from a wombat she had cared for when it was young. By choice, 'Birubi' had relocated himself into the forest but on cold nights he liked to come and sit by the fire. It seemed that the movement of Birubi between the forest and the house illustrated the relationship Plumwood had established with her surroundings.

Plumwood moved into her forest home in 1974 when her husband Richard Routley got a job at the Australian National University. Living so close to the forest helped to nurture their philosophical work, yet soon after they moved into the area they discovered, to their horror, that there were plans to turn nearly all the nearby tableland forests into pine plantations. In

Val Plumwood with her good friend Birubi the wombat, circa 1987
(Val Plumwood)

trying to draw attention to the folly of what was planned, the Routleys (as Val was still named at the time) wrote an article for the *Current Affairs Bulletin* that was mentioned in a parliamentary debate by Tom Uren, a minister in the federal Labor government of the time. After some more research they then wrote a book called *The Fight for the Forests*. Plumwood told the authors that the title of the book was only something that came to her a week before the book went into print and was a bit misleading because the book was 'actually a very academic piece' that offered an 'early critique of the forestry industry'. To the surprise of the authors, the book went through three editions in as many years. When it first appeared, its views got only limited support from a conservation movement that tended to be sympathetic to the forestry industry. According to Plumwood, Milo Dunphy at the Total Environment Centre in Sydney stood out as the conservationist who was willing to support their views, although they also heard that Judith Wright was sympathetic. By the time the conservation movement had developed what Plumwood called a 'critical edge on forestry' the plans for the southern tablelands had been shelved for economic reasons. However, the book by the Routleys had been influential enough to cause the federal government to withdraw funding assistance for the creation of pine plantations.

In campaigning against the destruction of native forests Plumwood had become a conservation activist. But the issue also gave her an 'abiding intellectual interest'. She saw the desire to construct artificial forests with straight lines of trees that could be easily harvested by machines as a product of rationalism; an attempt to impose 'order' on nature. To better understand this urge to rationalise the forests she began her journey into the philosophical roots of rationalism.

Philosophy has been Plumwood's passion since her early days at Sydney University in the late 1950s. Despite doing very well at school she had to defy her parents to go to university and they could see no merit at all in her interest in philosophy. Her mother worried that she might become too scholarly and, consequently, unattractive to men. The idea that scholarly women would end up being poor and lonely was widespread at the time and was a major barrier for women wanting to enter tertiary education. In Plumwood's case, her parents wanted her to escape the poverty of her childhood and they did not think that an interest in philosophy would achieve that aim.

Plumwood was born in 1939 when her parents lived in a shack with hessian walls painted with cement. They had come to Sydney from the country in search of work and put their names down for a free land grant at Terrey Hills, then on the edge of the city. Her father worked as a hod-carrier (a labourer in the building industry). When the war came he went into the army. After the war he managed to get a little poultry farm going at the Terrey Hills property but it was a real struggle. As a young girl Plumwood was without toys. However, what she did have was ready access to beautiful sandstone country on the edge of Kuringai Chase National Park. The bush was her playground and she acquired an abiding love for the Sydney region wild flowers and the octopus-like *Angopheras*. When she was ten the family had to give up the farm because they could not compete with the emerging battery farms ('a foretaste of a cruel future'). Her father was able to get a low-level job in the public service interviewing other struggling farmers and they moved closer to his work. First they lived at Collaroy ('that wasn't too bad') and then they moved to Kogarah ('Arr, hated that. I missed the bush terribly. It was as if a big part of me had been silenced').

> It's important to say that my parents were very thoughtful people and they encouraged me to think. They only had a few years of education between them, but my father was a published writer and our house was full of good literature. I learnt to read very early. My mother was my first teacher and, even though she left school at 12, she nurtured my love of learning. When I was sent to school at the age of eight I was shocked to see students being punished for making mistakes. I had never experienced that with my mother and I did well academically; well enough to win a Commonwealth Scholarship to go on to university.

Plumwood's parents were divided about her going to university. Her mother saw no use in it at all, while her father thought it would be useful provided she did something 'practical'. As a compromise she passed up the Commonwealth Scholarship for a less competitive Teacher's Scholarship that would virtually guarantee her a job as a teacher once she had completed her studies. However, within a very short time she had fallen in love with philosophy and abandoned all thoughts of becoming a teacher. At 18 she married a fellow student in a 'shotgun marriage' and by 21 she had 'two children and a divorce'. Now her mother agreed that university study could help her find a more fulfilling life and so she agreed to look after the children so that Plumwood could resume her studies. In her second stint at Sydney University she encountered the famed 'Sydney Push' – a circle of intellectuals and artists seeking to break the shackles of conservative Australian society. One of her English lecturers was the 'Push' identity and later feminist author Germaine Greer.

At the end of her undergraduate studies, Plumwood married fellow philosopher Richard Routley and they followed the convention of the time in heading off for an extended working holiday to 'see the world'. They travelled through the Middle East and spent time living in Britain where they could pursue their interests in philosophy. They spent nearly a year living in Scotland. Plumwood says that travelling in the Middle East made a strong impression on her because she learnt that forests had been long ago turned into deserts. In Scotland she lived near a pretty little beech forest. But the charm of that forest began to wear off after a year and she longed to return to the Australian bush. She remembered that as a child she had visited the homes of older relatives who had images of European scenes on their walls because they thought the Australian bush was very boring by comparison. By contrast, she had grown up playing in the Australian bush and after a long absence she began to miss it badly.

Returning to Australia in the late 1960s, Plumwood experienced a powerful prejudice against married women working, which even permeated the universities. On one occasion she was told that a younger, less qualified man had been given a job she had applied for because he had a family to look after. It was not until the early 1980s that Australian universities began to open up for women and then Plumwood was able to get a position at the Australian National University. From there she moved to another position at Macquarie University in Sydney, where she developed subjects in environmental philosophy and feminism, building on ideas she had pioneered in the 1970s.

Between 1973 and 1982 Val and Richard Routley published a series of important papers – both separately and together – that initiated a discussion of environmental philosophy in Australia. At the same time that the Norwegian philosopher Arne Naess was developing the concept of 'deep ecology' in Europe, Richard Routley was attracting the notice of fellow philosophers for his strong critique of human supremacism.[3] Val's independent contribution to the critique of human-centredness began with her 1975 review of John Passmore's *Man's Responsibility for Nature*. Another article published in 1978 attracted a lot of attention and an article she wrote about rainforests that was published in *The Ecologist* in 1981 was still being cited in the late 1990s. The Routleys established an important and successful collaboration yet, as a man in an established academic position, Richard was given most of the credit for their joint work. Articles that were written jointly were often cited in his name only.[4] Val became the victim of the tendency to devalue women's work. Her collaboration with Richard was doomed.

Plumwood's experience in having her voice silenced made her think more deeply about the suppression of agency.

> Women – playing roles like wife, secretary, nurse, tea-maker and fundraiser – have been given no credit for their work in supporting male 'achievers'. Their contributions have been systematically erased; their agency denied. Men deny their dependency on supportive women. The women are seen as being inessential. Hegemonic accounts of agency erase the contribution of the 'inessential'. The same thing happens with nature. We treat it as non-agent – defined as the absence of human agency – and by making it inessential we deny our dependency on nature.

> I had been thinking about all these things before my encounter with the crocodile. But that gave me a much more powerful experience of the agency of nature and of my own vulnerability. My experience with the crocodile made me think more deeply about ideas I had been developing as a feminist philosopher.

As mentioned earlier, Plumwood told the authors that one bizarre thing that happened during the crocodile attack was that the thought popped into her head that she now had a good excuse for being late with a review article she had promised for a special 1985 women's edition of *The Australian Journal of Philosophy*. The journal's guest editors Genevieve Lloyd and Jan Thompson had asked her to write a review of the emerging litera-ture on 'ecofeminism'. The term was not yet well established and the literature was limited, but Plumwood found a strong interest in what she read. She had been particularly impressed with Rosemary Ruether's book *New Woman, New Earth*, partly because she envied the greater freedom that theologians tend to have compared to philosophers. Philosophy in Australia was dominated by 'hard-headed traditions' that were rather hostile to 'free thinking', so Plumwood saw Ruether's work as fresh ('it helped me to understand the legacy of the Greeks').

Plumwood had also been influenced by other feminist philosophers – like Genevieve Lloyd in *Man of Reason* (1984) – who were looking critically at the western conception of reason and its relationship with constructions of masculinity. She saw an opportunity to draw the threads together to give ecofeminism a stronger philosophical basis and so she made this the subject of her PhD thesis. She used the thesis to prepare a book, published under the title *Feminism and the Mastery of Nature* in 1993. In this book, Plumwood traces the dominance of rationalism back to Plato and shows how this mode of thinking produces a polarisation of competing concepts, or 'dualisms', such as reason versus intuition, mind versus body, culture versus nature. In this process differences are converted into value judgements, allowing an hegemony of culturally dominant concepts (such as reason over intuition).

Both Plumwood and another Australian, the sociologist Ariel Salleh, have made significant contributions to the development of ecofeminist thought internationally. While Plumwood has explored the philosophical roots of western, patriarchal, cultures, Salleh has explored the social structures that create power hierarchies – from 'God' through men to women, children and nature. Plumwood's interests in seeking a stronger emotional connection with non-human nature is also reminiscent of Charles Birch's call for humans to 'regain compassion' for other living things (see chapter 7). Birch reached this conclusion as a result of practical and theoretical work as a 'rational' scientist and through his interest in organicist philosophy. Plumwood reached a similar view (although she criticises Birch for wanting to extend human attributes to non-human nature) through a philosophical critique of rationalism and through her own sensuous experiences with nature.

Plumwood believes that a sensuous experience of nature is essential for an understanding that is not locked in the constraints of rationalism. She has welcomed the debate in Australia about whether or not the term 'wilderness' betrays a colonial mentality that saw Australia as being an empty continent before the arrival of the Europeans.[5] She agrees that it is important to 'decolonise' the western concept of wilderness by recognising the long history of occupation by indigenous people. But she also argues that the problem is broader than that because terms like 'wilderness' reflect a perception of nature as being other than human or defined as human absence. Hence, the solution is not as easy as deleting particular words, like 'wilderness'. People need to have sensuous experiences of nature to become more aware of its presence and agency. When we experience the presence of nature sensuously and directly we can extend our empathy to places we have never visited. She began an article on the wilderness debate by saying that she keeps a poster on her wall of a remote place in Tasmania that she is never likely to see for herself but, for her, it's enough to know it is there, hopefully safe from 'developers'. Because our culture is so destructive of nature it can only be fully present when humans are not – i.e. in a wilderness. If a culture is more nurturing of nature then human presence and natural presence can be compatible.

Plumwood's empathy for nature is constantly renewed by direct experiences in her 'enchanted forest' at Plumwood Mountain. In an essay titled 'Forest Lover, Live Forever!' she wrote:

> At Plumwood Mountain, the enchanted forest still plays out for me its passionate and mysterious dramas, but I am no longer as content as I was when young just to take in the wonder and vigour of it all, to watch, to listen and be filled by it. Now I aspire to decipher a little of the overall plot, to discern the complex connections ordering the myriad individual life stories, hoping to be given a glimpse of their larger meanings. (Plumwood, 1999, p. 67)

As Plumwood learnt so painfully on that fateful day in Kakadu, not all dramas in nature have happy endings. But if they did they would be much less enthralling. Later in the same essay she recounted one drama she was privileged to observe:

> In late dusk, that magical between time when shapes grow fuzzy and green light suffuses the world, I sit down on a fallen treefern trunk to listen to midsummer cricket madrigals. A cricket tunes up its trilling drone, a deafening, melodious jack-hammer, right next to me. I admire the pink and green sky, pondering the meaning of the trilling. Unexpectedly, my question is immediately answered. Right at my feet a large insect lands. It is another cricket, presumably one of the opposite sex attracted by the first cricket's love song, just as it was meant to be. My heart starts to warm – a real Romeo and Juliet story is being played out here in the dusk. But Marcus, one of the small water skinks that inhabit this spot (he is familiar, I see him regularly basking on this treefern, hence the name) has anticipated this little romance. I watch transfixed, torn between laughter and tears, as Marcus rushes forward from his hiding place under the treefern trunk, and – seizing the radiant bride in his jaws – carries her off as a meal. There is food for thought as well as body here. Marcus knows, the cricket knows, and I know, that it will *not* do to think of this world as a benign sphere of love, harmony and peace. (Plumwood, 1999, p. 67)

A happy ending for Marcus but not for the 'radiant bride'. Pleasure and pain coexist and danger is always lurking.

A passionate interest in the landscapes surrounding Braidwood is something Plumwood has shared with a famous resident of the area, Judith Wright. Plumwood has written[6] that she first got to know Wright by letter and telephone in the mid-1970s because of her work on forests. At this time Wright was a prominent figure in the Australian Conservation Foundation and Plumwood was impressed by the fact that she was both accessible and generous with her time. However, when they both found themselves living in the same area their acquaintance deepened and Wright liked to quiz her younger friend about what she was teaching at Macquarie University. In her tribute to Wright, published after the poet's death, Plumwood wrote:

> Judith wrote passionate poetry that expressed sophisticated philosophical
> ideas as well as intense experience of and love for living nature. In an age
> when both philosophy and poetry are academicised, she integrated her
> poetry with the key social concerns of her age, and her philosophy with the
> way she lived her life, allowing each most admirably to inform the other.
> (Plumwood, 2000a)

In Plumwood's experience, Wright was always a person who was prepared
to say what she thought even if this was confronting to those who looked
up to her. Yet she had tremendous empathy for non-human life, demonstrating the 'complexity and poignancy of each of the earth's lives'.
Plumwood told the authors that one of her most treasured possessions is a
gift from Wright – an autographed copy of her volume of poems *Birds*.

A common theme in the work of Wright and Plumwood is the idea
that we need to find imaginative ways to reconnect with nature. In her 1968
essay *Conservation as a Concept*, Wright suggested a need for 'a renewed
humility and a revival of imaginative participation in a life process which
includes us'.[7] In Plumwood's terms we need to create ways to overcome
the human/nature dualism. Plumwood believes that Wright helped
Australians to see stories embedded in particular landscapes (as in the New
England poems) and she relates this to the concept of 'storied residence' that
she attributes to North American writers Jim Cheney and Holmes Rolston. In
the article 'Forest Lover, Live Forever!', Plumwood wrote that storied
residence requires 'discerning through long-attentive engagement with a
place how its myriad unfolding histories weave a web of subtle connections.'
This sort of engagement with places can lead us into a communicative relationship with nature. In Plumwood's view it can produce a 'communicative
ethic' that helps us overcome our erstwhile alienation from nature.

Of course, Plumwood has the advantage of being able to walk into
the surrounding forest to witness a myriad of little dramas unfolding.
However, she rejects the idea that this makes her privileged:

> We all have to find our own ways of experiencing nature. It was a real
> struggle to establish a home here. I've never had much money and used
> basic materials and my own labour to build the house. The forest is only
> here because my being here has stopped someone else clearing it for money.
> For me the house has been about survival not indulgence. In our culture it
> takes great effort to reconnect with nature. You have to be pretty deter-
> mined. At the same time I think that wherever people live they can start to
> rethink their dependency on nature. It's not just about beauty and romantic
> sensibility, but survival. We have to enter a communication with nature to
> work out something that is to our mutual benefit. Communication is the
> basis of negotiation and accommodation. If we recognise that nature has
> agency the communication becomes dialogical rather than monological.
> This is the basis for ecological rationality.

Plumwood has suggested a 'renaming project' as a way of getting many more people to think about their current relationships with places. The precise origin of many existing place names may be largely forgotten. However, Plumwood has written:

> ... the naming practices of non-indigenous Australians tend to be eurocentric and to register a monological or non-interactive relationship with a land conceived as passive and silent ... [while] ... Aboriginal narrative patterns can help to show us possibilities for a richer dialogical relationship which our dominant conceptual schemes obscure. (Plumwood, 2000b, p. 99)

At one level, the problem with Australian place names is that they have been so arbitrary and Plumwood cites as an example the use of dictionaries and prescribed formulas to name suburban streets. At another level, a name may sound appropriate yet on further investigation it become clear that the reason for the name is not known and Plumwood cites the example of her father calling his Terry Hills farm 'Wyeera' because, he said, it was an Aboriginal word that meant to 'dig the soil' but, when pressed, he did not know which tribe might have used such a word nor from whence they came. However, worse than this is that place names reflect the process of colonisation in which a 'grid of foreign names'[8] was imposed on the land to celebrate its 'conquest'. Worst of all is that some places were named after people who were directly responsible for the massacre of the indigenous people, like the Stirling who gave his name to the Stirling Highway which runs through Perth. Plumwood told the authors that she was largely put on this line of thought by something Frederick Turner had written about naming practices adopted by Christopher Columbus when he 'discovered' the 'new world' and the realisation that something very similar had been done in Australia. Turner wrote:

> To each bit of land he saw [Columbus] brought the mental map of Europe, with which he had sailed. Anciently place names arose like rocks or trees out of the contours and colours of the lands themselves. As a group took up residence in an area, that area would be dotted with names commemorating events that took place in it. Where one tribal group supplanted another, it too would respond to the land, its shapes, moods, and to tribal experience had there. Now came these newest arrivals, but the first names by which they designated the islands were in no way appropriate to the islands themselves. Instead the Admiral scattered nomenclature of all Christianity over these lands, firing his familiar names like cannonballs against the unresisting new world ... An enamoured Adam in this naked garden, he established dominion by naming. (Turner, 1986, p. 131)

Whereas Columbus used the names of Christian saints, Australian colonisers liked to honour distant 'figures in the British Colonial Office' with names like Sydney, Melbourne, and Brisbane. Other names reflected a desire to transform Australian landscapes into something resembling a European Arcadian dream, using existing British place names such as Perth, Ipswich, or Penrith. Just as 'acclimatisation societies' were set up to introduce European plants and animals thought to be more valuable than the indigenous ones, European names were seen as adding value to the 'empty land'. In Plumwood's view we have ended up with a set of 'feral' names that compete with some pre-existing ones.

> An empty and highly conventionalised naming practice is both a symptom and a partial cause of an empty relationship to the land. If we want meaningful relationships with the land that expresses a healthier pattern than the colonial one, we have to look at naming it in meaningful terms that acknowledge its agency and narrative depth. So I want to propose a renaming project as a project of cultural convergence, cross-fertilisation, reconciliation and decolonisation. (Plumwood, 2000b, p. 103)

Plumwood suggested that a starting point for such a project could be to replace the most offensive names – replacing, for example, the name of the Stirling Highway with something like the Jack Davis Highway.[9] The project, however, would achieve little if it meant that another set of names was imposed from above, without local debate and involvement, because in many places the conclusion might be reached that the existing name is quite relevant.

> However, the point is they could only reach that conclusion by doing some research and learning about the hidden history of the place. At the local level it would be part of a reconciliation exercise in that white Australians would be seeking to emulate the Aborigines in establishing a narrative and storied connection with the land. It should involve extensive consultation with Aboriginal people, but the aim would not be to restore Aboriginal names but rather to think of names that are appropriate for existing people and identities. A good way to start would be to invite storytellers and people with creative interests to create place narratives that might help to uncover embedded names.

As a philosopher, Plumwood has sought to explain the roots of our alienation from nature. As an alienated citizen herself she has sought to intervene in political processes when the opportunities have arisen. For her, philosophy and activism have been brought together in the form of ecofeminism.

THE CARTOONIST: MICHAEL LEUNIG

Alienation has also been a major theme in the work of the popular Melbourne-based cartoonist Michael Leunig. Best known for his depictions of lonely, isolated people who do not fit comfortably into modern, technocratic society, he has also taken up the related theme of human alienation from nature. For example, a classic cartoon from 1974[10] shows a family watching a sunrise on television while a real sunrise is happening, unnoticed, outside their window.

Leunig was born in 1945 into a working class family in the Melbourne suburb of Maidstone. He has lived most of his life in Melbourne and most of his cartoons have an urban theme. But he has come to the view that cities are 'not my spiritual home' because they are 'isolating and confusing'. He made these comments in a short film, *Central Australian tracks in the sand* (1989),[11] which shows him making a kind of pilgrimage to the central Australian desert in search of a deeper connection with the 'soul' of Australian landscapes. In explaining the reason for undertaking the journey, Leunig says that there is 'power and mystery' in the desert and an opportunity to observe the way that Aboriginal people relate to ancient landscapes. Given that much of his work has been a *cri de coeur* on behalf of lonely and isolated people, the desert would seem to be an unlikely spiritual home for him. However, in the film he interviews a woman, Molly Clark, who has lived on her own at a station on the edge of the Simpson Desert since her husband died many years earlier. Outside occasional trips to 'town' and visits from people passing by, her only contact with other people is by radio. When they talk about loneliness Molly says that she does sometimes miss the company of other people but she says that she does not feel isolated because she has a strong fascination for the country she lives in. She doubts if she could ever feel as much at home anywhere else. By contrast, many of Leunig's cartoon characters live in cities in which they are surrounded by other people, yet they feel profoundly isolated.

Leunig has made the point that cities were less intimidating places to live when he was young and that his childhood was not an unhappy one.[12] There were lots of open spaces not far from the family home and it was still common to see horses and carts rattling along the streets (dispensing milk, collecting bottles, etc.). There was a sense that the countryside was not far away. Most children had pets; Leunig rates an early influence as being the love of his dog Dinah.[13] Many houses had chooks. The Maribyrnong River was within walking distance.

> I always had animals. We would shop at the market for our weekly food and once I bought a duckling and it imprinted on me, used to follow me around everywhere. It would try to follow me to school. Coming home, I'd turn the corner and the duck would see me and come running – not a dog but a duck! So I always got ducks after that. I like their conformation, they are playful and good-humoured and innocent with those rounded beaks and all; they aren't as predatory as chooks.[14]

Sunrise by Michael Leunig (1974)

This explains why ducks regularly appear in Leunig cartoons; a symbol of unalienated innocence.

When Leunig was about eight he went off with a cousin to one of the more risky places in the neighbourhood – the rubbish tip. 'I'm walking along this path in the sunshine, happy as a bird, never been happier,' he has written (Lennox, 1998). 'I think this was the first significant time my mother had let me go alone to a slightly risky place . . .' Then, without warning, he slid into a pit full of burning coals and tangled wire, suffering extensive and serious burns to his legs. After emergency surgery he had to lie, swathed in bandages, in a darkened room for months while he could hear children playing outside. When the burns finally healed he had to learn to walk again and by this time he felt socially isolated. As a result, he became something of a loner at school. It did not help that he had no aptitude for, or interest in, sport – the arena in which boyhood reputations and friends are often made. Even worse, he saw a funny side to the intensity with which sport is played and watched. Thrown into a school football team (Australian Rules) he once found himself with the ball in hand in front of an open goal. It would have been an easy matter to score six points but that seemed too obvious to Leunig so, instead, he simply ran between the posts to score a single point. Across the line he doubled up in laughter but when he turned around he discovered that no one else had got the joke (Lennox, 1998).

Leunig has said that he turned to drawing because it was a 'poor boy's way of doing things'.[15] 'In my father's day the way up might have been becoming a boxer. Drawing was what I could do.' For a long time he harboured hopes of becoming a writer but when he eventually got to university he felt that his education had not equipped him to mix with the literary set so he went back to his 'funny little drawings'. Leunig started out with a rather simple drawing style, but he always had a strong sense of the absurd and a capacity to reveal human emotions. While at Monash University he managed to get some of his cartoons published in the student newspaper *Lot's Wife* and after that he contributed to various 'underground' publications during the Vietnam war protest days in the 1960s. He was surprised and delighted to find that some of this work had come to the attention of the editors of *The Age*. They decided that Leunig's irreverent style might be just the thing for a new evening newspaper, *Newsday*, they were launching, called and they offered him his first job as a political cartoonist. *Newsday* lasted only about six months but that was long enough for Leunig to enhance his reputation. When *Newsday* folded he was able to walk into a plum job at the popular and irreverent newspaper *Nation Review*.

Leunig was at *Nation Review* when his famous and ubiquitous character Mr Curly was born. He remembers playing around with a figure and putting a curl on top of his head

> I liked the look, the funniness, no explanation, and out of that I thought, 'Yeah, maybe this little fellow with a curl on his head lives in a world where there are other curly things.' So I started to put this together in a picture and it's one of my favourite drawings to this day: Mr Curly comes home to this little idyllic cottage with a curl on its roof, to his wife and children with little curls on their heads and a dog with a curly tail. He comes home to a curly world . . . (Lennox, 1998)

In Mr Curly, Leunig had found a way to represent people who do not fit in with straight lines and 'normal' expectations. He had created his classic anti-hero, his own alter ego. However, a representation of the pain of alienation was not enough, even if Mr Curly could retreat to a different, imaginary world. Leunig wanted to engage in a more positive search for meaning. Given that he forged his reputation in the secular and cynical world of political commentary, it was virtually another act of rebellion by Leunig to find a new outlet in the exploration of religious language and iconography. In 1989 he contributed a cartoon to the *Sunday Age* that took the format of a prayer and just as his surprised fans decided that this was a new form of irony he surprised them even more by showing them that he was genuinely interested in the concept of prayer (Debelle, 1990). 'My friends thought I'd become a born-again Christian! I hadn't, but I'd discovered the word "God".' Leunig had begun to notice how often people used the word God in everyday language (perhaps unconsciously) to express powerful emotions so he thought it might be a key to something. '[M]aybe it means many things, maybe when you say it, the heart hears it the same as the word "love".' But for Leunig words that make cynics feel uncomfortable are worth exploring. 'I want to bring these words back into my language and use them comfortably.'

Leunig's exploration of religious language has led him to the view that 'spirituality' is a quality that is sadly missing in 'modern' society:

> Spirituality is in a very sad, sorry and sick state and the language of it is disappearing. That wouldn't matter except that the effect of this is decline everywhere, it's apparent. We are suffering from an absence of love . . . The statistics are pretty horrific about the use of Valium, Mogadon, suicides, mysterious illnesses of the spirit . . . (Debelle, 1990)

Leunig's sister Mary, herself an accomplished cartoonist, has worried about her brother's willingness to bare his soul in public. 'I think Michael's pretty brave to get so personal', she has said. 'I sometimes think maybe he's revealing too much about himself . . .' But perhaps the bravery has paid off in very direct ways as Leunig has attracted an even broader circle of admirers interested in his deeper thoughts. As David Tacey (1993, pp. 161–162) has said Leunig has single-handedly broken the mould of cartoonists as satirical commentators on ephemeral political events and has 'won back for popular culture the more subtle realities of the soul.'

Empathy with nature has long been a feature of Leunig's work. For example, a 1974 cartoon depicted Mr Curly perched in a tree crowing proudly like a bird. But it was Leunig's pursuit of soul that led him to take the pilgrimage to Central Australia, where he accompanied a group of Aboriginal elders on a journey into their country. It has been his interest in spirituality that has given him a stronger empathy with Aboriginal art:

> I feel I know this [A]boriginal art, there's a lot of hype going on, exploitation . . . but within that, I think there is something that will pass into the minds of children . . . It might draw them in, it might be this little window through which they peer and find another world through which they explore knowledge.[16]

At the end of the film *Central Australian tracks in the sand* we see Leunig mapping out an enormous depiction of Mr Curly in the desert sand; a new home for a lost soul.

THE 'HIPPIE DROPOUT': JOHN SEED

One year after Michael Leunig was born in Melbourne, Janos Kampfner was born in Hungary.[17] His Jewish parents had managed to survive the Nazi Holocaust but had lost close relatives in the death camps. His father, Jancsi Kampfner, had spent three years of the war in a forced labour camp. In 1951 he brought his family to Australia where Janos' mother, Judy Cassab, would emerge as one of the country's foremost portrait and landscape painters.

Janos Kampfner returned to Europe, to live in England, in 1968, where he worked as a systems engineer for IBM. In London he found soulmates who felt that modern society had become spiritually bankrupt and they sparked his lifelong interest in eastern philosophy and religions. Not long after he returned to Australia in 1971, he and his partner Greta Nelson headed for the north coast of New South Wales where people interested in 'alternative lifestyles' were starting to congregate. They got involved with the Tuntable Falls commune, helped organise the 1974 Aquarius Festival at Nimbin, and then helped to build a very successful community at Bodhi Farm. Janos Kampfner became John Seed.

In 1979 Seed was fully occupied with life at Bodhi Farm; helping run its Buddhist meditation centre, growing organic vegetables, learning to be a father, and helping deliver babies.[18] Less than 10 kilometres down the road a remnant area of the once mighty Big Scrub rainforest was about to be logged but Seed was so engrossed in his 'alternative' life that he had never even ventured into the beautiful Terania Creek forest. However, one day, while he was shopping at The Channon markets he came across some people he knew appealing for others to come the next day for a protest against the commencement of the logging. His first reaction was to want to avoid involvement because 'I was a Buddhist – someone who watched as things rose and passed away' (in Stone, 1989). However, like a lot of other people from the 'hippie' communities in the area, he went along to see what was going to happen. The atmosphere was quite jovial. Local police knew about the protest and advised the loggers to delay their work for the day. The event got a lot of media attention and many participants thought that might be the end of the story. However, people who, like Seed, had not been in the forest before, were stunned by its beauty and outraged by the idea that the ancient trees were about to be felled. By the end of the day, a decision was taken to resume the protest the next day. This time the protesters came ready to camp in the forest and two days turned into two weeks and then two months. Among the 'hippies' was a large number of professional people who had dropped out of mainstream society and they now used their old skills to organise the camp and do the media work. Many of those who had migrated to the region surrounding Nimbin were trying to escape the rapacious nature of 'mainstream' society, but an aspect of that society had followed them into their retreats and threatened its extraordinary natural beauty. For Seed, his first visit to the rainforest was like an epiphany that would seize his soul. He had come to love the regrowth forest around Bodhi Farm, but this was something entirely different and he knew he had to commit himself totally to its preservation.

So Seed became a very active member of the protest camp that was dedicated to using non-violent tactics to try to convince the loggers and, through the media, the state government that the forest was too precious to be logged. The protesters went out of their way to be nice to the local police, even offering head massages at the end of long and stressful days, and to convince the loggers that the protests were not aimed against them personally. However, as the stalemate continued, more highly trained police were brought in from places like Newcastle and, with them holding the protesters at bay, the loggers started attacking the forest giants. The conflict became more intense with protesters climbing threatened trees and sitting in front of bulldozers, while the loggers set out to fell as many trees as they could as quickly as they could. Just when it looked like the protest was going to fail, the state government, concerned about the intense media interest in the protest, announced that the logging would be suspended to allow for an inquiry into the granting of the licence. While the inquiry found that logging should proceed, Premier Neville Wran ignored its findings and announced a total ban on logging in the forest.

The Terania Creek protest was unlike any forest protest previously seen in Australia. With a large group of 'hippies' camping in the forest, playing musical instruments and singing, making props and devising theatrical forms of protest it was very colourful and 'media-friendly'. One thing that undoubtedly influenced Wran was a public opinion poll conducted at the height of the media coverage of the protest that indicated that 70% of people in the state opposed the logging. A close bond grew up among those in the camp and after they had secured the protection of Terania Creek they made plans for further protests to prevent logging in the surrounding Nightcap Ranges. A further campaign of civil disobedience resulted in hundreds of arrests but it also resulted in a decision by the government to create six new national parks, including one that covered the whole Nightcap area. John Seed was elated at the protesters' 'victory' but he also wanted to give credit to the forests themselves. In later years he became famous for the comment 'We did not save the forest, we were part of the forest saving itself.' For him personally he said that he first went to Terania Creek thinking he was going to save a forest and came out thinking the forest had saved him (Trenowyth, 1995).

At the height of the forest protests, in 1981, Seed set up a Rainforest Information Centre in a small shopfront in Lismore and set about learning all he could about rainforest ecology and about the fate of rainforests all around the world. He began to produce *World Rainforest Report* that reflected his own intense learning. Before long he moved from Bodhi Farm to a house in Lismore that became a kind of mecca for forest protesters from around the world. In 1982 he organised a group of protest veterans to travel with him to join the Franklin blockade in Tasmania (see chapter 10). In 1983, he helped organise an intense campaign in marginal electorates to help support the election of the Hawke Labor Government on the basis of its undertaking to save the Franklin (which it did). After that, attention turned to efforts to block the building of a road through the Daintree Forest north of Cairns in Queensland. This time the protesters lost but they fought a well-publicised battle and Seed and his colleagues had now created a national network of experienced campaigners who continued their work in many parts of the country.

Meanwhile, Seed's *World Rainforest Report* was starting to get international distribution and so the Rainforest Information Centre turned its attention to protests in other parts of the world, including Malaysia and Brazil. In 1984 Seed put together a Rainforest Roadshow, which featured a combination of film, music and talk and took it to the United States. The success of this venture led to the formation of the US Rainforest Action Network (RAN), which managed to establish a presence in 50 cities by the end of the 1980s. Seed also began travelling to other parts of the world in which rainforests were in danger. With a broad base of support in both Australia and the US he was able to raise a significant amount of money to publicise the plight of forest-dwelling communities in places such as

Sarawak (Malaysia) and Papua New Guinea. He also began financing alternative, community-based economic development projects that provided alternatives to continued logging. These included a large reforestation and sustainable agriculture project with the Awa people living on the periphery of a large rainforest in Ecuador and support for the promotion of Walkabout Sawmills in Papua New Guinea and the Solomon Islands that encouraged small, community-based logging operations rather than 'industrial logging'.

Seed has said that at the end of the campaign to save the forest at Terania Creek he looked for international organisations working on rainforest issues and found none; just isolated scientists making dire, but largely unheard, predictions about likely consequences of forest clearance and disturbance. Through the distribution of his newsletter and by travelling extensively he was able to put together the sort of network he had been looking for. In the late 1980s that network was able to demonstrate its power by mounting a highly effective consumer boycott of the hamburger chain Burger King, which was providing incentives for rainforest clearance in Costa Rica by importing cheap beef from that country to the United States. The campaign was run mainly in the US with support actions in some other countries. It involved the widespread picketing of Burger King restaurants in cities across the US, combined with some effective street theatre and media work. After a year of bad publicity the Burger King operation announced that it would switch to beef produced on established farms in the US and, as a result, beef exports from Costa Rica fell by about 80 per cent the following year.

In 1987 the respected international journal *Christian Science Monitor* featured an article on Seed that described him as 'the town crier in the global village' (August 24–30, 1987). The article also cited the world-renowned US ecologist Paul Ehrlich as saying that Seed 'is a tremendous example of what a person without strong professional training in ecology can do by becoming informed – and then becoming an activist'.

During one of his trips to the US in the mid-1980s, Seed attended a workshop on 'Despair and Empowerment in the Nuclear Age' run by the Buddhist scholar and anti-nuclear activist Joanna Macy. By then he had already come across the concept of 'deep ecology' as pioneered by Arne Naess, but he was highly impressed with Macy's ability to make these ideas seem of immediate, practical relevance. He was impressed with the way she was able to reinvigorate 'burnt out' activists. At the end of the workshop, Seed talked to Macy about the work he had been doing for rainforests and invited her to visit the Rainforest Information Centre in Lismore. Macy accepted the invitation and during a walk in the rainforest that the protesters had protected, she and Seed began to talk about ways of getting people to extend their sense of identity to include nature. Out of this conversation grew the ideas for a new kind of empowerment workshops that became known as the Council of All Beings and in 1988 Seed collaborated with Macy, Pat Fleming, and Arne Naess to produce a book about applying deep ecology ideas in this way. Borrowing a line from Aldo Leopold's *A Sand Country Almanac*, the book was *Thinking Like a Mountain*.[19] In the following years, Seed travelled extensively, sometimes with Macy but often alone, running Council of All Beings workshop in Australia and the US. After paying for his own living expenses the surplus funds went to projects being run by the Rainforest Information Centre.

Seed (1993) has said that the article by Naess in *Thinking Like a Mountain* provides a very good explanation of the key motivation for the work he does with deep ecology concepts. According to Naess, people working for the conservation of nature have to recognise that an appeal to a sense of duty or responsibility is a 'treacherous basis' for their work. Rather than hopeful appeals to a spirit of altruism, conservationists should try to help people establish a much broader sense of self because:

> If we can IDENTIFY with the Earth we don't need altruism, If we have the EXPERIENCE of ourselves, not merely as isolated, separate, skin encapsulated egos, but also as part of the larger body of the Earth, then the defense of nature becomes merely self defense and this does not require a highly elevated moral stature. (Seed, 1993)

Seed cites Naess as saying:

> The requisite care flows naturally if the self is widened and deepened so that protection of free nature is felt and conceived of as protection of our very selves. (Seed, 1993)

John Seed with his mother Judy Cassab (Judy Cassab)

In working with this idea, Seed began to see the Council of All Beings workshops as being a kind of therapy that could help people find their own path towards greater self-fulfilment. However, while travelling in the south-west United States he had the privilege of watching an ancient Hopi ritual that included more elaborate versions of some of the exercises that he and Macy had devised. This made him rethink the idea of the workshops as therapy, because, whereas therapy can be seen as a one-off event, rituals are continually repeated as a way of reinforcing and celebrating ways of being in the world. He started to think that what he was doing was addressing a concern identified by writers such as Joseph Campbell, who have said that the loss of ritual and myth in western society has been a major source of anxiety. So Seed began to see his work as a form of ritual that was aimed at cultural revival rather than personal therapy.

Of course deep ecologists were not the first scholars to promote a concept of 'enlightened self-interest' (see Hayward, 1998) as a motivation for social change. However, they have based it on a broad, ecological concept of the self that is new and innovative. Seed's early comment about becoming part of a 'forest saving itself' led him on his own journey towards an ecological sense of self. During a 1995 interview with the Sydney writer Samantha Trenoweth (1995), he said:

> When I tell you that I represent the earth, I mean that, when you don't represent your ego and when you don't represent Australia and when you don't represent Christianity and when you don't represent humanism and when you don't represent anything else, then that's automatically what you represent. I don't think of it as something I've constructed so much as, when you pull all the other things away, that's what's left because that's always been underneath it all. The sense of separation from nature, which everybody suffers, is an illusion.

> I'm not talking, here, about anything more mysterious than trying to hold your breath for three minutes. The fact that we can't hold our breath for three minutes totally demonstrates our interdependence. That's our relationship with the air. Then there's the water and the soil and all the other earth cycles which move through us and which define us. The psyche grows out of this soil and out of this water and out of this air. (Trenoweth, 1995, p. 25)

In talking about the 'dangerous illusion' that we are superior to nature, Seed likes to quote James Lovelock, one of the key promoters of the 'Gaia hypothesis', who likened this situation to that of the brain deciding to mine the liver. This would suggest the brain is stupid rather than intelligent because it has no existence outside the body as a whole. Seed adds in his own inimical way:

It's like a tree with a billion leaves and one leaf thinks that the sole purpose of the tree is to be a place for that one leaf to grow . . . If this attitude is lived out with our current technology . . . we start to destroy the tree on which we're growing. We believe in our independence, even when ecology clearly shows us our extraordinary interdependence. (Trenoweth, 1995, p. 27)

Although Val Plumwood has critiqued deep ecology for promoting an identity with nature rather than respecting difference, it is clear that Seed's very organic philosophy is consistent with her call for a stronger 'communicative ethic' in our dealings with nature. Seed also shares with Leunig a deep desire to understand the loss of 'soul' in our 'modern' culture. In Plumwood, Leunig, and Seed we find three Australians who have worked in different fields, using different languages and modes of communication, but with a similar message about the root cause of our alienation from non-human life processes and the consequent impoverishment of our sense of self. In different ways, they argue for a less 'rational', more sensuous, more imaginative, and more engaging, and more responsible relationships with nature.

CHAPTER 12 | CONCLUSIONS

The story of Australia's ecological pioneers is an emergent one. We began our exploration of it because we were attracted by what was for us a very interesting paradox. This was that, from the perspective of environmental care, the colonisation of Australia began disastrously yet, during its second century of existence, Australian settler society managed to produce a surprisingly high number of internationally significant ecological thinkers and practitioners. Most of these 'ecological pioneers' were individuals who recognised, long before the rest of the population, that because our settlement practices were not informed by either an appreciation of the unique qualities of Australian landscapes or an understanding of ecological processes, they would have long-term costs for both the people and the land. In diverse ways all of these pioneers have made significant contributions to the development of values and practices that reflect an attitude of caring and responsible action towards our still largely taken-for-granted and misunderstood landscapes. For many of the pioneers profiled in this book, their starting point was a critique of current practices and a demand for more conserving methods of resource management. In many cases, however, this seemed to naturally evolve into the much more radical quest of seeking a more dialogical relationship with the land, in which there is an evolving integration of the 'wisdoms' of both humans and nature.

In the process of both researching and writing this book, we have also experienced a similar progression. Although we started by seeking to highlight the main contributions of each of the selected pioneers to our understanding of Australian landscapes, we found ourselves looking afresh at such landscapes through their eyes. Rather than just recording the stories of these exceptional people, we found ourselves in deep dialogue with them, and at moments even 'becoming' some of them. One of our hopes is that our readers have been able to join in this dialogue and be as inspired by our subjects as we have been. We chose to tell their stories as social histories, rather than focusing on only the intellectual and practical contributions of them as individuals. As noted in the Introduction, we wanted to be able to tell the stories of these individuals in ways that maximise their power to inspire the reader, so we did not feel bound by exact chronological ordering or systematic rational argument. These are, therefore, not tidy, sanitised stories, but hopefully more alive representations of Australia's unfolding ecological history. For the sake of coherence, however, we endeavoured to weave the diverse narratives into a meta-narrative – particularly by using the introductions to each chapter to draw together the various connecting threads. Our main hope has been that the reader can relate these stories to his or her own experiences and understandings and, in the process, deepen their appreciation of Australian landscapes, thereby permitting them to further develop their own storied and dialogic relationship with those landscapes.

In the very early stages of this book's conception we were inspired by Donald Worster's account of the history of ecological ideas in Europe and North America (1994). We concur with his observation that ecology had a history long before it had a name. With Worster, we have also found it useful to distinguish between the development of an 'imperial' tradition of ecology, which in Australia we have taken to start with the arrival from England of Joseph Banks, and a contrasting 'Arcadian' or romantic ecology tradition, which is apparent in the works of so many of Australia's artists, writers, nature-lovers and philosophers. Although this Arcadian vision had to be rethought in the context of Australian landscapes and ecosystems, there is little doubt that much of our present ecological understanding has been influenced by an emerging aesthetic appreciation of the land. Because of this, we have taken a much broader view than Worster of what constitutes pioneering ecological thought and action, and we have focused much less than he did on the imperial tradition. Consequently, we have included many people who probably would never have thought of themselves as being ecological pioneers. Had it been possible for all the pioneers discussed in this book to feel united in a quest for deeper ecological understanding there could have been much more cross-fertilisation among their ideas, and their common feelings of being isolated and marginalised might have been considerably reduced. It is not too late for us to explore the synergy in the work of these ecological pioneers, but it is sadly too late for many of them to see how influential their work could be.

At the same time, we have argued that being in the margins has some advantages for those who want to challenge what is going on in 'the centre'. Throughout history, new ideas have almost invariably had their origins in the margins of 'mainstream' thought. In a parallel way, we note that in nature biodiversity is highest and ecological interactions and evolutionary processes most intense at edges, such as between land and sea. From a study of 'natural' ecosystems, therefore, we learn of the importance of 'edge effects'. However, most pioneer thinkers and practitioners who find themselves in the margins have experienced feeling undervalued, ignored and ridiculed; some even feeling that their lives were under threat. Such marginalisation may be partially self-imposed, particularly when the message is perceived to be too radical or too dogmatic. More often, however, it can be traced to the narrow-minded resistance to change so common in societies (and individuals) that are uncritically 'wedded' to the *status quo,* or extrapolations of it. We shall return to this phenomenon later.

Tim Bonyhady (2000) has effectively argued that, from the earliest days of European settlement in Australia, there have always been some people who were interested in nature conservation. Consequently, it is not useful to try to date the birth of our conservation movement. He acknowledges, however, that whereas the early conservationists were primarily motivated by a desire to enjoy observing nature as a recreational activity, it was not until the 1860s and 1870s, firstly in the state of Victoria, that a more significant shift in public awareness took place that was to have much more profound personal and political implications. In this book we have recognised this shift as an important starting point for the evolving relationship of non-indigenous Australians with Australian landscapes. Furthermore, unlike Bonyhady, we have found it helpful to link the changing level of ecological awareness to a social movement that has grown over time, with recognisable and overlapping phases in its development. As discussed above, we have tried to avoid a linear approach in our exploration of ideas that have emerged in uneven ways. As in nature, the evolutionary process has involved two complementary processes: on the one hand an almost imperceptible ongoing gradual process of change, complemented by periodic bursts triggered by unexpected dramatic events. Some of the pioneers profiled in this book were responsible for quantum leaps in the development of theory and practice by combining long and patient work with an ability to seize the opportunities that presented themselves.

We have noted that the earliest pioneers – artists, writers and nature-lovers – displayed an extraordinary sensibility in being willing and able to 'listen to' and learn from the land. As their dialogical relationship with nature evolved, most of them experienced a deepening engagement with the non-human world (and probably also with their inner psyches). In its early stages, this engagement was most evident as a 'stoical' response to the refusal of landscapes to be conquered (as illustrated, for example, in the works of Henry Lawson and Russell Drysdale). From Lawson through to Patrick White (in *Voss*), many landscape-sensitive Australian writers have challenged coast-hugging white Australians to dare to deeply contemplate that which they have feared the most – the arid inland, Australia's 'dead heart'. Such writers recognised that for a sense of belonging to emerge, white Australians would firstly have to acknowledge and confront their deep sense of not belonging. In the context of a 'frontier society', this new ecological sensibility may also be recognised as a growing empathy for what lay beyond the frontier. People like Eleanor Dark, Judith Wright and Russell Drysdale were far ahead of their time in recognising that the settler society had paid a high price for its failure to acknowledge and learn from the ecological wisdom of Australia's Aboriginal people. The concept of *terra nullius* not only obliterated the indigenous people as active agents, it also led to a profound devaluing of most of Australia's phenomenally rich flora and fauna, which were commonly regarded as having no utilitarian value. Most plant communities in Australia were perceived, for example, as 'rubbish vegetation' that had to be cleared to make way for the more familiar 'useful' exotics. The nature preservation movement in Australia developed as a response to this, and it owes its existence to people – like Myles Dunphy – who had profound empathy for the remnant tracts of Australian nature that had somehow avoided colonisation. Their concept of 'pristine nature', however, still reflected the legacy of *terra nullius* in that it failed to acknowledge the long-term engagement of the indigenous people with *all* of Australia's diverse landscapes.

At the opposite end of the spectrum to the preservationists have been pioneers like P. A. Yeomans, who sought to develop ways to better *manage* Australian landscapes and intervene to design better outcomes that would benefit both people and nature. For many, this represents a perpetuation of a 'softer', but still potentially harmful, version of the myth of human 'mastery' over nature. However, Yeomans was a powerful advocate of the need to learn from nature and, in many ways, his approach shared much with the Aboriginal notion of 'caring for country' (discussed in chapter 9). Rather than trying to referee disputes between the advocates of preservation and of ecological management of the land, we believe that the two traditions can be complementary and appropriate in different circumstances. In looking for complementarity between diverse traditions we must, however, remain conscious of the consequences of our dualistic western legacy of human 'mastery' over, and simultaneous separation from, nature. The exploration of the overlaps of various traditions may lead to some new, emergent insights and the diagram (opposite) suggests a way of integrating different approaches to engagement with the land around the notion of 'ecological citizenship'.

As strategies for changing dominant social attitudes and practices the three approaches all have some strengths and weaknesses, as suggested in the table opposite.

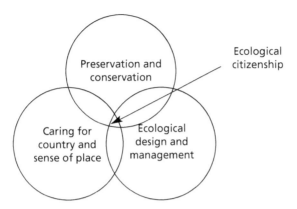

If we acknowledge that all three traditions have both strengths and weaknesses we may be more inclined to draw ideas from each. Unfortunately, each of the traditions has sometimes been defended with such zeal as to give rise to territorial wars between competing 'ecological worldviews'. A more rewarding, if more difficult, task is to seek an integration of their diverse ecological insights. Perhaps the notion of 'ecological citizenship' can facilitate such integration. However, there is still much work to do to advance this idea and the authors regard this as very important work for deepening our notions of ecological sustainability.

Perhaps the most difficult integration of all is to make use of the ecological insights emerging from such radically different cosmologies as those

	Strengths	Weaknesses
Preservation and conservation	Fosters respect for nature. Recognitionof non-utilitarian rights of other species and the need to maintain their habitats.	Fosters a separation of people and nature and leaves the 'weaker' category of nature in a vulnerable position. Only slows down the rate of loss of biodiversity.
Caring for country and a sense of place	Deep integration with the 'other' leading to paradoxical insights into practical caring.	Dependent on a high level of local knowledge and engagement, and a spiritual attitude to nature, which is lacking in most of the population.
Ecological design and management.	Maximally uses scientific knowledge and practical experience to understand and work with nature.	Too limited by quantifiable data and proof thereby neglecting much intuitive 'ecological wisdom'. Can sustain the illusion that humans are in control.

of the indigenous Australians and Australian settler society. On the one hand, the ecological wisdom of the indigenous people is often dismissed on the grounds that they lived in pre-industrial cultures. On the other hand, some well-meaning non-indigenous Australians want to appropriate the beliefs and values of indigenous cultures that evolved over many thousands of years in fundamentally different circumstances to their own. It is not easy to strike an appropriate balance between engagement and non-engagement with Aboriginal beliefs – avoiding the extremes of arrogant dismissal of the indigenous cultures and shallow, tokenistic appropriation of selected beliefs. What we have noted in this book, however, is that a mutual interest in wanting to understand Australian landscapes has led to some important convergences; starting first, perhaps, in the field of art. Some white Australian artists, like Sidney Nolan, developed an Aboriginal-like capacity to perceive the land 'from the inside out' and, from the opposite direction, Aboriginal artists came to experiment with new technologies to move beyond the iconography of their traditional art forms. We also noted that white Australian writers joined the artists in the conscious development of a mythology surrounding relationships between people and the land; again emulating Aboriginal people in emphasising the role that mythology can play in forging a sense of belonging to the land.

Some of those we count as ecological pioneers were among the first to realise that white Australians have much to learn from the people who have lived in this land for more than 60,000 years. However, for most Australians the legacy of *terra nullius* only came to be challenged by the emergence of the 'modern' Aboriginal land rights movement and the consequent ruling of the High Court in the so-called Mabo case (see chapter 9). In challenging the settler concept of land ownership and showing how people can belong to the land, the pioneers of the Aboriginal land rights movement certainly shocked many white Australians by pointing out how their relationship to the land was shallow and irresponsible. Instead of the word 'landscape', Aboriginal people have chosen to use the English word 'country' to suggest a deeper sense of belonging to what Deborah Bird Rose (1996) has called 'nourishing terrains'. In a similar way, Aboriginal people have also borrowed and enriched the English word 'dreaming' to refer to ways in which they engage with the creative spirit of the land that nurtures them. Where once it was common to dismiss the spiritual beliefs of Aboriginal people as being the product of the 'primitive mind', some white Australians have suggested that we need to work on our own 'whitefella dreaming' in order to re-enchant our relationships with the non-human world (for a discussion of this concept, see Mulligan, 2000). In so doing we might move from relatively abstract notions of space and place to a greater appreciation of the uniqueness of every single place.

As Val Plumwood has pointed out (see chapter 11) our ecologically inappropriate concepts of place are reflected in naming practices that continue to celebrate our colonisation of the land and its original inhabitants. Such inappropriate naming practices blind us to the rich diversity of stories that are embedded in particular places and landscapes. Aboriginal notions of 'storied landscapes' (discussed in chapter 9) can help us to understand how dialogical, rather than monological, relationships with the land can be built. They show us how our 'dreaming' can be built on the many layers within these stories, from the personal and local to the national and cultural. For us 'whitefellas', many stories generated by science will be a part of our dreaming; helping to build a dreaming that can also inspire indigenous people working with western cultures.

John Seed is one of those who have called for the development of a new, ecological sense of self (see chapter 11). This would include an extension of empathy for the non-human into what might be called a self-interested exploration of our connectedness with everything we depend on for the continuity of life. Such an ecological sensitivity has much in common with the Aboriginal notion of caring for country and the capacity to value the uniqueness of every place. With such ideas in mind we might talk of a movement from empathy (for that which is different) to 'ecological empathy' (emphasising complex and mutualistic relationships across differences).

Empathy has sometimes manifested itself in surprising ways. For example, scientists like Francis Ratcliffe, Bert Andrewartha and Charles Birch were employed to find ways to control 'pests' that were regarded as the 'enemies' of agriculture. However, in their engagements with the 'enemy' they began to perceive the world from the perspective of the organisms in question, and they started to search for ways in which humans and other organisms might co-exist rather than compete. Even with good intentions such peaceful coexistence can be difficult to achieve. However, an ecological reframing of the problem made it much easier to consider, and this reframing was driven by empathy.

A fairly simple exercise we recommend to all readers is to try to step outside the experience of being human for a while and imagine how the world might appear to you if you were looking at it from the perspective of another organism. How might you assess your needs and 'desires'? What would it take, for example, for you to feel contented as, say, an ant living in a very particular place? Of course, such a movement can never be complete; we can never fully know what it would be like to be another species; however in the process of trying to imagine it we can deepen our understanding of others and our empathy for them. We also invite you to experience 'becoming' one or more of the ecological pioneers profiled in this book. Pay attention to how that makes you feel. What would you do if society seemed uninterested in your gifts of understanding? Who are likely to be your greatest supporters (and enemies)? How will you continue in your 'mission'?

As Val Plumwood has pointed out, a movement towards what the authors are calling 'ecological empathy' would be a movement beyond identifying with the non-human to a deeper awareness of entities that will always be 'other than' human. We might start with identity work, as May Gibbs and the other pioneers of 'bush stories' for children did. However, a deeper engagement with the non-human world will soon lead us to the realisation that our ability to truly empathise with non-human experiences will always be limited. This should be a cause of celebration, however, rather than despair because it reminds us that there is a world of infinite complexity and richness for us to explore – endlessly. Plumwood suggests the need for a 'communicative ethic' that encourages us to continually watch, listen, learn and share.

It comes as little surprise that pioneer thinkers and practitioners tend to have a rather robust self-belief. Some of those discussed in this book – for example, Sidney Nolan and Judith Wright – have eventually won widespread acceptance and, even, praise. Fortunately, this did not stop them being critical and outspoken – Judith Wright, for example, was notorious for 'speaking her mind', even to those who were her strongest supporters. Others – like Philip Crosbie Morrison and Jack Mundey – have experienced a less enduring notoriety. Others still – like P. A. Yeomans and Charles Birch – have attracted some attention at particular moments in their lives, but have yet to receive the recognition they deserve. Many of our pioneers have had considerable experience of having to work from within the margins of society, some even being vilified as being harmful zealots. To maintain their critical voices, they have needed considerable patience, as well as a strong belief in both themselves and their ideas. They have needed an attentiveness and willingness to learn, especially through patient observation of nature, and they have also needed a willingness to act resolutely when action has been called for.

Rarely have they offered generalised, 'quick-fix' solutions to the problems that have emerged from inappropriate settlement practices. Rather, their approaches have demanded attentiveness and patience, and they have often warned that we will continue to make mistakes if we insist on forcing the pace of nature. Paradoxically, Yeomans demonstrated that natural processes can actually build topsoil much more quickly than the impatient, interventionist methods that prevail in agriculture; demonstrating that the appeal of the 'quick-fix' approach may be an illusion, even in terms of short-term efficiency. However, in all such endeavours it is essential that we shift the emphasis from short-term gain to long-term sustainability. The Bradley sisters demonstrated this so clearly. They found that if bush regeneration is carried out at a pace dictated by the capacity of the bush to regenerate itself then the results will be much more enduring than if interventions reflect an impatient human pace. Similarly, Permaculture promotes the idea that attention to ecological design in the early stages will result in less maintenance work in the longer term. Being attentive and patient is not the same as simply slowing down, but it does encourage more contemplation and reflection. Val Plumwood, for example, advocates a patient, communicative, process for building new relationships with the non-human world. This message about patience can be uncomfortable for those who are in a hurry to rectify the wrongs of the past. However, quick-fix solutions invariably exacerbate the problems. The message is clear – we need to work on much longer-term changes and move in time with nature's processes.

One thing that stands out clearly from our review of Australia's ecological pioneers is that they have mostly been driven by an all-consuming *passion* – a passion for learning about nature and a commitment to act on this learning and share it with others. From the early naturalists like Donald Macdonald and Charles Barrett, to scientists like Charles Birch and Stephen Boyden, to conservationists like Judith Wright and John Sinclair, to activists like Bob Brown and John Seed, we find that all have worked with a passionate intensity, often at considerable personal cost. Of course, passion sometimes needs to be tempered by other personal qualities. In the pioneers we profiled we found two particular qualities that seemed to temper passion as a driving force, as illustrated in this diagram:

Although Australia has produced some outstanding ecological thinkers and practitioners, it would be foolish to exaggerate the influence they have had on 'mainstream' Australian beliefs and practices. There have been major shifts in areas, such as aesthetic appreciation of Australian landscapes, the desire to preserve our natural and cultural heritage, the ongoing discourse around land and identity, and an acknowledgement that we have much to learn from the indigenous people. However, it has been much more difficult to make significant shifts in areas such as resource and land management. Because of the resistance to change in these areas, it has been common for people like David Holmgren, Ted Trainer, Bob Brown, and Val Plumwood to be either marginalised or treated as impractical zealots. Ecological thinking continues to present a very radical challenge to a society that remains obsessed with economic 'growth', affluence, and consumerism. However, that is also why ecological thinking retains its important 'cutting edge', and it is why a book like this is as relevant for visioning and planning the future as it is for understanding the past.

Radicalism shares some important qualities with diversity. In natural systems too much diversity can be harmful; what is needed is 'functional' or 'requisite' diversity. Similarly, self-imposed and self-perpetuating radicalism can become nothing more than an irritating distraction. Radical thinkers need to find ways to keep their ideas fresh and relevant – constantly subjecting them to critical review. To be effective in bringing about social change, it is necessary to help people identify small, personally relevant, steps they can carry through to completion in order to sustainably reorient their beliefs and practices.[1] We need more tolerance and understanding at both ends of the dialogue in our society, between the radical voices and those who feel trapped by the *status quo*.

We also need to remain mindful that the *status quo* has an infinite capacity to blunt radical voices by either ridiculing them (as when forest campaigners are labeled as 'tree-huggers') or by absorbing and institutionalising part of their message (as in the promotion of 'green consumerism' or the subversion and dilution of the notion of environmentally sustainable development). Sharon Beder (2000), for example, has produced a disturbing analysis of the many ways in which the corporate world has responded to the 'threat' of environmentalism globally. However, it is too easy to externalise the problem by pinning all the blame on the corporate world, and neglect to reflect deeply on our own attitudes, beliefs and practices. Radical voices sometimes make us feel personally uncomfortable, but if we are willing to engage with such 'irritating' ideas, the process may free us from our conceptual prisons and facilitate progressive change.

Michael Marien (1982) has suggested that succumbing to what he called the 'sandbox syndrome' has often effectively sidelined those who advocate a 'green' message. Sandboxes are very convenient for parents because they enable them to keep their children preoccupied, 'out of their hair', and making no demands on their attention. Many children readily accept the rules of the game by staying in the sandbox. In a similar way, Marien argues, the green challenge to the values of 'hyper-industrialized' society can also be quarantined into a sandbox if the 'children' who want to play there are encouraged to create, or imagine, boundaries that they should not cross. An enemy of the green movement might encourage the eco-children to construct their own sandbox by doing things like creating their own jargon, developing a sense of equality *within* the sandbox and a sense of superiority towards those outside, and confusing goals and enthusiasm with results. If, however, our subversive 'enemy' actually became convinced of the green message, he/she would turn around and advise these same people to do the reverse of everything he/she had previously advised; i.e., use simple language, collaborate across differences in order to work together for common goals, and measure success against solid outcomes.

As Michel Foucault might say, we tend to accept forms of self-discipline that discourage critical thought. There are two sides to the problem – a tendency towards self-perpetuating and self-limiting radicalism and an intolerance of destablising ideas and voices. We need to be vigilant in keeping ourselves open to destabiling ideas and, at the same time, subject them all to critical assessment.

In the Introduction we presented a table that compared 'ecological understandings' with 'prevailing assumptions/practices' as a way of highlighting the radical nature of ecological thinking (see page 9). It is now appropriate to reflect on how the stories of our ecological pioneers have provided illustrations of some of those ecological concepts. For example:

BEING RESPONSIVE TO EARLY INDICATORS RATHER THAN
WAITING FOR CRISES
The thinkers and practitioners we have highlighted were pioneers precisely because they were ahead of their times in seeing the nature and scope of problems emerging in the relationships between people and the environment. We have argued that further problems and crises could have been avoided if our settler society had been more receptive to the ideas of such pioneers, many of whom were confined to the margins of society.

BEING INTERESTED IN CYCLICAL REGENERATIVE RELATIONSHIPS
RATHER THAN LINEAR CONCEPTS OF GROWTH AND 'PROGRESS'
No one has better expressed an understanding of nature's regenerative cycles of life and death than Judith Wright in her poem *Rainforest* (see page 103). Russell Drysdale began working on this theme in his drought series of 1944. Those who have sought to emulate nature in ecological design and management (see chapter 8) have all emphasised the importance of working with cycles; with perhaps Ted Trainer being the strongest critic of linear concepts of growth. Regenerative cycles are central to Aboriginal cosmology and to the extent that there have been some convergences between

the ecological understandings of indigenous and white Australians, this theme has been central to that convergence.

SEEING A NEED TO USE RESOURCES FOR MAINTENANCE RATHER THAN JUST PRODUCTION

This, of course, has been a guiding principle of the nature conservation movement, which seeks to preserve biodiversity and leave, for future generations of people, a rich natural legacy. The theme has been picked up in the extension of conservation politics into green politics, through the promotion of a concern for building and maintaining cultural as well as natural capital. Again, this is also a guiding principle of the work of those involved in ecological design and management.

MUTUALISM FAVOURED OVER COMPETITION

The discussion above about the movement from sensibility to empathy and on to an 'ecological empathy' has been driven by the desire to overcome a competitive 'othering' of the natural world. Only by overcoming the western separation of humans from nature can we overcome the competitive illusion of human mastery and find ways to work with – rather than against – nature.

FUNCTIONAL DIVERSITY, COMPLEXITY AND STABILITY FAVOURED OVER CONTROLLED SIMPLIFICATION

Once again this has been a driving principle for those engaged in ecological design and management. It was an important principle illuminated by the work of ecological scientists like Francis Ratcliffe, Bert Andrewartha and Charles Birch (see chapter 7). As discussed above, the 'edge effect', manifested in natural systems, is evident in the ways in which pioneers have frequently operated in the margins of dominant human systems, subjecting mainstream ideas and practices to an ecological critique. We also noted above the distinction between simple diversity and functional diversity in regard to the contribution that can be made by radical voices.

SEEING A NEED FOR BOTH SPECIALISTS AND GENERALISTS RATHER THAN JUST SPECIALISTS

Many of the pioneers profiled in this book have been bold thinkers who have dared to step outside a more usual narrow specialisation. At the same time many have paradoxically been able to make more significant contributions within their area(s) of concern. Hopefully we have also modelled this principle in the way we have constructed this book; drawing together stories from a wide range of fields that all relate to a general interest in ecological thought and action.

VALUING A MUCH WIDER RANGE OF ROLES AND NICHES WITHIN A COMMUNITY

Again this relates to the marginalisation of people who question prevailing value judgements about who and what is important in our society. A classic manifestation of this may be in the way that research funding is allocated; with projects aimed at perpetuating linear growth-oriented, input-output, technology-dependent processes being strongly favoured over projects like Keyline farming and Permaculture, with their emphasis on cycles, complex relationships, maintenance, long-term benefits, and minimal human interventions once the systems are established. Concepts like Permaculture have found a niche in our society, but they have not yet received the institutional support and attention they deserve.

SEEING THE IMPORTANCE OF THE UNIQUENESS OF TIME AND PLACE

This may be one of the most important things that non-indigenous Australians can learn from the cosmologies of the indigenous people; engendering a much deeper sense of belonging to the landscapes that we tend to pass through, always on the way to 'somewhere else'. It was also the basis on which important perceptual breakthroughs were made in understanding the uniqueness of Australian landscapes, especially in art. The Heidelberg painters, for example, used techniques borrowed from time- and place-sensitive European impressionism to begin exploring Australian landscapes 'from the inside out'.

GRADUAL CO-EVOLUTIONARY CHANGE WITH BURSTS OF EMERGENT CREATIVITY RATHER THAN A FORCED PACE OF CHANGE RESULTING IN UNEXPECTED CONSEQUENCES

Working at the pace dictated by nature itself was the key lesson that united the work of people like P. A. Yeomans, David Holmgren and the Bradley sisters. As discussed above, we have also modelled this emergent conception of enduring and enriching change in the way we have constructed this book.

Perhaps that brings us to the end of a cycle we began in the Introduction when we first compared 'ecological understandings' with 'prevailing assumptions and practices'. Such a comparison can lead to an articulation of guiding principles for the development of viable ecological alternatives, and this is a higher order of understanding than trying to articulate specific policies that might flow from such principles. The scope of our work has been too great to end it with a discussion about such specific policies. However, we were pleased to find that our complex, emergent story did provide support for the principles with which we set out. At the same time, it gave us a deeper insight into how these principles operate in practice. This book has been both a confirmation of these ecological principles and an exploration of how they have been expressed in the life and work of our chosen ecological pioneers.

What we have created in this work will resonate with readers in ways that we cannot even envisage, and that is the main thing we set out to achieve. As researchers and authors we too have learnt more from our journey than we could hope to articulate here, and we are left with questions and new ideas that we will continue to explore in greater depth. Indeed we have reached the end of our text with more question marks than exclamation marks. This seems entirely appropriate, however, because ecological thinking encourages us to grapple with complexity and to constantly engage with the unknown, rather than seek neat closures and tidy, rational, solutions.

NOTES

1 INTRODUCTION

1 He made this point in a documentary series titled *Beyond the Fatal Shore* screened by ABC Television in 2000.
2 A term introduced by the ecofeminist philosopher Val Plumwood, see chapter 11.
3 An example used by Robert Hughes in *Beyond the Fatal Shore*.
4 This theme has been well covered in Tim Flannery's popular book *The Future Eaters: An Ecological History of the Australasian Lands and People* (Reed Books, Melbourne, 1994).
5 See Worster's (1994) account of Darwin's life and work.
6 According to which, it is considered unwise to manipulate natural systems until one is confident that the outcomes will be positive and involve minimal risk.

2 THE COLONISATION OF AUSTRALIAN NATURE AND EARLY ECOLOGICAL THOUGHT

1 In his highly acclaimed history of ecology, *Nature's Economy: A History of Ecological Ideas* (1994, second edition), Donald Worster has said that Linnaeus and his disciples began an *imperial* tradition in natural history (which Worster saw as a precursor of ecology). He contrasts this with an *arcadian* tradition pioneered by English natural historian Gilbert White in the late eighteenth century. Worster called the Linnaean tradition 'imperial' because it aimed to assert the mastery of humans over nature. Banks effectively extended this notion of imperial mastery by arguing that the imperial powers of Europe – especially Britain – should use science to assert their mastery over nature in the far-flung colonies.
2 See *The Voyage of the Beagle* by Charles Darwin, abridged and edited by Millicent Selsam (1966).
3 In the nineteenth century this term implied land that had not been 'improved' for human use. See Margaret Robertson, Kevin Vang and A. J. Brown (1992).
4 The poem, by a writer using the pen name 'Raupo' and simply titled 'The Bush', is reproduced in *The Poets' Discovery* edited by Jordan and Pierce (1990), pp. 243–245.
5 Writings about Gordon, even as late as Sladen's work of 1934, blur the distinction between his life and work in building the legend of the man and his poetry.
6 According to research presented by Nigel Padbury at a workshop at the Aboriginal Studies Association national conference (2000).

3 SEEING THE LAND IN A NEW LIGHT: PEOPLE AND LANDSCAPES IN AUSTRALIAN ART

1 According to Spate (1972), these two artists made Roberts more interested in the representation of light and colour but they did not have an immediate impact on his own technique.

2 William Moore's 1934 account of *The Story of Australian Art* includes a valuable, detailed account of the Heidelberg camps. See pp. 69–84.

3 A three-part series titled *A Brush With Landscape*, transcript from the Australian Broadcasting Corporation, Sydney.

4 As quoted in an interpretation of one of his paintings in London's Tate Gallery, 1998.

5 This point is borne out by some of Glover's Tasmanian paintings in which he depicts Aboriginal people happily recreating in rather Arcadian landscapes (with native plants). Yet these paintings were made after most of the island's indigenous inhabitants had been killed or rounded up and sent off to islands in Bass Strait. Glover's paintings were, then, a romanticised interpretation of an ideal.

6 For a discussion of the evolution of the term 'wilderness' see Margaret Robertson, Kevin Vang and A. J. Brown (1992).

7 See, for example, an article on Russell Drysdale by Janet Hawley in *The Good Weekend* (1997).

8 The term 'hideous blank' was used by the editor of the Melbourne paper *The Argus* in 1858, as cited by Haynes (1999).

9 For an interesting discussion about the use of landscape images in promotional campaigns, see *The Lie of the Land*, edited by Annette Shiell and Ann Stephen (1992).

10 He also had a house in Suffolk in England and often alternated between England and Australia.

11 As explained by a tour guide at a Drysdale retrospective at the National Gallery in Melbourne in December 1997.

12 Phrase used by a tour guide at the Drysdale retrospective in Melbourne in December 1997.

13 This comment was made in the same year that the Yolngu people of north-eastern Arnhem Land effectively launched the 'modern' Aboriginal land rights movement by sending their bark petition to Canberra (see chapter 9).

14 Information about Henry Gold has been taken from past editions of the newsletter of the Colong Foundation, as provided to the authors by Alex Colley.

4 OF DROVERS' WIVES AND A TIMELESS LAND: LAND AND IDENTITY IN AUSTRALIAN LITERATURE

1 This term is used quite widely. See, for example, Dutton (1985) and *The Penguin New Literary History of Australia* (1988).

2 Such criticisms are discussed, for example, in essays by Bruce Clunies Ross and Shirley Walker in *The Penguin New Literary History of Australia*, edited by Laurie Hergenham (1988).

3 The point about the unique character of the built environment is made strongly in the Boyer Lectures of David Malouf (1998).

4 According to Lawson's own recollections in an autobiographical story written in 1908, it was actually quite a substantial structure with semi-permanent walls added.

5 Biographies of Henry Lawson abound. An interesting one is *In Search of Henry Lawson* by Manning Clark (1978), later republished as *Henry Lawson: the Man and the Legend* (1985). Professor Colin Roderick became an acknowledged expert on the life and work of Lawson and he produced two biographies in 1982 and 1991.

6 A habit of Lawson's often referred to in tributes published at the time of his death.

7 See an article by her in Roderick (1972).

8 See the text of an address by Miles Franklin in Roderick (1972).

9 In Roderick (1972), p. 269.

10 Eleanor Dark's life and work have been covered in biographies by A. Grove Day (1976) and Barbara Brooks with Judith Clark (1998).

11 They moved there in 1922 and later built a much larger house on it with a studio for Eleanor in the garden. It now operates as a Writers' Centre, offering accommodation for writers who need some space for their work.

12 Having made a special study of the anthropological work of writers such as A. P. Elkin and Herbert Basedow and W. B. Spencer and F. J. Gillen – see Brooks and Clark (1998), p. 351. These anthropologists had not studied coastal Aborigines and Dark's understanding was entirely second-hand. But she was the first writer/historian to try to understand historical events from the 'other side'.

13 From her famous poem 'South of My Days' published in *The Moving Image*.

14 *Half a Lifetime* (1999), edited by Patricia Clarke, pp. 32–33.

15 From an essay titled 'The Upside-Down Hut' in *Australian Letters*, 3.4 (1961), pp. 30–34.

16 An example was R. F. Brissenden, writing in 1966, who said that Wright had allowed her political concerns to 'weigh too heavily on her' (Brady, 2001).

17 From 'Patrick White and the Story of Australia', in *Going On Talking* (1992), p. 52.

18 For biographical information about White, see Kiernan (1986); Bliss (1991); and Marr (1991).

19 For a summary and analysis of White's work, see Kiernan (1980) and Bliss (1991).

5 TAKING NATURE TO THE PUBLIC: JOURNALISTS, BROADCASTERS AND WRITERS AS EDUCATORS

1 Information provided by Dakin's long-time research assistant Isabel Bennett in an interview in Sydney in 2000.

2 He so enjoyed a visit to Yarrabah Aboriginal community near Cairns that he even joked about the possibility of 'chucking' science in order to become a missionary, Pizzey, 1992, p. 51.

3 He was a member of several clubs, including the Savage Club, which counted Robert Menzies among its members.

4 Information about the life and career of Vincent Serventy comes from the interview mentioned in the text and from a special tribute published by the Wild Life Preservation Society in 1996 to mark his fifty years of conservation work.

5 Direct quotes are taken from the interview mentioned above.

6 Doug Lowe survived for nearly four years as premier but was dumped by the party in favour of a more conservative leader, Harry Holgate.

7 The account of the contact between Serventy and Butler was provided to the authors by Serventy.

8 A good example has been Tim Flannery's book *The Future Eaters: An Ecological History of the Australian Lands and People* (1994), which attracted much controversy and discussion, as Flannery had hoped.

6 TOWARDS A CONSERVATION ETHIC: BIRTH OF THE CONSERVATION MOVEMENT

1 This account of how the Blue Gum Forest was saved has been recounted in a number of places. See, for example, *Myles Dunphy: Selected Writings*, compiled and annotated by Patrick Thompson (1986); and Macqueen (1997).

2 The story of this campaign is also covered in *Myles Dunphy: Selected Writings*.

3 For a discussion of the work of Collins and Romeo Lahey and the establishment of the first national parks in Queensland, see Hutton and Connors (1999), pp. 33–34.

4 Information about the life and work of Romeo Lahey is available from Binna Burra Lodge in Queensland (http://www.binnaburralodge.com.au/Inp.htm).

5 Named after a Queensland governor who also gave his name to a famous Australian cake.

6 Partly due to the work of organisations like the Wilderness Society, but largely due to the proliferation of nature programs on television (see chapter 6).

7 Sections of these diaries have been included in *Myles Dunphy: Selected Writings*, edited by Thompson (1986) and this was the source for information about Dunphy's adventurous life.

8 The authors interviewed Dot Butler and Alex Colley in Sydney in August 1996.

9 See Meredith interviews broadcast on *Ockham's Razor* in 1998, Australian Broadcasting Commission, transcript, p. 3.

10 Prominent in making this argument has been Professor Marcia Langton, an outspoken Aboriginal academic, whose views on the matter were summarised in an article published in *Wilderness News* in the Summer 1995/96 edition, The Wilderness Society, Melbourne.

11 For two sides of this debate see Flannery (1994) and Horton (2000).

12 Eric Rolls has written about changes that have taken place in the Piliga Forest in New South Wales since Aboriginal people were pushed out.

13 Such sentiments are repeated a number of times in his diaries, as cited by Patrick Thompson (1986).

14 As well as campaigning earlier for national parks, Lahey formed a National Parks Association two years before Dunphy formed his National Parks and Primitive Areas Council.

15 An account of this is included in Brady (1998). Wright has also written her own account of the struggle in *The Coral Battleground* (1977).

16 For a more detailed account of the Great Barrier Reef campaign see Hutton and Connors (1999).

17 Campaigns against sand-mining at Cooloola and Fraser Island are well covered by Hutton and Connors (1999).

18 She was interviewed by Helen Rusden and tapes and transcripts of their discussions are available from the library.

19 First made in a paper delivered at an Ecopolitics Conference at the Northern Territory University in Darwin in 1995.

20 For another account of the Fraser Island campaign, see Hutton and Connors (1999).

21 Such as Peter Christoff in his review of the book by Hutton and Connors in *Ecopolitics: Thought and Action*, 1.1 (2000).

7 WORKING IN THE BORDERLANDS: AUSTRALIAN INNOVATIONS IN ECOLOGICAL SCIENCE

1 In delivering the Templeton Lecture at Sydney University in 1999, the world-famous North American ecologist Paul Ehrlich suggested that 'Per capita Australia has the best group of ecologists in the world.'

2 Birch's views are taken from an interview conducted with him in Sydney in May 1997.

3 This argument is developed in his 1993 book *Regaining Compassion for Humanity and Nature*.

4 In an interview conducted at her Sydney home in October 2000.

5 Details of Dakin's life and career were provided by Bennett from her extensive and well-kept records.

6 Robert May (eventually Sir Robert May) went on to become Chief Scientist in Britain, but at this time he was a young physicist at Sydney University.

7 Information about Francis Ratcliffe is taken from an essay by Libby Robin 'Ecology: a science of empire?' in Griffiths and Robins (1997).

8 A compromise was that Boyden could not use capital letters in the name of the new unit because it was not an established discipline.

9 Conducted in Canberra in 1997.

10 For an explanation see *Western Civilization in Biological Perspective: Patterns in Biohistory* (1987).

11 For an explanation of chaos and complexity theory see Waldrop (1992).

8 THINKING LIKE AN ECOSYSTEM: AUSTRALIAN INNOVATIONS IN LAND AND RESOURCE MANAGEMENT

1 Stuart Hill interviewed Neville Yeomans in July 1998.

2 Percival A. Yeomans (1954, 1958 and 1965). *Water for Every Farm* was republished in 1978, incorporating *The Keyline Plan* (1954). Ken Yeomans, youngest son of P. A. Yeomans, republished it in 1993 with extensive revisions.

3 This plough has been further improved by Yeomans' son Allan, who manages Yeomans Plow Co in Southport, Queensland.

4 Stuart Hill interviewed David Holmgren in March 1999.

5 Keyline Designs, PO Box 3289, Southport, Queensland 4215.

6 Based on Stuart Hill's telephone conversation with Allan Yeomans on May 4, 2001. The manuscript of his unpublished book 'Green Pawns and Global Warming' (1998) can be found at http://www.yeomanskeyline.com/chapters_index.htm

7 Unreferenced information about David Holmgren's life and work is based on Stuart Hill's interview in March 1999.

8 From the interview with Stuart Hill in March 1999.

9 From an article Holmgren wrote about the plans for this project in the Victorian magazine *Green Connections*, 2.2 (1996), pp. 20–21; see also his website (http://www.spacountry. net.au/holmgren).

10 Unreferenced information about Ted Trainer's life and work is based on an interview with Stuart Hill in April 2000.

9 CHALLENGING *TERRA NULLIUS* VIEWS: THE ABORIGINAL LAND RIGHTS MOVEMENT

1 We use the term 'modern' here because Aboriginal people point out that a struggle over rights to land was at the heart of all major conflicts between settlers and the indigenous people all through the process of colonisation.

2 Hardy (1968, p. 143) recites this story, as told to him by Lupgna Giari.

3 Rick Farley quoted in an article in the Spectrum section of the *Sydney Morning Herald* by Tony Stephens on October 25, 1997.

4 From Deborah Bird Rose (1992). In this book, Rose examines the land tenure and management systems of the Ngarinmen people in the northern part of the Victoria River country.

5 Speaking in Darwin in July 1997.

6 Despite the fact that Gatjil Djerkura did not emerge well from this dispute, the Howard government saw fit to appoint him as chairman of ATSIC in 1997 once it assumed the power to appoint a person to this position (previously the ATSIC Board had elected its own chairperson). The promotion of Djerkura was widely seen as an attempt to undermine the authority of the Northern Land Council.

7 Text of this speech published in *Land Rights News*, March 1997, Northern Land Council, Darwin.

10 GREEN POLITICS IN THE WIDE BROWN LAND: WILDERNESS POLITICS AND SOCIAL JUSTICE AGENDAS

1 The idea that Tasmania could become an industrial power because it had a greater potential than any other state to generate hydro-electricity. The pursuit of this strategy meant that the HEC became one of the most powerful organisations in the state.

2 Tasmania has a unique electoral system for state elections, the Hare–Clark system, which is based on proportional representation, a little like election to the Australian senate. This means a candidate winning 12.5% of the vote can win a seat representing a seven-member electorate.

3 The Lake Pedder Restoration Committee, which launched a campaign called *Pedder 2000*, with David Bellamy as patron. Details from the Wilderness Society, Hobart.

4 Ideas promoted by the UTG have been taken from a range of documents made available by Patsy Jones, widow of Richard Jones. The most important summary of these ideas was *The New Ethic*, adopted in 1972 and recently reprinted in *The Daily Planet: Tasmania's Green Magazine*, No. 48, May–June 1997, published by the Tasmanian Greens, Hobart, p. 19.

5 The term 'Laborials' was first used by Des Shield and later picked up by Dr Norm Sanders, an environmental activist who was elected to the Tasmanian Parliament as a

member of the Australian Democrats in 1979. It was used frequently by Richard Jones and the UTG. See Pamela Walker (1989), p. 165.

6 In an interview conducted by Stuart Hill in Hobart in 1996, Patsy Jones said it was still listed in the Hobart telephone book.

7 As confirmed by Patsy Jones at interview in 1996.

8 Information about Jones' life comes from the 1996 interview with Patsy Jones.

9 As an environmental activist in the US, Sanders himself had once made the media by insisting on his legal right to fish from a three-metre rowboat anchored in front of a joint oil rig that was to be towed out to a new well site, Peter Thompson (1984), p. 114.

10 When the authors interviewed Brown in Canberra in 1997 he responded to a question about whether or not he had any mentors he could mention by name the 'young Pedder activists who were all younger than me'.

11 Information about Brown's life and career come from Peter Thompson (1984) or from an interview with Brown conducted by the authors in Canberra in 1997.

12 Karen Alexander, personal communication, 2001.

13 As he told the authors in the interview conducted in Canberra in 1997.

14 This and following quotes are taken from the interview the authors conducted with Brown in Canberra in 1997.

15 This gathering was initiated by Queensland's Drew Hutton.

16 Information about Mundey's international contacts and his international reputation is taken from the book by Meredith Burgmann and Verity Burgmann (1999).

17 Unless otherwise stated, information about Jack Mundey is taken from his book *Jack Mundey: Green Bans and Beyond* (1981).

18 For a full account of where bans were applied see Burgmann and Burgmann (1999).

19 At Victoria Street in Kings Cross, the property developer hired thugs to smash their way into houses being protected by a green ban and a prominent supporter of that particular ban – Juanita Nielsen – disappeared without trace (see Burgmann and Burgmann, 1999). It has been estimated that in two years the bans held up $3 billion worth of development (Burgmann and Burgmann, 1999, p. 278).

20 See the *Bulletin* of April 7, 1992, pp. 56–57.

21 In a letter to the authors of September 8, 1999.

22 As cited in Mundey's letter to the authors.

23 As cited by Mundey in a contribution to the book *The Battlers for Kelly's Bush* edited by Pip Kalajzich (1996).

24 In his letter to the authors.

25 The Text Publishing Company Melbourne, 1996.

26 All achieved during the period of the Labor–Green Accord of 1989–92.

11 TOWARDS A COMMUNICATIVE ETHIC: AUSTRALIAN CONTRIBUTIONS TO ECOPHILOSOPHY

1 The story of Plumwood's experience in Kakadu is based on an account she published in *Terra Nova* 1.3 (1996), pp. 33–44. It was also a subject of discussion when the authors interviewed her for this book at her home near Braidwood, New South Wales, in December 1997.

2 Passmore's book, *Man's Responsibility for Nature*, was published in 1974 (Duckworth, London).

3 Routley, who changed his name to Sylvan, established an international reputation for his work on environmental ethics. Along with Plumwood, his work was acknowledged by the US philosopher J. Baird Callicott in his 1994 book *Earth's Insights: a survey of ecological ethics from the Mediterranean basin to the Australian outback*.

4 Plumwood mentioned the example of a recent book that gave Routley the exclusive credit for an article in which she had been the senior author.

5 This debate was sparked by the Aboriginal academic Professor Marcia Langton who has written that conservationists who believe in people-free 'wilderness' deny the long history of Aboriginal occupation of such areas.

6 In a tribute to Wright published after the poet's death in *Arena*, 49 (2000), p. 44.

7 See 'Conservation as a Concept', *Quadrant*, 12.1 (1968), pp. 29–33.

8 Plumwood attributes this phrase to Aboriginal linguist Eve Feisl.

9 In honour of the Perth-based Aboriginal poet and human rights activist Jack Davis.

10 In *The Penguin Leunig* (1974).

11 *Central Australian tracks in the sand* is a film by David Morgan and David Greig, Kestrel Films, Richmond Victoria.

12 Leunig has discussed his life experiences in many published interviews. See, for example, articles in *Good Weekend* (John Fairfax and Sons, Sydney) on November 10, 1990 and June 13, 1998; *Temenos: Australian Jungian and Cultural Review*, Issue 1, 1995 (Canberra); *Fetish* (Anthropology and Sociology Club, Monash University), Volume 2, September 1990.

13 See preface to *A Bunch of Poesy* published by Angus and Robertson, Sydney.

14 From an essay on Michael Leunig in the book *Seven Remarkable Men* by Gina Lennox (1998).

15 Comments about Leunig's becoming a cartoonist and his views on spirituality are taken from an interview published in *St Mark's Review*, Summer 1995, St Mark's Institute of Theology, Canberra.

16 From an interview published in *Fetish* (Anthropology and Sociology Club, Monash University) September 1990.

17 Information about Janos Kampfner/John Seed's life is taken from his mother's published diaries, *Judy Cassab Diaries* (1995).

18 Information about Seed's life and work as an environmentalist is taken from the following sources: *The Future of God* by Samantha Trenoweth (1995); an article by Pat Stone in the May/June 1989 edition of *Mother Earth* journal; an article by Rick Fields in the November/December 1989 edition of *Yoga Journal*; and a paper given by Seed at an Ecopsychology Symposium held as part of the 28th Annual Conference of the Australian Psychological Society, October 1993.

19 New Society Publishers, Philadelphia, USA, 1988.

12 CONCLUSIONS

1 The author Stuart Hill (1999a, 1999b, 1999c) has written extensively about the role of small initiatives in the process of change. See also the recent book *The Tipping Point: How Little Things Can Make a Big Difference* by Malcolm Gladwell (2000).

BIBLIOGRAPHY

Abram, David (1997) *The Spell of the Sensuous: Perceptions and Language in a More-Than-Human World*. Vintage Books, New York.

Albrecht, William A. (1975) *The Albrecht Papers I and II*. Acres USA, Rayton, MO.

Allen, Christopher (1997) *Art in Australia: From Colonization to Postmodernism*. Thames and Hudson, London.

Andrewartha, H. G. and L.C. Birch (1954) *The Distribution and Abundance of Animals*. University of Chicago Press, Chicago.

Andrewartha, H.G. and L.C. Birch, (1984) *The Ecology Web: More on the Distribution and Abundance of Animals*. University of Chicago Press, Chicago.

Astbury, Leigh (1985) *City Bushmen: the Heidelberg School and the Rural Mythology*. Oxford University Press, Melbourne.

Ashton, Peter (1999) *Endeavour: The Story of Captain Cook's First Great Epic Voyage*. Windrush Press, UK.

Balfour, Lady Eve (1975) *The Living Soul and the Haughley Experiment*. Faber and Faber, London.

Bardon, Geoffrey (1979) *Aboriginal Art of the Western Desert*. Rigby, Adelaide.

Beder, Sharon (2000, 2nd edition) *Global Spin: The Corporate Assault on Environmentalism*. Scribe Publications, Melbourne.

Bennett, Hugh H. (1939) *Soil Conservation*. McGraw-Hill, New York.

Biehl, Janet and Murray Bookchin (1998) *The Politics of Social Ecology: Libertarian Municipalism*. Black Rose Books, New York.

Birch, Charles (1993) *Regaining Compassion for Humanity and Nature*. University of New South Wales Press, Sydney.

Birch, Charles (1999) *Biology and the Riddle of Life*. University of New South Wales Press, Sydney.

Bliss, Carolyn (1986) *Patrick White's Fiction: the Paradox of Fortunate Failure*. Macmillan, London.

Bonyhady, Tim (1984) *Images in Opposition*. Oxford University Press, Melbourne.

Bonyhady, Tim (1991) *Burke and Wills: From Melbourne to Myth*. David Ell Press, Sydney.

Bonyhady, Tim (1993) A Different Streeton. *Art Monthly*, 61, pp. 8–12.

Bonyhady, Tim (1994) 'The Art of the Wilderness'. In Barton, Will (ed.) *Wilderness: The Future*. Envirobooks, Sydney.

Bonyhady, Tim (2000a) 'Fine Art of Activism'. In Spectrum section of the *Sydney Morning Herald*, p. 4. July 15. John Fairfax and Sons, Sydney.

Bonyhady, Tim (2000b) *The Colonial Earth*. Melbourne University Press.

Boucher, Douglas H. (1982) 'The Ecology of Mutualism'. In *Annual Review of Ecological Systems*, 13, pp. 315–347.

Boyden, Stephen (1987) *Western Civilization in Biological Perspective: Patterns in Biohistory*. Clarendon Press, Oxford.

Bradley, Joan (1971) *Bush Regeneration*. Mosman Parklands and Ashton Park Association, Spit Junction, NSW.

Bradley, Joan (edited by Joan Larkin, Audrey Lenning and Jean Walker) (1988, reprinted in 1997) *Bringing Back the Bush: The Bradley Method of Bush Regeneration*. Lansdowne Press, Sydney.

Brady, Veronica (1998) *South of My Days: A Biography of Judith Wright*. Angus and Robertson, Sydney.

Brady, Veronica (2001) 'Judith Wright: The Politics of Poetics'. In *Southerly*, 61.1, p. 82. Halstead Press, Sydney.

Bramwell, Anna (1989) *Ecology in the 20th Century: A History*. Yale University Press, New Haven, CT.

Brent, Peter (1981) *Charles Darwin: 'A Man of Enlarged Curiosity'*. Heinemann, London.

Bromfield, Louis (1947) *Malabar Farm*. Ballantine Books, New York.

Brooks, Barbara and Judith Clark (1998) *Eleanor Dark: A Writer's Life*. Macmillan, Sydney.

Brown, Bob and Peter Singer (1996) *The Greens*. Text Publishing, Melbourne.

Burgmann, Meredith and Verity Burgmann (1999) *Green Bans, Red Union: Environmental Activism and the New South Wales Builders Labourers' Federation*. University of New South Wales Press, Sydney.

Butel, Elizabeth (1985) *Margaret Preston*. Imprint Books, Sydney.

Butler, Dot (1992) *The Barefoot Bushwalker*. ABC Enterprises, Sydney.

Callicott, J. Baird (1994) *Earth's insights: a survey of ecological ethics from the Mediterranean basin to the Australian outback*. University of California Press, Berkeley, CA.

Campbell, Andrew (1994) *Landcare: Communities Shaping the Land and the Future*. Allen and Unwin, Sydney.

Cassab, Judy (1995) *Judy Cassab Diaries*. Alfred A. Knopf, Sydney.

Christoff, Peter (2000) 'Review of *A History of the Australian Environmental Movement* by Drew Hutton and Libby Connors'. In *Ecopolitics: Thought & Action*, 1.1. Social Ecology Research Group, University of Western Sydney–Hawkesbury, Sydney.

Clark, Manning (1978) *In Search of Henry Lawson*. Macmillan, Melbourne.

Crowley, Kate (2000) 'Parliamentary Experiences of the Tasmanian Greens: The Politics of the Periphery'. In *Ecopolitics: Thought & Action*, 1.1, Social Ecololgy Research Group, University of Western Sydney–Hawkesbury, Sydney.

Daly, Herman and John Cobb (1994, 2nd edition) *For the Common Good*. Beacon, Boston, MA.

Darwin, Charles (1882) *The Formation of Vegetable Mould Through the Action of Worms*. John Murray, London.

Darwin, Charles *The Voyage of the Beagle*, abridged and edited by Millicent Selsam (1996), The World's Work, UK.

Day, A. Grove (1976) *Eleanor Dark*. Twayne Publishers, Boston, MA.

Debelle, Penelope (1990) *Leunig*. In *Good Weekend*, November 10, 1990 John Fairfax & Sons, Sydney, pp. 53–58.

Doherty, Darren (1996) *Underground. Green Connections*, 2.1, pp. 21–22.

Dunphy, Myles (1986) *Myles Dunphy: Selected Writings*. Selected and annotated by Patrick Thompson, Ballagirin, Sydney.

Dutton, Geoffrey (1964) *Russell Drysdale*. Thames and Hudson, London.

Dutton, Geoffrey (1974a) *Sir Russell Drysdale's Australia*. Ure Smith, Sydney.

Dutton, Geoffrey (1974b) *White on Black: The Australian Aborigine Portrayed in Art*. Macmillan, Melbourne.

Dutton, Geoffrey (ed.) (1985) *The Literature of Australia*. Penguin Books, Melbourne.

Dutton, Geoffrey (1991) *The Picador Henry Lawson*. Pan Macmillan, Sydney.

Dutton, Geoffrey (1992) *Artists' Portraits: Selected and Introduced by Geoffrey Dutton*. National Library, Canberra.

Ehrlich, Paul (1968) *The Population Bomb*. Ballantine Books, New York.

Farwell, George and F. Johnson (1949) *This Land of Ours*. Angus and Robertson, Sydney.

Faulkner, Edward H. (1943) *Ploughman's Folly*. Grosset & Dunlap, New York.

Figgis, Penelope (1999) *Australia's National Parks and Protected Areas: A Discussion Paper*. Australian Committee for the IUCN, Sydney.

Flannery, Tim (1994) *The Future Eaters: An Ecological History of the Australasian Lands and People*. Reed Books, Melbourne.

Fox, Stephen (1981) *John Muir and his Legacy: The American Conservation Movement*. Little, Brown and Co, Boston, MA.

Friend, Dick (1997) 'A Movement Grows'. In *The Daily Planet*, 48, The Tasmanian Greens, Hobart.

Geiger, Rudolf (1965) *The Climate Near the Ground*. Harvard University Press, Cambridge, MA.

Ghimire, Krishna and Michel Pimbert (1997) *Social Change and Conservation*. Earthscan, London.

Gladwell, Malcolm (2000) *The Tipping Point: How Little Things Can Make a Big Difference*. Little, Brown, London.

Griffiths, Tom (1996) *Hunters and Collectors: The Antiquarian Imagination in Australia*. Cambridge University Press, Cambridge, UK.

Griffiths, Tom and Libby Robbins (eds.) (1997) *Ecology and Empire: Environmental History of Settler Societies*. Melbourne University Press.

Grove, Richard (1995) *Green Imperialism: Colonial Expansion, Tropical Island Edens and the Origins of Environmentalism*. Cambridge University Press, New York.

Hagmaier, S., J. Kommmerall, M. Stengil and M. Wurfel (2000) *Eurotopia: Directory of Intentional Communities and Eco-villages in Europe 2000/2001*. Poppau, Okodorf Seiden Linden, Germany.

Hardy, Frank (1968) *The Unlucky Australians*. Nelson, Melbourne.

Hardy, Jane, J. V. S. Megaw and M. Ruth Megaw (1992) *The Heritage of Namatjira: The Watercolourists of Central Australia*. William Heinemann, Melbourne.

Hawley, Janet (1997) 'Man in a Landscape'. In *The Good Weekend*, December 13, John Fairfax and Sons, Sydney.

Hawley, Janet (1999) 'Elusive Genius of the Gentle Pixie'. In *Sydney Morning Herald*, April 4. John Fairfax and Sons, Sydney.

Hay, Peter, Robyn Eckersley and Geoff Holloway (eds.) (1989) *Environmental Politics in Australia and New Zealand*. Centre for Environmental Studies, University of Tasmania, Hobart.

Haynes, Roslynn (1998) *Seeing the Centre: The Australian Desert in Literature, Art and Film*. Cambridge University Press, Melbourne.

Haynes, Roslynn (1999) 'Images of the Desert in Literature, Visual Arts and Film'. Paper delivered at the Fenner Conference at the Australian National University, Canberra.

Hayward, Tim (1998) *Political Theory and Ecological Values*. Polity Press, Cambridge, UK.

Hergenham, Laurie (ed.) (1988) *The Penguin New Literary History of Australia*. Penguin, Melbourne.[5]

Hicks, Sir Stanton (1955) 'Keyline Farming and the Australian Future'. *Keyline*, 1.1, pp. 2–7.

Hill, Stuart B. (1982) 'A Global and Agriculture Policy for Western Countries: Laying the Foundations'. In *Nutrition and Health*, 1.2, AB Academic Publishers, UK.

Hill, Stuart B. (1999a) 'Social Ecology as Future Stories'. In *A Social Ecology Journal*, 1, pp. 197–208. School of Social Ecology and Lifelong Learning, University of Western Sydney.

Hill, Stuart (1999b) *Landcare: A Multi-stakeholder Approach to Agricultural Sustainability in Australia*. In A.K. Dragun and C. Tisdell (eds.) *Sustainable Agriculture and Environment: Globalisation and the Impact of Trade Liberalisation*. Edward Elgar, Cheltenham, UK, pp. 125–134.

Hill, Stuart (1999c) *Conservation Challenges and Opportunities for the Future*. In *New Solutions for Sustainability – A Conference on Integrated Natural Resource Management*. Nature Conservation Council, Sydney pp. 145–148.

Hodge, Bob and Vijay Mishra (1990) *Dark Side of the Dream: Australian Literature and the Postcolonial Mind*. Allen and Unwin, Sydney.

Holmes, J. Macdonald. (1960) *The Geographical Basis of Keyline*. Angus and Robertson, Sydney.

Holmgren, David (1985, second edition 1993) *Permaculture in the Bush*. Holmgren Design Services, Hepburn Springs, Victoria.

Holmgren, David (1994) *Trees on the Treeless Plains: Revegetation Manual for the Volcanic Landscapes of Central Victoria*. Holmgren Design Services, Hepburn, Victoria.

Holmgren, David (1995) 'The Permaculture Movement and Education'. In *Goldfields Permaculture & Landcarers*, 3, pp. 14–16.

Holmgren, David (1995) *Hepburn Permaculture Gardens: 10 Years of Sustainable Living*. Holmgren Design Services, Hepburn, Victoria.

Holmgren, David (1996) 'Fryers Forest Village'. In *Green Connections*, 2.2, pp. 20–21.

Holmgren, David (1997) 'Getting Started'. In *Green Connections*, 10, pp. 28–31.

Hope, A. D. (1975) *Judith Wright*. Oxford University Press, Melbourne.

Horton, David (2000) *The Pure State of Nature: Sacred Cows, Destructive Myths and the Environment*. Allen and Unwin, Sydney.

Howard, Sir Albert (1940) *An Agricultural Testament*. Oxford University Press, Oxford, UK.

Howard, Sir Albert (1947) *The Soil and Health*. Schoken Books, New York.

Hughes, Robert (1970) *The Art of Australia*. Pelican Books, Melbourne.

Hughes, Robert (2000) *Beyond the Fatal Shore*. Television documentary series screened by the Australian Broadcasting Corporation, Sydney.

Hulsberg, Werner (1988) *The German Greens: A Social and Political Profile*. Verso, London.

Humphries, Edith and Douglas Sladen (1912) *Adam Lindsay Gordon and His Friends in England and Australia*. Constable and Company, London.

Hunter, Dale, Anne Bailey and Bill Taylor (1997) *Co-operacy: A New Way of Being at Work*. Tandem, Birkenhead, New Zealand.

Hutton, Drew and Libby Connors (1999) *A History of the Australian Environmental Movement*. Cambridge University Press, Melbourne.

Jordan, Richard and Peter Pierce (eds.) (1990) *The Poets' Discovery*. Melbourne University Press.

Joyce, Clayton (1991) *Patrick White: A Tribute*. Angus and Robertson, Sydney.

Kalajzich, Pip (ed.) (1996) *The Battlers for Kelly's Bush*. Cercus, Sydney.

Kiernan, Brian (1980) *Patrick White*. Macmillan, London.

King, Franklin H. (1911) *Farmers of Forty Centuries*. Reprinted by Rodale Press, Emmaus, PA.

Kormondy, Edward and Frank McCormick (1981) *Handbook of Contemporary Developments in World Ecology*. Greenwood Press, Westport, CN.

Kramer, Leonie (ed.) (1986) *The Greats: The 50 Men and Women Who Most Helped to Shape Modern Australia*. Angus and Robertson, Sydney.

Law, Margie (1997) 'From Tiny Beginnings'. In *The Daily Planet*, 48, The Tasmanian Greens, Hobart.

Lennox, Gina (1998) *Seven Remarkable Men*. Allen and Unwin, Sydney.

Leunig, Michael (1974) *The Penguin Leunig*. Penguin, Melbourne.

Leunig, Michael (1998) *A Bunch of Poesy*. Angus and Robertson, Sydney.

Lines, William (1991) *Taming the Great South Land: A History of the Conquest of Nature in Australia*. Allen and Unwin, Sydney.

Lines, William (1994) *An All Consuming Passion: Origins, Modernity, and the Australian Life of Georgiana Molloy*. University of California Press, Berkeley, CA.

Linton, Michael and T. Greco (1990) 'LETS: The Local Exchange Trading System'. In Van

Andrauss et al. (eds.) *Home!: A Bioregional Reader*. New Society Publishers, Gabriola Is, BC, Canada.

Lloyd, Genevieve (1984) *The Man of Reason: 'Male' and 'Female' in Western Philosophy*. Methuen, London.

McCarrison, Sir Robert (1944) *Nutrition and National Health*. Faber and Faber, London.

MacIntosh, Robert (1985) *The Background of Ecology: Concept and Theory*. Cambridge University Press, Cambridge, UK.

Mackay, David (1985) *In the Wake of Cook: Exploration, Science and Empire 1780–1801*. Victoria University Press, Wellington, New Zealand.

McConvell, Patrick and Rod Hagen (1981) 'A Traditional Land Claim by the Gurindji to Daguragu Station'. Central Lands Council, Alice Springs, Northern Territory.

McLaren, John (1981) *Xavier Herbert's 'Capricornia' and 'Poor Fellow My Country'*. Shillington House, Melbourne.

McLean, Alastair (1972) *Captain Cook*. Collins, London.

Macqueen, Andy (1997) *Back from the Brink: Blue Gum Forest and the Grose Wilderness*. Self-published, Springwood, New South Wales.

McQueen, Humphrey (1996) *Tom Roberts*. Macmillan, Sydney.

Malouf, David (1998) *A Spirit of Play: The Making of Australian Consciousness*. Boyer Lectures, Australian Broadcasting Corporation, Sydney.

Marien, Michael (1982) 'The "Transformation" As a Sandbox Syndrome'. In *Journal of Humanistic Inquiry*, Winter Edition, Association for Humanistic Psychology, San Francisco, CA.

Marr, David (1991) *Patrick White: A Life*. Random House, Sydney.

Mellor, Bill (1990) 'Gourmet of the Grub'. In *Time Australia*, 6.34, pp. 60–62.

Meredith, Peter (1999) *Myles and Milo*. Allen and Unwin, Sydney.

Mollison, Bill (1988) *Permaculture: A Designer's Manual*. Tagari Publishing, Tyalgum, New South Wales.

Mollison, Bill (1996) *Travels in Dream*. Tagari Publishing, Tyalgum, New South Wales.

Mollison, Bill and David Holmgren (1978) *Permaculture One: A Perennial Agriculture for Human Settlements*. Transworld, Melbourne.

Moore, William (1934) *The Story of Australian Art*. Angus and Robertson, Sydney.

Mulligan, Martin (2000) 'Towards a "Whitefella" Dreaming: Re-enchanting Nature Conservation "work"'. In *Overland*, 161, Summer 2000, pp. 58–61, OL Society, Melbourne.

Mulligan, Martin (2001a) 'Re-enchanting Conservation Work: Reflections on the Australian Experience'. In *Environmental Values*, 10.1, pp. 19–33, White Horse Press, Cambridge UK.

Mulligan, Martin (2001b) 'Rethinking Romantic Ecology'. In *Social Ecology Journal*, 2, School of Social Ecology and Lifelong Learning, University of Western Sydney.

Mumford, Lewis (1934) *Technics and Civilisation*. Harcourt, Brace & Co., New York.

Mundey, Jack (1981) *Jack Mundey: Green Bans and Beyond*. Angus and Robertson, Sydney.

Murray, Les (1999) *The Quality of Sprawl: Thoughts about Australia*. Duffy and Snellgrove, Sydney.

Norgaard, Richard (1994) *Development Betrayed*. Routledge, New York.

Neilsen, Philip (1996) *Imagined Lives: A Study of David Malouf*. University of Queensland Press, Brisbane.

Odum, Howard T. (1971) *Environment, Power and Society*. John Wiley, New York.

Palmer, Vance (1954) *The Legend of the Nineties*. Melbourne University Press.

Passmore, John (1974) *Man's Responsibility for Nature*. Duckworth, London.

Pearce, Barry (1983) *A Century of Australian Landscape: Mood and Moment*. The Beagle Press, Sydney.

Pfeiffer, Erhenfried (1943) *Bio-Dynamic Farming and Gardening, Soil Fertility, Renewal and Preservation*. Anthroposophic Printers, Spring Valley, New York.

Pizzey, Graham (1992) *Crosbie Morrison: Voice of Nature*. Victoria Press, Melbourne.

Philips, Franz (1967) *Arthur Boyd*. Thames and Hudson, London.

Plumwood, Val (1993) *Feminism and the Mastery of Nature*. Routledge, London.

Plumwood, Val (1996) 'Being Prey'. In *Terra Nova*, 1.3, pp. 33–44. MIT Press, US.

Plumwood, Val (1998) *Wilderness Skepticism and Wilderness Dualism* in J.B. Callicott and M. Nelson (eds.) *The Great New Wilderness Debate*. University of Georgia Press, pp. 652–90.

Plumwood, Val (1999) *Forest Lover, Live Forever – A Philosophy of the Forests* in G. Borschmann (ed.) *The People's Forest: A Living History of the Australian Bush*. People's Forest Press, Canberra pp. 61–70.

Plumwood, Val (2000a) 'Remembering Judith Wright'. In *Arena Magazine* no. 49, Arena Printing and Publishing, Melbourne, p. 44.

Plumwood, Val (2000b) 'Belonging, Naming and Decolonisation'. In *Ecopolitics: Thought & Action*, 1.1, Social Ecology Reserch Group, University of Western Sydney–Hawkesbury, Sydney.

Prineas, Peter (1998) *National Parks: New Visions for a New Century*. Nature Conservation Council, Sydney.

Pybus, Cassandra and Richard Flanagan (eds.) (1990) *The Rest of the World is Watching: Tasmania and the Greens*. Pan Macmillan, Sydney.

Reynolds, Henry (1996) *Aboriginal Sovereignty: Reflections on Race, State and Nation*. Allen and Unwin, Sydney.

Reynolds, Henry (1998) *This Whispering in our Hearts*. Allen and Unwin, Sydney.

Robertson, Margaret, Kevin Vang and A. J. Brown (1992) *Wilderness in Australia: Issues and Options*. Australian Heritage Commission, Canberra.

Roderick, Colin (ed.) (1972) *Henry Lawson: Criticism 1894–1971*. Angus and Robertson, Sydney.

Roderick, Colin (1982a) *The Real Henry Lawson*. Rigby. Adelaide.

Roderick, Colin (1982b) *Miles Franklin: Her Brilliant Career*. Lansdowne, Sydney.

Roderick, Colin (1991) *Henry Lawson: A Life*. Angus and Robertson, Sydney.

Roderick, Colin (1993) *Banjo Paterson: Poet by Accident*. Allen and Unwin, Sydney.

Rolls, Eric (1981) *A Million Wild Acres*. Thomas Nelson, Melbourne.

Rose, Deborah Bird (1992) *Dingo Makes us Human: Life and Land in an Australian Aboriginal Culture*. Cambridge University Press, Melbourne.

Rose, Deborah Bird (1996) *Nourishing Terrains: Australian Aboriginal Views of Landscapes and Wilderness*. Australian Heritage Commission, Canberra.

Ross, Robert L. (ed.) (1991) *International Literature in English: Essays on the Major Writers*. St James Press Chicago, IL.

Salleh, Ariel (1997) *Ecofeminism as Politics: Nature, Marx and the Postmodern*. Zed Books, London.

Seddon, George (1995) *Searching for the Snowy: an Environmental History*. Allen and Unwin, Sydney.

Seddon, George (1997) *Landprints: Reflections on Place and Landscape*. Cambridge University Press, Melbourne.

Seed, John et al. (1988) *Thinking Like a Mountain: Towards a Council of All Beings*. New Society Publishers, Philadelphia, USA.

Semmler, Clement (1974) *The Banjo of the Bush: The Life and Times of A. B. Paterson*. University of Queensland Press, Brisbane.

Serventy, Vincent (1999) *Vincent Serventy: An Australian Life; Memoirs of a Naturalist, Conservationist, Traveller and Writer*. Fremantle Arts Centre Press, Fremantle.

Shiell, Annette and Ann Stephen (1992) *The Lie of the Land*, National Centre for Australian Studies, Monash University, Melbourne.

Sinclair, John and Peter Corris (1994) *Fighting for Fraser Island: A Man and an Island*. Kerr Publishing, Sydney.

Sladen, Douglas (1934) *Adam Lindsay Gordon: The Life and Best Poems of the Poet of Australia*. Hutchinson and Co., London.

Smith, Bernard (1991) *Australian Painting 1788–1990*. Oxford University Press, Melbourne.

Smith, Geoffrey (1997) *Russell Drysdale 1912–81*. National Gallery of Victoria, Melbourne.

Smith, Russell (1977) *Tree Crops: A Permanent Agriculture*. Devlin-Adair, Old Greenwich, MA.

Spate, Virginia (1972) *Tom Roberts*. Lansdowne, Melbourne.

St Barbe Baker, Richard (1970, reprinted 1979) *My Life My Trees*. Findhorn, Forres, Scotland

Stephens, Tony (2000) 'Conscience of the Country Lives On'. In *Sydney Morning Herald*, July 27, p. 13. John Fairfax and Sons, Sydney.

Strauss, Jennifer (1995) *Judith Wright: Oxford Australian Writers*. Oxford University Press, Melbourne.

Sykes, Friend (1946) *Humus and the Farmer*. Faber and Faber, London.

Tacey, David (1995) *Edge of the Sacred*. Harper Collins, Melbourne.

Tané, Haikai (1995) *Ecography: Mapping and Modelling Landscape Ecosystems*. The Murray–Darling Basin Commission, Canberra.

Thompson, Peter (1984) *Bob Brown of the Franklin River*. Allen and Unwin, Sydney.

Thornhill, John (1992) *Making Australia: Exploring our National Conversation*. Millennium Books, Sydney.

Trainer, Ted (1995) *The Conserver Society: Alternatives for Sustainability*. Zed Books, London.

Trainer, Ted (1996) *Towards a Sustainable Economy*. Envirobooks, Sydney.

Trainer, Ted (2001) 'The Simpler Way: The Global Situation, The Sustainable Alternative Society and The Transition'. Unpublished manuscript (available at http://www.arts.unsw.edu.au/socialworl/trainer.html).

Trenoweth, Samantha (1995) *The Future of God*. Millennium Books, Sydney.

Turner, Frederick (1986) *Beyond Geography*. Rutgers University Press, New Brunswick, NJ.

Voisin, Andre and A. Lecomte (1962) *Rotational Grazing*. Crosby Lockwood, London.

Waldrop, M. Mitchell (1992) *Complexity: The Emerging Science at the Edge of Order and Chaos*. Penguin Books, New York.

Walker, Pamela (1996) 'The United Tasmania Group'. Unpublished thesis, Department of Political Science, University of Tasmania, Hobart.

Walsh, Maureen (1985) *May Gibbs: Mother of the Gumnuts*. Angus and Robertson, Sydney.

Williams, Nancy (1986) *The Yolngu and their Land: A System of Land Tenure and the Fight for its Recognition*. Australian Institute for Aboriginal Studies, Canberra.

Winton, Tim (1997) 'Time and Tide'. In *The Good Weekend*, November 8, John Fairfax and Sons, Sydney.

Worster, Donald (1994, 2nd edition) *Nature's Economy: A History of Ecological Ideas*. Cambridge University Press, Cambridge, UK.

Wright, Judith (1959) *Generations of Men*. Oxford University Press, Melbourne.

Wright, Judith (1961) 'The Upside-Down Hut'. In *Australian Letters*, 3.4.

Wright, Judith (1965) *Preoccupations in Australian Poetry*. Oxford University Press, Melbourne.

Wright, Judith (1977) *The Coral Battleground*. Thomas Nelson, Melbourne.

Wright, Judith (1992) *Going On Talking*. Butterfly Books, Springwood, New South Wales.

Wright, Judith (1994) *Judith Wright Collected Poems 1942–1985*. Angus and Robertson, Sydney.

Wright, Judith (with Patricia Clark) (1999) *Half a Lifetime*. Text Publishing, Melbourne.

Yeomans, Ken (1993) *Water for Every Farm: Yeomans Keyline Plan*. Keyline Designs, Southport, Queensland.

Yeomans, Percival A. (1954) *The Keyline Plan*. Self-published, Sydney.

Yeomans, Percival A. (1958) *The Challenge of the Landscape: The Development and Practices of Keyline*. Keyline Publishing, Sydney.

Yeomans, Percival A. (1965) *Water for Every Farm*. Murray Books, Sydney.

Yeomans, Percival A. (1971) *The City Forest: The Keyline Plan for the Human Environment Revolution*. Keyline Publishing, Sydney.

Yeomans, Percival A. (1978) *Water for Every Farm Using the Keyline Plan*. Murray Books, Sydney.

INDEX

Page references to photographs and illustrations appear in **bold** type.

Milton Keynes UK
Ingram Content Group UK Ltd.
UKHW021437181024
449640UK00022B/330

9 780521 009560